Emerging Digital Spaces in Contemporary Society

# Emerging Digital Spaces in Contemporary Society

## Properties of Technology

Edited by

Phillip Kalantzis-Cope
*PhD Student, The New School for Social Research*

and

Karim Gherab-Martín
*Visiting Research Scholar, Harvard University*

First published 2011 by
PALGRAVE MACMILLAN

Palgrave Macmillan in the UK is an imprint of Macmillan Publishers Limited,
registered in England, company number 785998, of Houndmills, Basingstoke,
Hampshire RG21 6XS.

Palgrave Macmillan in the US is a division of St Martin's Press LLC,
175 Fifth Avenue, New York, NY 10010.

Palgrave Macmillan is the global academic imprint of the above companies and
has companies and representatives throughout the world.

Palgrave® and Macmillan® are registered trademarks in the United States,
the United Kingdom, Europe and other countries.

ISBN  978–0–230–27346–7      hardback

This book is printed on paper suitable for recycling and made from fully managed
and sustained forest sources. Logging, pulping and manufacturing processes are
expected to conform to the environmental regulations of the country of origin.

A catalogue record for this book is available from the British Library.

A catalog record for this book is available from the Library of Congress.

10   9   8   7   6   5   4   3   2   1
20   19   18   17   16   15   14   13   12   11

Printed and bound in Great Britain by
CPI Antony Rowe, Chippenham and Eastbourne

# Contents

# List of Tables and Figures

**Tables**

**Figure**

# Notes on Contributors

## Part I   Digital Communication

**Robin Mansell** is Professor of New Media and the Internet and former Head of the Department of Media and Communications, London School of Economics and Political Science. She is past president of IAMCR—the International Association for Media and Communications Research. Her research focuses on the social, economic, and technical issues arising from new technologies, especially in the computer and telecommunication industries. She has a long-standing interest in the structure of telecommunications and media markets, the sources of regulatory effectiveness and failure, and issues around Internet governance.

**Ilhem Allagui** is currently an assistant professor at the Mass Communication Department, American University of Sharjah. Dr. Allagui holds a PhD in Communication Studies and a Master's degree in Communication Studies from the University of Montreal-Canada as well as a graduate degree in Marketing and Bachelor's degree in Business Administration. Dr. Allagui has contributed to a number of forums and congresses, she was invited by the United Nations for the Alliance of Civilizations Forum (Spain, 2008), by the German Ministry of Foreign Affairs for the Arab-German Media Dialogue (Amman, 2007), and has contributed to the Canadian National Conference (2001) "Information Deficit: Canadian Solutions: Canadian Content on the Internet," Calgary, Alberta. Allagui has served on an editorial board of an electronic journal and as a reviewer for the *International Journal of Technology, Knowledge and Society*, and the journalism research and education section of the IAMCR (International Association for Media and Communication Research). Prior to her academic career, Dr. Allagui held various professional experiences in the new media and advertising field in Montreal, among them experience in the development of virtual communities, experience in an advertising agency as well as in the marketing consultancy field.

**Alfonso Unceta** has a PhD in Political Science and Sociology, and is Head Professor of Sociology at the University of the Basque Country. He has participated in over 30 research projects and external contracts. He has authored more than 40 publications (books, collective books, and articles in magazines). His main lines of work are in innovation, education, and governance. Since 2007 he has been the Director of the Master's course in Management of Innovation and Knowledge. He is also Dean of the Faculty of Social Sciences and Communication. He was the Director of Universities and Deputy Councilor for Education of the Basque Government.

**Jennifer A. George-Palilonis** is the George & Frances Ball Distinguished Professor of Multimedia Journalism and journalism graphics sequence coordinator at Ball State University. There she teaches courses in graphics reporting, multimedia storytelling, and interactive media design. She began her career as a news designer for the Detroit Free Press in 1996. She went on to be the Deputy News Design Director at the *Chicago Sun-Times* in 1999. She has been teaching since 2001 and has spoken at more than 30 conferences and seminars. Her research interests include visual rhetoric, multimedia storytelling, media convergence, and digital publishing. She has a Master's degree from Ball State in Composition and Rhetoric and a Bachelor's degree in Journalism. She is also the author of *A Practical Guide to Graphics Reporting* (2006) and *Design Interactive* (2004), an electronic textbook on basic design principles.

**John Belcher** has taught electromagnetism at all levels at the Massachusetts Institute of Technology, from graduate courses to introductory first year physics courses. He is interested in visualization and was the lead in developing many physics visualizations that can be found at http://web.mit.edu/8.02t/www/802TEAL3D/. Professor Belcher's research interests are in the areas of space plasma physics, outer planet magnetospheres, solar wind in the outer heliosphere, and astrophysical plasmas. He was the principal investigator on the Voyager Plasma Science Experiment during the Voyager Nepture Encounter—the end of the Grand Tour. He is now a co-investigator on the Plasma Science Experiment on board the Voyager Interstellar Mission.

## Part II   Defining New Media

**Carlos Elias** is Full Professor in Journalism at Carlos III University of Madrid (Spain), and he was Visiting Fellow at London School of Economics (2005/6). Educated in Chemistry (BSc and MA) and Journalism (BSc, MA, and PhD), he worked as political journalist for EFE News Agency (the most important mass media in Spanish) and as scientific correspondent for *El Mundo* (one of the most influential newspapers in Spain). His main research focuses on the relationship between policy-making, science, mass media culture, and the future of mass media in digital society. He has published four books and several academic articles. His latest book is *Reason Strangled: The Science Crisis in the Contemporary Society* (2008).

**Reisa Levine** has been working in the realm of digital media since before the web existed, and is currently the Producer for CitizenShift, an online social-media platform. She began her artistic practice on stage as a member of the Narroway Theatre Troupe and from there went on to make her first films and videos. After receiving a BA in Film Production from Concordia University, she worked in Nicaragua as the director/producer of a series of animated films on health and nutrition. Reisa has a strong technical background with over 20 years of experience in film and video production as well as in web development. Over the past decade, Reisa has produced several web-based media projects as an employee of the National Film Board of Canada and as an independent producer. While raising a family and working full time, she also completed an MA in Educational Technology at Concordia University. Reisa has worked as a consultant, lecturer, radio host, and arts council jury member, and is the recent past president of StudioXX, Montreal's feminist digital arts centre.

**Nathan Angelo** is a doctoral candidate at The New School for Social Research. He is interested in racial politics, political rhetoric, and the American presidency. He has taught classes at The New School, Borough of Manhattan Community College, and Lehman College. He received a MA in Political Theory from New York University and an MA in Education from Lehman College.

**Seth Thompson** is an educator and arts journalist involved in documenting and interpreting art, design, and culture through print and online presentations. He is currently an assistant professor of design at the American University of Sharjah in the United Arab Emirates. He has written on the arts for such magazines as *Afterimage*, *Art Calendar*, *Bidoun* and *Dialogue*. Thompson's documentaries, *Evolving Traditions: Artists Working in New Media* and *Outside the Box: New Cinematic Experiences* have aired on such television stations as PBS 45 & 49, Northeast, Ohio; KDOL Channel 18, Oakland, California; DUTV, Philadelphia, PA; and Triangle Television, Auckland, New Zealand.

## Part III   The Texts of Digital Publishing

**John B. Thompson** is Professor of Sociology at the University of Cambridge and Fellow of Jesus College, Cambridge. His publications include *Ideology and Modern Culture* (1990), *The Media and Modernity* (1995), *Political Scandal* (2000), and *Books in the Digital Age* (2005); he was awarded the European Amalfi Prize for Sociology and the Social Sciences in 2001 for *Political Scandal*. His new book, *Merchants of Culture* (2010), examines the changing world of Anglo-American trade publishing and the making of bestsellers.

**John W. Warren** is Director of Marketing, Publications, at the RAND Corporation, a non-profit institution that helps improve policy and decision making through research and analysis. John has nearly two decades of experience in the publishing industry, with special focus on marketing and digital publishing. Previously, John managed marketing efforts for Mexican publisher Fondo de Cultura Económica, Sage Publications, and Sylvan Learning, Inc., and has provided consulting services to firms seeking to expand business in Mexico and South America. He has presented papers at major publishing conferences in the US and internationally. He has a Master's degree in International Management from the Graduate School of International Relations and Pacific Studies (IR/PS) at the University of California, San Diego. John is the winner of the Common Ground International Award for Excellence in the area of the book.

**Yasmin Ibrahim** is Reader in International Business and Communications at Queen Mary, University of London. Before this she worked as a senior lecturer at the University of Brighton. She is a visiting lecturer at Kingston University where she lectures on the postgraduate International Political Communication, Campaigning and Advocacy Programme. Her research is multi-disciplinary and weaves in different fields and disciplines in analyzing the role of communications and technologies in society and its consequences for culture and humanity. Beyond her interest in media and communications, she writes on Islam and modernity, memory studies, globalization, and visual culture.

**José Morillo-Velarde** is a professor of Library Science at the University San Pablo-CEU, Vice-president of the ADLUG – rhe user group of Amicus Dose/Libis and Director of Archives & Libraries of the University Foundation San Pablo-CEU.

## Part IV   The Digital Citizen

**Timothy W. Luke** is the University Distinguished Professor of Political Science, and Program Chair, Government and International Affairs, School of Public and International Affairs at Virginia Tech. His areas of research include environmental politics and cultural studies, comparative politics, international political economy, and modern critical social and political theory. He teaches courses in the history of political thought, contemporary political theory, and comparative and international politics. He serves on the editorial board of *Capitalism Nature Socialism, Culture and Politics: An International Journal of Theory, Environmental Communication, Fast Capitalism, International Political Sociology, International Journal of Innovation and Sustainable Development, Journal of Information Technology and Policy, Organization & Environment, New Political Science, Current Perspectives in Social Theory*, and *Telos*.

**Christopher Wilson** has researched human rights jurisprudence regarding transnational corporations, investigated corporate activity for the Norwegian Pension Fund's Council on Ethics, coordinated NGO training in new media for the Global Forum on Freedom of Expression, and now works as an information consultant for the UNDP, where he is designing a social media strategy and online community for activists who use democracy assessments to drive democratic reform in their countries.

**Julie Uldam** is a PhD fellow at the Department of Intercultural Communication and Management at Copenhagen Business School and the

Department of Media and Communications at London School of Economics. She is the Young Scholars' representative in ECREA's Communication & Democracy section. Her research addresses online media practices of social movement organizations in relation to the tensions that emerge at the intersection between management of visibility and fostering political engagement. She has previously worked as a PR and web consultant at Amnesty International. Her current research on global justice movement organizations is funded by the Danish Research Council.

**Mario Toboso** is a tenured scientist at the Department of Science, Technology and Society in the Institute of Philosophy of the Spanish National Research Council (CSIC). He received his Degree in Physics at the University of Salamanca, Spain, and PhD at the same university (Department of Philosophy, Logic and Philosophy of Science). He received his Master's degree in "Design for All and Accessibility to ICT" at the EOI Business School, Madrid. His research interests focus on studies on science, technology, and society; philosophy of science and philosophy of technology; social philosophy and civil rights for persons with disabilities; analysis of disability in Amartya Sen's approach; study of human agency and functionality under the point of view of functioning diversity; accessibility; and inclusive technological design. Mario is a member of the Independent Living Forum, Spain.

## Part V   Power, Knowledge, Surveillance

**David Lyon**'s research, writing, and teaching interests revolve around major social transformations in the modern world. Questions of the information society, globalization, secularization, surveillance, and postmodernity all feature prominently in his work. Best known for his work in surveillance studies, David Lyon's research and writing span several other areas as well. Starting with historical sociology in the 1970s, his early work was on secularization processes—and the critique of some key theories—in the modern world. Following this, his main research explores other forms of social transformation that are both characteristic and constitutive of modernity. From the mid-1980s, this prompted a critical examination of the much-hyped 'microelectronics revolution' that gave way to the so-called information society. In the 1990s Lyon focused his analyses on the social origins, incidence, and consequences of processing personal data, arguing that surveillance has become a major dimension of modernity in its own right. This aspect of his research has expanded considerably, especially since 9/11, and now includes investigations of airport security, urban surveillance infrastructures, and national identification systems. As Director of the Surveillance Studies Centre, Lyon works with a multidisciplinary and international team on several related initiatives from primary research to theoretical development as well as associated media, policy, and advocacy activities.

**Brian Jefferson** is a PhD candidate in the Department of Politics at the New School for Social Research in New York City. His research interests include American politics, police studies, and processes of political subjectivity. He currently lectures at Brooklyn College, City University of New York.

**Joseph Ferenbok** is a post-doctoral fellow with the Identity, Privacy and Security Institute (IPSI) at the University of Toronto. Joseph received his BA (English and Biology) and MA in Humanities Computing with a focus on visual information design, both from the University of Alberta. He received his PhD from the University of Toronto and his thesis looks at a theoretical framework for understanding the techno-political development of "the face" as a strategy of institutional identification, surveillance, and empowerment. His research interests focus on themes of privacy, accountability, and transparency in (video) surveillance and how "the face" may play a positive role in redressing some of the one-sided views of surveillance. His research includes the augmentation of identification documents with technologies, such as radio frequency ID (RFID) chips and biometrics like face recognition/comparison software and how these technologies change relationships of power between individuals and institutions. He is also interested in the promise of privacy enhancing technologies (PETs) and how these may be integrated into surveillance systems to meet regulatory compliance and enhance the potential of supporting autonomy, dignity, due process, empowerment, and social equity.

## Part VI   Digital Property

**Phillip Kalantzis-Cope** is a teaching fellow at The New School for General Studies, and a PhD candidate in the Department of Political Science, The New School for Social Research in New York City. He is a board member of *The International Journal of Technology, Knowledge and Society*, and founding editor of *The International Journal of the Image*.

**Roberto Feltrero** is a lecturer at the Department of Logic, History and Philosophy of Science at the UNED (Spanish Open University). His research interests include cognitive technologies, epistemology and technology, user interface design and evaluation, ethics and computing, philosophy and free software. He has published articles on open access and the philosophy of free software in *Isegoría* and *Arbor*. He is also the author of books including *El Software Libre y la construcción ética de la Sociedad del Conocimiento* (Barcelona: Icaria 2007).

**Mikkel Flyverbom** is an associate professor at the Copenhagen School of Business. His research focuses on organizational and communicative relations among corporations, international organizations, and civil society groups, with a particular interest in the relationship between organizational processes and emerging issues in global politics and governance. Currently he is working on three different projects. One is how the Internet is constructed as an object of global governance in the context of the UN and at the intersection of a range of private and public actors. The second is an ethnographic study of the ways in which companies manage globalization and uncertainties in their everyday work. Finally, he is involved in a collaborative research project which explores how corporations, international institutions, and NGOs collectively seek to advance transparency and anti-corruption strategies in the private sector. Theoretically and methodologically, Flyverbom is broadly interested in organizational communication, governance, sociology, and international relations, with a particular focus on practice-oriented, relational approaches, such as actor-network theory, governmentality studies, and global ethnography.

**Rubeena Aliar** is a researcher with interests in fields of environment, culture, health, gender issues, disaster management, and sustainable development. After graduating with a Master's degree in Applied Geography from the University of Madras, she completed her doctoral research in applied geography on local health traditions and complementary medicine in Kerala, India. On completion of her doctoral research, she worked as Project Fellow at the Kerala State Land Use Board on the degradation of wetlands in Kerala, later joining the Institute of Land and Disaster Management (which is the state-run administrative training institute for disaster management in Kerala) as assistant professor.

## Part VII   The Digital Commons

**John Willinsky** is currently on the faculty of the Stanford University School of Education. Until 2007 he was the Pacific Press Professor of Literacy and Technology and Distinguished University Scholar in the Department of Language and Literacy Education at the University of British Columbia (UBC). He is a fellow of the Royal Society of Canada.

**Aloka Parasher-Sen** graduated with History Honors (1973) from Delhi University and did her graduate (1974) and doctoral studies (1978) from the University of London. Her teaching career spans three decades primarily located at the Department of History, University of Hyderabad, India. She was DAAD Fellow (1986/7) and occupant of the Rotating Chair in India Studies (2007/8) at the South Asia Institute of the University of Heidelberg, Germany, and Fulbright Scholar and Visiting Professor (1992) at the University of California, Berkeley, US. She was awarded the Career Award in Social Sciences and Humanities by the University Grants Commission, New Delhi (1989–91), and nominated Member Indian Council for Historical Research, New Delhi, by the Government of India (1994–7). Her major publications are in the main area of her interest in social history, namely, early Indian attitudes towards foreigners, tribes, and excluded castes. Her other work is on different aspects of the history and archaeology of Peninsular India. She has delivered invited lectures and presented papers at conferences in Canada, France, Germany, Hong Kong, Japan, the UK, and the US She is currently Saroj and Prem Singhmar Chair in Classical Indian Polity and Society, Department of History and Classics, University of Alberta, Canada.

**David Yates** is an assistant professor of Computer Information Systems at Bentley University. Yates's research areas include computer networking, data communications, sensor networks, embedded systems, operating systems, and computer architecture. Before joining Bentley, he held research and academic positions at the University of Massachusetts and Boston University. In the corporate arena, he was a co-founder and vice president of software development at InfoLibria – a startup that grew to become a leading provider of hardware and software for building content distribution and delivery networks before it was acquired. He holds several US patents for processes and systems related to computer networking, content management, and mobile computing. He holds a PhD and MSc from the University of Massachusetts and a BSc from Tufts University.

**Anas Tawileh** has a BSc in Engineering from Damascus University, and a Master's degree in Information Systems Engineering from Cardiff University. He is a consultant for IT solutions, networking, and information security for many companies and organizations. He founded the Syrian GNU/Linux Users Group, and contributed to the development of many free and open source software (FOSS) projects. He was an invited speaker in many events; the topics he covered include FOSS advocacy, Arabization, FOSS licensing and security. Tawileh is the co-founder of the Internet Society—Syria Chapter, and the author of *Open Source, Unlimited Opportunities* (2009). He has many published papers and interviews in the Arab and international media; he is a member of the IEEE, ACM, NaSPA, ASSIST, and USENIX. He is currently working on his PhD at Cardiff University, and his main areas of interest include FOSS, systems design, information security, human–computer interaction, and multi-agent systems. He launched the Nosstia Centre of Excellence in Damascus, an NGO aiming to become the pioneer competency centre for FOSS-related activities in the Middle East, and designed several training programs which were delivered in the centre.

**Elena Moschini** is a senior lecturer in Digital Media and Communications Technology and the MA Digital Media course leader at London Metropolitan University, UK. Before joining the university, she has worked in the multimedia industry in Switzerland, Italy, and the UK, managing and developing a number of interactive projects. Her research interests include e-learning and game-based learning, learning in virtual worlds, digital literacy, game design, game audiences, the game industry, social media and virtual worlds, and digital media industries. She is a member of the Higher Education Academy, UK; the Learning Technology Research Institute, London Metropolitan University; the International Game Developer Association; and The British Interactive Media Association.

## Part VIII   New Infrastructures of Science

**Karim Gherab-Martín** is a theoretical physicist and has a PhD in Philosophy of Science and Technology. He has taught at Harvard University, Universidad Autónoma in Madrid, University of the Basque Country in Bilbao, and San Pablo CEU University. He has worked as an IT consultant in Spain and Latin America, and has written strategic reports on digital libraries for the Government of Madrid. Some of his latest publications include *The New Temple of Knowledge: Towards a Universal Digital Library* (2009) and *Science and Culture on the Web* (2009).

**Khosrow Farahbakhsh** is an associate professor at the School of Engineering, University of Guelph, Ontario, Canada. Dr. Farahbakhsh has nearly 20 years of design, consulting, and academic experience in environmental engineering addressing issues ranging from air pollution control, water, waste water and rainwater management, soil and groundwater remediation, pollution prevention, and resource recovery and energy from waste. His current research interest includes integrated water resources management, water reuse, water safety in First Nations communities, rainwater harvesting, microbial fuel cells, resource recovery from wastewater and sustainability, and the second law of thermodynamics.

**Benjamin Kelly** is a PhD candidate with the Sociology Department at McMaster University in Canada. His interests include environmental sociology, science studies, and ethnographic research. He is currently studying social interaction and knowledge production among multi-stakeholder learning platforms within pollution prevention arenas.

**Julio E. Rubio** is the Director for Research and Graduate Studies, Vicepresidency for Research and Technological Development, Tecnologico de Monterrey Cuernavaca, Morelos, Mexico. He holds a PhD in Philosophy of Science, University of Valencia, Spain; Master's degree in Philosophy of Science, Autonomous Metropolitan University, Mexico; and Master's degree in Science in Physics, University of Texas at El Paso. He is Electrical Engineer, Tecnologico de Monterrey, Mexico; Member of National Researchers System, Mexico, 2000–6; President, Tecnologico de Monterrey Campus Santa Fe, 2004–6; and Research and Graduate Studies Director since 2006. He has written articles and book chapters in epistemology of natural sciences and social studies of science.

**Manuel González Villa** is a member of the Mathematics Department at the University of Illinois (Chicago).

## Part IX   Digital Aesthetics

**Sean Cubitt** is Professor of Global Media and Communications at Winchester School of Art, University of Southampton, and Professorial Fellow in Media and Communications at the University of Melbourne. His publications include *Timeshift: On Video Culture* (1991), *Videography: Video Media as Art and Culture* (1993), *Digital Aesthetics* (1998), *Simulation and Social Theory* (2001), *The Cinema Effect* (2005), and *EcoMedia* (2005). He is the series editor for Leonardo Books at MIT Press. His current research is on public screens and the transformation of public space and on genealogies of digital light technologies.

**John Byrne** is a lecturer at Liverpool School of Art and Design. He is also currently the Manager and Curator of the Site Project at Liverpool School of Art and Design. The aim of the Site Project is to interface the work of students and staff at Liverpool School of Art and Design with the work of national and international artists and designers and to bring this work into contact with local, national, and international audiences and publics. Over the last 15 years Byrne has published internationally on issues surrounding contemporary art and its relationships to technology and popular culture. During this period Byrne has also worked collaboratively on the curation and production of a range of events, shows, projects, and films. Most recently, Byrne has co-produced "Damien Hirst: One Night Only" for the Lewis Glucksman Gallery, Cork, Ireland: co-curated and managed the "Martha Rosler Library" at Liverpool School of Art and Design (in collaboration with e-Flux and Liverpool Biennial); produced and curated "CreamTVfields" at the Creamfields 10 Festival in August 2008; and co-produced "Noodle Bar" (Static Gallery/Liverpool Biennial 08). Byrne has recently completed his first book *High Art: Low Life: Art, Technology and Popular Culture in the Last Century*, which was expected to be published by Rebelo Publications and Cambridge University Press in spring 2009. Byrne curated The Centre of Attention's project "Darling," in collaboration with FACT, in Liverpool's new Art and Design Academy in spring 2009 and is working in collaboration with Charles Esche (Director of the Van Abbemusuem), Annie Fletcher (Programmes Curator at Van Abbemuseum), and Juan Cruz (Head of Art at Liverpool School of Art and Design) on the "Autonomy" research project.

**Melissa D. Milton-Smith** after graduating from The University of Western Australia with First Class Honours in History, completed a Master's degree in Architecture and the Moving Image at Cambridge University, UK. This hybrid degree addressed the interplay between architectural theory and digital media through the production of practical and theoretical theses. After completing her Master's degree in 2001, Melissa returned to Australia to complete a PhD funded by an Australian Postgraduate Award. Her thesis "A Conversation on Globalisation and Digital Art" was completed at The University of Western Australia. Melissa was previously employed as the Multimedia Manager at the Museum of Australian Democracy in Canberra, and is now the Head of Communications and Media at the University of Notre Dame Australia.

**Tamsyn Gilbert** is a PhD student in Sociology at The New School for Social Research. She has worked as a pre-screener for Documentary Fortnight at The Museum of Modern Art (MoMA) in 2009 and 2010. She has also completed internships at MoMA and at Anthology Film Archives. Her current research looks at the intersection of cultural institutions (archives, museums, libraries), the digital life, and everyday life.

## Part X   Digital Labor

**Eran Fisher** completed his PhD at the New School for Social Research in New York in 2008. He is interested in critical social theory, technology, and capitalism. His articles on the intersection of capitalism and network technology appeared in the *European Journal of Social Theory*, *Fast Capitalism*, and *Working USA: The Journal of Labor and Society*. His book *Media and New Capitalism in the Digital Age: The Spirit of Networks* (2010) explores the ideological transformations involved in the rise of contemporary techno-capitalism.

**Owen Darbishire,** after reading PPE (politics, philosophy, and economics) at Balliol College, Oxford, UK, between 1984 and 1987, worked as an economist and subsequently taught economics. Between 1990 and 1996 he attended the School of Industrial and Labor Relations at Cornell University, during which time he was also a Visiting Scholar at the IAAEG in Trier, Germany. In 1996 he returned to Oxford as a University Lecturer in Organisational Behaviour and Industrial Relations, and as a Fellow of Pembroke College where he is currently Dean of Visiting Students.

**Madeline Carr** is a PhD candidate at the Australian National University. Her thesis is titled "A Political History of the Internet: Implications for US Power." Her research interests include American foreign policy, international relations theory, the philosophy of technology, and the nexus between the media and politics. She tutors for AusAID in the Graduate Studies in International Affairs program at ANU.

**Suriyani Muhamad** is Senior Lecturer in Department of Economics, University of Malaysia Terengganu, Malaysia. She obtained her Bachelor of Economics from University of Malaya, Kuala Lumpur in 1999; Master's degree in Local Economic Development in 2001 from the London School of Economics, UK; and Doctoral degree from the University of Manchester, UK in 2007. Her specializations are in labour economics, human resource development, and economic development.

## Part XI   Technology, Culture, and Society

**David Hakken** is a Professor of Informatics and Director of Social Informatics Program at Indiana University, Bloomington. He studied cultural anthropology at Stanford, Chicago, and the American University in Washington, DC in the 1960s and 1970s. The abiding concerns of his research career have been the complex ways in which social change, culture, and technology, especially automated information and communication technology, co-construct each other. This has led him to study worker education, public policy, and workplace use of new information technology in Britain and the US; software development in Britain, the Nordic countries, the US, and Malaysia; social service and technology (for example, assistive technology) in the US; and techno-science in Chinese and Malaysian scholarship and higher education.

**Maurizio Teli** received his PhD in Sociology and Social Research from University of Trento with a dissertation titled "Freedom and Practices: Free and Open Source Software in the OpenSolaris Case" in 2008. Since then, he has been working at Museo Tridentino di Scienze Naturali as a researcher and scientific coordinator of the European project "My Ideal City." His research interests are connected to the relations between information technologies and organisation, with a particular focus on the political dimensions of the use of information technologies as organizing devices.

**Anjali Gera Roy** is Professor in the Department of Humanities and Social Sciences at the Indian Institute of Technology, Kharagpur. She has published several essays on postcolonial literature and theory in scholarly journals. She has co-edited a special issue of *The Literary Criterion* on "New Directions in African Writing." Her books include *Three Great African Novelists* (2001) and *Wole Soyinka: An Anthology of Recent Criticism* (2006).

**Amareswar Galla** was educated in south and north India, including the prestigious Jawaharlal Nehru University in New Delhi. He is the first Professor of Museum Studies in Australia at the University of Queensland. He provides strategic cultural leadership in Australia and the Asia Pacific Region as the founding Director of the UNESCO Pacific Asia Observatory for Cultural Diversity in Human Development. He has been Vice President of the International Council of Museums (ICOM), Paris, since 2004, and is the founding Chairperson of the ICOM Cross Cultural Task Force. Amareswar Galla is one of the leading experts in the world on museums, sustainable heritage development, and poverty alleviation through culture, and has worked extensively in Vietnam, South Africa, Iraq, the Pacific, Europe, Asia, and Australia. He has been a key advisor for the UNESCO World Commission for Culture and Development, and was UNESCO technical advisor and guest curator of international projects in Vietnam responsible for the development of World Heritage sites in Hoi An and Ha Long Bay, working on the development of the world's first floating museum, the Cua Van Cultural Centre in Ha Long Bay.

**William James Stover** is Professor of Political Science and International Relations at Santa Clara University, where he teaches world politics and international law. He has an MA in International Relations from the American University (Washington), School of International Service, and a PhD in political science from the State University of New York (Buffalo). He has written widely on peace and security and information technology in world affairs. Most recently, he was named a Senior Fulbright Specialist on Information Technology and Conflict Resolution by the Council for International Exchange of Scholars where he makes available open source software for use in conflict resolution simulations.

## Part XII  Digital Identities

**Marcus Breen** is Associate Professor in the Department of Communication Studies at Northeastern University, Boston. He was born and educated in Australia, where he worked as a journalist, researcher, and consultant before becoming an academic. In 1996 he moved to the US. His research and teaching involves explorations of digital technology, cultural studies, and cultural policy studies and their intersection with industrial formations. He has worked and published research on music, film and information, communication, and telecommunications. His most recent book is *Rock Dogs: Politics and the Australian Music Industry* (2006). He is currently writing *Uprising: The Internet's Unintended Consequences*.

**Emily D. Arthur** recently completed her Master's Degree in Sociology at the University of Victoria. Her thesis project sought to explore the role of digital environments in facilitating and coordinating knowledge production through narrative-based dialogue. She will be exploring new areas of digital space in her PhD studies. Her research interests include social media, digital culture, human–technology interactions, sexuality, identity work, social (re)organizations of knowledge, and radical uses of the potentialities of online space.

**Jan Lüdert** is a PhD student in the department of Political Science at the University of British Columbia. He holds a Masters of Arts in International Relations from the Australian National University and a Bachelor of Business Administration from Hamburg University for Economics and Politics. He has studied in Tanzania focusing on sociology and economics and before his graduate studies was involved in grassroots development programs in Botswana for two years. This work in the non-governmental sector of international development has spurred his interest in indigenous peoples' advocacy, which ultimately brought him to the University of British Columbia. He intends in his PhD dissertation to trace the connection between international indigenous elites and experts and their dialogical relationship with local indigenous groups. Here, he is in particular interested to analyze the construction of local identities that have been influenced by global actors and structures of the international indigenous movement. He hopes that his research will help to better our understanding of international politics and will further support the indigenous activism that has been established within the last decades. For more information, please visit his website at jl.luedert.com.

**Verónica Sanz** is a PhD candidate at the Department of Logic and Philosophy of Science at the University Complutense of Madrid (Spain). She graduated in Philosophy at the University of Valladolid (1999) and got her Master's Degree at University Complutense within the graduate program "Between Science and Philosophy" (2002). Her main interests are philosophy of technology and science and technology studies, particularly information and communication technologies and artificial intelligence. During 2003/4 she received a scholarship from University of California and University Complutense to attend during two semesters the Graduate Program of the Office for the History of Science and Technology (History Department) at University of California at Berkeley.

From 2005 to 2007 she was an assistant researcher at the Department of Science, Technology and Society in the National Council of Scientific Research (CSIC). During that period, she was the coordinator of the permanent Seminar of Junior Researches (SIJI) and the Seminar of Science, Technology and Gender (STG). She has participated in five research projects, including "Practis: Philosophy of Social and Human Technosciences," "Social Perception of Technologies of the Body," and "Civil Society and Governance of Science and Technology in Spain." She has presented papers in 24 national and international conferences and workshops. In 2008 she was a fellow researcher at the Institute of Advanced Studies on Science, Technology and Society (IAS-STS) in Graz, Austria. Currently she is a visiting graduate student at the European Centre for Soft Computing in Mieres, Asturias.

## Part XIII   Information Globalsim

**Jan Nederveen Pieterse** is Mellichamp Professor of Global Studies and Sociology in the Global and International Studies Program at the University of California, Santa Barbara. He specializes in globalization, development studies, and cultural studies. He focuses on new trends in twenty-first-century globalization and the implications of economic crisis. He has been visiting professor in Brazil, China, Germany, India, Indonesia, Japan, Pakistan, South Africa, Sri Lanka, Sweden, and Thailand. He is an editor of Clarity Press and associate editor of *Futures, Globalizations, Encounters, European Journal of Social Theory, Ethnicities, Third Text*, and *Journal of Social Affairs*.

**Robert M. Bichler** is lecturer at the Shanghai International Studies University (SISU) and member of the Unified Theory of Information (UTI) Research Group, Association for the Advancement of Information Sciences. Previously he was a research fellow and lecturer at the ICT&S (Information and Communication Technologies & Society) Center at the University of Salzburg. He recently co-published a special issue of the journal *Information, Communication & Society* on the topic "Sustainable Development and ICTs."

**Mili Kalia** completed her MSc Research Methodology in Sociology at St Antony's College, University of Oxford, UK. She is currently a PhD student in Sociology at the University of Virginia.

**Charles C. Chiemeke** is presently an assistant professor of Economics at the American University of Kuwait and teaches microeconomics, macroeconomics, and econometrics courses mostly to students majoring in economics. Previously, he worked as an assistant professor of Economics and Finance at the Maastricht School of Management MBA outreach programme in Kuwait, taught economics and financial resources management courses to MBA and pre-MBA students and supervised students writing their thesis. In addition, he has a Master's degree in International Business Administration from the University of Vienna and a PhD in Economics from the Vienna University of Economics and Business Administration, Austria.

## Part XIV   Reading Machines

**Jean-Claude Guédon** is Professor of Comparative Literature at the Université de Montréal. He has been much involved with the emergence of digital culture, electronic publishing, and open access. Former member of the Open Society Institute Information Program Sub-board, and of the Board of Electronic Information for Libraries, and former Vice President of the Canadian Federation for the Humanities and Social Sciences, he is presently a member of the editorial board of interdisciplinary science reviews, of the Scientific Advisory Board for the European project OAPEN, and a member of the steering committee of Open Humanities Press.

**José Luis González-Quirós** has a PhD in Philosophy (Universidad Complutense, Madrid). He has worked as Philosophy Professor at the Universidad Complutense and the Spanish National Research Council (CSIC, Madrid), and also as Vice Director of the Universidad Complutense summer courses. He has taught courses and seminars at the universities of Wyoming (US),

Lund (Sweden), Loyola (Chicago, US), and Veracruz (Mexico). He held the positions of Secretary General at the Spanish Institute of Emigration, Director of Research at the Official Institute of Radio and Television, and Secretary General at Fundesco. He is now a tenured scientist at the Rey Juan Carlos University (Madrid), and is a member of the editorial staff for the magazines *Revista de Libros*, *Nueva Revista*, *Dendra Medica*, and *Revista Hispano-Cubana*. He also regularly contributes articles to the newspapers *Gaceta de los Negocios* and *El Confidencial*. He is the founder and first director of the magazine *Cuadernos de pensamiento político*.

**José Antonio Millán** has maintained a website on reading and electronic media (librosybitios.com) since 1999. He has published the following books on this subject: *La lectura en España. Informe 2008. Leer para aprender* (Reading in Spain: 2008 Report. Reading to Learn) (as editor) and *Edición electrónica y multimedia* (Electronic and Multimedia Publishing) (1996). He has contributed chapters in books such as *El lector en red* (The Internet Reader) in Inés Miret and Cristina Armendano (coordinators), *Lectura y bibliotecas escolares* (Reading and School Libraries) (2009); and his articles include "Leer sin papel" (Reading without Paper) in *El País* (2009), and "La era de las máquinas lectoras" (The Age of Reading Machines)] in *Arbor* (2009). He has directed seminars and taught courses on this subject. He is a member of the Ministry of Culture's "Reading and Books Observatory" and "Electronic Book Work Group" (2009–) and is also on the advisory committee of various journals and institutions.

# Introduction

# Properties of Technology

*Phillip Kalantzis-Cope*

Our aim in bringing together this collection of papers is to uncover ways in which "the digital" is at once encroaching, reformulating and creating social spaces. Indeed, at times the digital may even reconfigure what it means to be social. In order to capture the complex dimensions of this digital shift we have included a comprehensive range of disciplinary fields—politics, sociology, science, philosophy, informatics, public policy, communications and media studies.

The notion of "properties" serves as the central conceptual tool of the collection, the binding element that ties together the disparate disciplinary perspectives and empirical contents of this book. There are three dimensions to the notion of "properties." The first sense of "property" is reflected in the book's structure. Each thematic area represents one aspect of the social transformations occurring around digital technologies, identified in order to present a comprehensive account of the spectrum of social issues arising from digital technologies. Within this context we have selected a mix of big picture theoretical analysis for the lead anchor chapters in each section, followed by detailed examples of the diverse effects of each general transformation in the accompanying case studies. The second sense of "property" is embodied in the expositions of the emergent and unique "properties" of digital technologies in general. The third sense of "property" emanates from the framing of digital technology within a longer continuum of action and thought associated with technological development in general. Each of our contributors, in their own particular way, addresses one or more dimensions of these properties.

Robin Mansell leads the first part on Digital Communication. Her chapter provides a thorough exploration of whether or not, as is often argued, information communication technologies (ICTs) can be tools for emancipatory social change. This chapter challenges our thinking to consider whether a global one-size-fits-all approach to technology is appropriate for the socially disparate web or for people who share and communicate through these technologies. Mansell also asks us how, and in what ways, the new technologies themselves communicate a social vision. Three case studies then provide striking and differing insights into the forces shaping digital "communication" technologies. Ilhem Allagui, in his snapshot of Arabic websites, illuminates how the nation-state continues to play a role in shaping digital content. Alfonso Unceta introduces us to e-participation in the Basque country of Spain and the ways digital networks may or many not aid political, social or economic participation. Completing this part, Jennifer A. George-Palilonis and John Belcher present the ways ICTs can facilitate collaborative learning processes.

Part II, "Defining New Media," is anchored by Carlos Elias, who frames the current digital shifts within a broader structural transformation in the dominant hierarchical order of knowledge creation. He focuses on how Web 2.0 technologies and the rise of a "convergence culture" are transforming the profession of journalism. He questions the nature of "credible" news, and the emerging boundaries between amateur and professional reporting. The case studies that follow illustrate the general issues raised by Elias in a variety of ways. Reisa Levine presents us with a glimpse of community engagement in digital production within the Canadian film industry. Nathan Angelo investigates Barack Obama's use of Twitter and the way in which new media technologies reflect the changing nature of political communication. Seth Thompson explores how digital technologies are reshaping the relationship between "object and audience" in the museum.

Part III, "The Texts of Digital Publishing," begins with John B. Thompson's chapter on the ways digital technologies are shaping the future of academic publishing and university presses, hitherto stalwarts in the knowledge production industry. This detailed overview of the publishing industry provides the base from which we can examine the digital transformation of "information" production, workflow and dissemination. John W. Warren's case study throws light on the emergence of free online textbooks, and the benefits of the open access publishing business model. Yasmin Ibrahim highlights the emerging cultural and political economy of digital book production. José Morillo-Velarde's case study reveals a temporal shift in the production of text, from periodical to instantaneous artifact.

In Part IV, "Digital Citizenship," Timothy W. Luke leads the discussion, focusing on the ways digital networks are encroaching and reshaping traditional boundaries of "rights" and "duties." This, he argues, suggests a transformation of both the definition of "citizenship" and the spatial boundaries of political membership—the "what," "where" and "who" of "digital citizenship." How asks whether our connectedness shapes the way we participate. The case studies that follow demonstrate the "political" uses of digital technologies and networks. Christopher Wilson highlights the use of Facebook in political activism in Egypt. Julie Uldam explores how social networking sites such as Facebook and YouTube can be used to facilitate global grassroots political movements. And Mario Toboso shows us how

ICTs can help people to cross boundaries of political participation, facilitating a "functional diversity" within digital society.

David Lyon anchors Part V, "Power, Knowledge, Surveillance." In this chapter Lyon details how our individual and collective digital information trails are not simply used to track personal preferences within social and economic networks, but are also representative of a broader logic of surveillance in the information age. In the first of the case studies, by focusing on the tactics of the New York Police Department (NYPD), Brian Jefferson demonstrates how digital technology maps criminality and political subjectivity. Yasmin Ibrahim maps the range of "participants" in what has been called the "surveillance society"—from CCTVs to citizen journalists. Finally, Joseph Ferenbok illuminates how even our face becomes a site of digital identification (digital signature).

In Part VI, "Digital Property," I contextualize the unfolding debates and visions for social justice arising from "digital" intellectual property rights, and the broader question of the ownership of intellectual goods in general. The case studies that follow present reference points for understanding the global dynamics of this digitally propelled social transformation. Roberto Feltrero points to the emergence of the free and open source software movement and peer-to-peer production as "ethical" statements within the digital domain. Mikkel Flyverbom explains the intersection of intellectual property rights and Internet governance in order to illuminate the emerging modalities of global governance. And Rubeena Aliar, through a case study of health traditions in rural India, provides insights into how property regimes that privilege the individual owner displace cultural forms based on collective ownership.

John Willinsky anchors Part VII, "The Digital Commons." In this chapter he explores open source collaborative knowledge production—the fundamental logic of the digital commons—through a study of Wikipedia. He presents us with a clear example of the ways in which such approaches to digital technology can promote "openness" by pointing to the diverse make up of participatory actors in the commons. Aloka Parasher-Sen follows, with a case study of a digital mapping exercise in Southern India, exploring the new ways of mapping space, culture and identity. David Yates and Anas Tawileh draw our attention to the digital divide in rural Africa and urban North America, arguing that this divide is not simply a case of the developed versus the developing world, but one that creates a new cartography based on differing access patterns, thereby forcing us to think about what is "common" about the "commons." Elena Moschini illuminates the demands and possibilities of "information literacy," based on the emerging new "common" language of the digital.

Karim Gherab-Martín anchors Part VIII, "New Infrastructures of Science," exploring the intersection of Web 2.0 technologies, the open source movement and the specific character of scientific knowledge. He presents us with a way to navigate the challenges of "digital" scientific endeavor based on the idea of "open reuse." The case studies that follow provide grounded reference points for understanding the contours of technology and scientific endeavor. In their example of approaches to water management, Khosrow Farahbakhsh and Benjamin Kelly force us to think outside the "digital box," in order to grasp the cultural dynamics of scientific development in general. Julio E. Rubio, through an examination of the issues surrounding transgenic corn in Mexico, illuminates the need to contextualize the digital within the broader transformative effect of scientific research—for example bioengineering and the interactions among stakeholders in the validation of scientific endeavor. Finally in this section, Manuel González Villa highlights the displacement at times of traditional models of scientific knowledge validation in digital knowledge communities, specifically the mathematical community.

Part IX, "Digital Aesthetics," is framed by Sean Cubitt's chapter, which uses the concept of "fabrication" as an historical and social process, and examines the relationship between the aesthetic experience of digital technologies and the material world. John Byrne then asks what the digital reveals about "radical art." What pushes the boundaries of the aesthetic experience? Melissa D. Milton-Smith examines the boundaries of "art" and "new media" and Tamsyn Gilbert presents us with the i-phone application Art Beat, which demonstrates the relationship between art and function in new digital technologies.

Part X, "Digital Labor," is anchored by Eran Fisher's analysis of the emergence of "prosumption," a hybrid of production and consumption, a mode of work characterized by networked "digital" labor practices. The meta-argument of this chapter is that this shift in labor practices reveals a "new spirit of capitalism" throwing into question the emancipatory potential of digital network technologies. The three case studies then provide us with differing examples of the spaces, practices and dynamics of "digital" labor. Owen Darbishire illuminates the emergence of an economy of servicing the digital with the UK telecommunications industry. Madeline Carr details how the US government has approached its digital labor force and its relationship to "national interest," and Suriyani Muhamad sheds light on the ICT job market in Malaysia.

David Hakken and Maurizio Teli anchor Part XI, "Technology, Culture, and Society." Their chapter examines the balance between a purely technological base for computing studies, and what they call a "socially conscious computing" or "technosocial" approach. At the core of this chapter is the question of who participates in the processes of technological development. The three case studies then explore this question in various ways. Anjali Gera Roy presents the global digital network of Bhangra production and dissemination, making the case that the digital, rather than simply being the conduit for the voice of the "mainstream," also can "globaliz[e] the voice of the margin." Amareswar Galla deals with the digitization of museum artifacts, focusing on how we collect, protect and open access to intangible heritage, balancing questions of whose voice is included, who is collecting and who retains ownership. William James Stover presents a classroom setting where virtual modeling and simulation can encourage critical thinking by taking political questions and "digitally" abstracting them, in order to solve real world issues.

Part XII, "Digital Identities," is anchored by Marcus Breen, who analyzes Internet pornography, exploring the ways in which the formation of identity on the Internet is a reflection of deeper social transformations within our shared cultural landscape. This chapter compels us to think about the forces and boundaries of "digital" identity formation. The three case studies that follow provide insights into the very different ways in which "identity" is formed, advocated and distributed on the Internet. Emily D. Arthur illuminates bisexual chat rooms and personal web diaries as forums to explore sexuality, presenting them as windows to the world and into one's own experience. Jan Lüdert unveils the development of the "UsMob" website, as a dialogue between an indigenous Australian community and the "digital social"—indigenous and non-indigenous people. Finally, Verónica Sanz sheds light on the gender disparities of those receiving computer science degrees.

Jan Nederveen Pieterse anchors Part XIII, "Information Globalism." This chapter takes a critical look at the logic of "information communication technology for development" and its relationship to the spread of "digital capitalism." The three case studies then delve into the issues of digital global interconnectedness. Robert Bichler takes us to Malawi, where the cost of a telephone line is triple the cost of the Internet service. Mili Kalia's study on dot.com marriages in India illuminates not just the effects of a material class order, but also a status order within Indian society that is being recreated in emerging "virtual matrimonial" networks. Charles C. Chiemeke, while noting the effects of the digital divide, provides an example of a Bangladesh Village Voice project "that provides loans to women borrowers for phone equipment and subscriptions to cellular services."

In Part XIV, "Reading Machines," Jean-Claude Guédon's anchor chapter outlines the social and technological ecologies around e-readers, focusing on the flow of documents and information across information communication technologies—specifically through the iPad. In the first case study, José Luis González-Quirós draws our attention to the cultural dynamics of the e-book market in Spain. José Antonio Millán critically addresses the future of the e-reader and John W. Warren outlines the distinct affordances of the "digital novel."

Through its excursions into a broad range of digital spaces and social practices, this book aims to identify the specificities of the digital moment—the emergent and unique "properties" of digital technologies in particular, and the novelties and continuities of these technologies in the wider frame of reference of technological development in general. In the analyses of the book we find that solutions to "virtual problems" are windows onto broader social contestations. Untangling the digital requires an untangling of the social, too. In other words, it is not possible to separate meaningfully the digital from the non-digital space.

Digital technologies can "open" new modalities of social, economic and political practice. However, at the same time the social can "close" emergent social transformations through the matrix of social practices that are built on foundations laid in digital networks. So, the argument that historic relations of power around race, gender, class and identity might be in some ways transcended in digital spaces might in fact be a utopian hope, painfully unrealizable as the digital becomes a new space for their reproduction. In this perspective, nothing changes fundamentally with the advent of the digital. In another perspective, though, as we are drawn together in new, digitally mediated social spaces, an inevitable increase in participation could emerge, which in turn might be taken to constitute enhanced "civic" belonging. Furthermore, in such a context, the integration of "civic" and "citizenship" produces a feedback loop of expectations and practices built on consent and participation. This may have flow-on effects into the material world. The new openings for participation in digital spaces, however, suggest that parallel openings are required in the world of social and material resources.

In order for there to be a productive conceptual and practical interaction between the "digital" and the "material," we need to take into account the spectrum of social transformations underway, as reflected in the "properties" of this book. For example arguments around labor practices need to be in conversation with culture, surveillance with participation in open source production, the digital divide with notions of citizenship, and so on. We also need to critically address emergent "properties" of digital technologies, as conduits for social transformation. Then we need to contextualize these "hopes" and "visions" of transformation within a broader historical and technological vision.

This book seeks to propose that we need to think more critically about the properties of new technologies, the cultural content that flows through them, and the social forms new technologies both reproduce and disrupt. This is the concretely social of our "digital future," the effects of technologies-in-use. The trap of technological determinism, at a fundamental level, is born out of misrecognition of the social nature of technological development and should therefore be eschewed. Likewise, technological development cannot be regarded as unidirectional, acultural or apolitical. Rather, the evolution of new technologies provides us with insights into the multifaceted, historically located, transformative dynamics of social agency.

Our aim in this book is to explore the properties of digital technologies in order to discern a range of cultural, social, economic and political trajectories or explanations of effects. In addition, we want to prompt readers to think constructively and strategically about the alternative pathways and transformative possibilities that the properties of technology allow for our social futures. We hope that this collection provides an overview of recent thinking about these matters, and that this in turn might suggest an agenda for new intellectual work and provide inspiration for practical action.

# Part I
# Digital Communication

# 1
# Technology, Innovation, Power, and Social Consequence

*Robin Mansell*

## Introduction[1]

There are many claims and counterclaims in the academic literature on innovations in information and communication technologies (ICTs) about their relationship to power. Different disciplinary perspectives privilege various assumptions about the social consequences that are likely to accompany the innovation process. In this chapter, some of these competing analytical perspectives are considered briefly. This is followed by an assessment of some of the issues that are deserving of deeper investigation. Although some analysts envisage a relatively smooth progression towards equitable access and use of these technologies in ways that, on balance, are empowering for citizens and consumers, others do not. In many instances claims about the nature of this relationship are supported by weak empirical evidence or underpinned by a disavowal of the notion that technologies are political. In this contribution, my aim is to set out the foundation for the assertion that the ground is very flimsy for the claim that innovation in ICTs inevitably favors decentralization, the flattening of hierarchy, or the automatic empowerment of human beings.

## Analytical perspectives in contention

ICTs always have been entwined with changes in society, an association that has been examined in detail by historians.[2] In some cases these technologies have been characterized as being revolutionary because of the new opportunities they offer for mediated relationships. Digital technologies are often depicted in the consultancy literature as calling into existence a new inclusive, social, and economic order (Le Guyader 2009), but a more measured response typically indicates that these technologies become woven into society in very complex ways. We can regard ICTs as being neither necessarily transformational nor completely malleable in the hands of their users (Mansell et al. 2007; Mansell and Silverstone 1996b). We have

13

several decades of scholarly research that consistently confirms this observation.[3]

Nevertheless, even today, in the academy, the most predominant analytical approach in the social sciences to innovations in ICTs remains diffusion theory, a theory that focuses principally on technology and individual behavior. In *The Diffusion of Innovations*, Rogers (Rogers 1962, 1995) aimed to explain how to inculcate awareness and enthusiasm for technical innovations such that even those individuals most resistant to their adoption might do so. By 1995, when the fourth edition of his book was published, he had modified his theory to account for many of the contextual factors that influence the diffusion of new technologies. Even so, the central concern in this theoretical model remained explaining the rate and direction of adoption of new technologies such as ICTs.[4] Research in the diffusion of innovations tradition has continued to develop and inform new generations of scholarly work on the patterning and scaling of networking—both technical and social (Monge and Contractor 2003)—as well as in the burgeoning fields of advertizing and marketing that are linked to behavioral theories of consumer decision making (Egan 2008; Van den Bulte and Joshi 2007). Most studies in this tradition presume that consumption is a matter of individual choice and that all such consumption is desirable. The social implications of consumption, the ethical issues raised by privacy intrusions and the personalization of ICT-based services, and the sustainability of intensely networked and mediated environments supported by ICTs are simply not considered.

In line with the predominance of the diffusionist tradition, there are substantial efforts to measure the ICT diffusion process. This work started in the US with the early contributions of Machlup (1962) and Porat and Rubin (1977), but it has since become a global ambition to benchmark progress toward the information society, knowledge economy, or whatever label is used to capture the phenomenon whereby information (or communication) services make a growing contribution to economic activity and information-related occupations as compared to industrial output or agriculture (Bell 1973; Ito 1991). Work in this tradition proceeds through the development of indicators and surveys that enable comparisons of a country's or region's investment performance and use of ICTs (OECD 2005; UNCTAD 2009), the assumption being that 'more' is always better.

The diffusion of innovations tradition is complemented by economic analysis of characteristics of information. Economic analysis has been brought to bear on the market exchange of information (Brousseau and Curien 2007). As Brousseau and Curien point out, "while ICTs seem to provide the technical support which should favor the efficient performance of an ultracompetitive market economy, they make information a public good, thus sowing the seeds of a cooperative economy" (Brousseau and Curien 2007: 21). This means that while some analysts are very enthusiastic about the growth and potential profitability of markets for information, others stress that traditional intellectual property rights protection should not be used to hinder the growth of an open, sharing, more cooperative environment that fosters artistic, educational, and scientific endeavor.[5] They argue that the market exchange of information needs to be complemented by analysis of the benefits and costs of information exchange, less encumbered by the costs of negotiating property rights (Benkler and Nissenbaum 2006; David 1995, 2005; Lessig 2001; Mansell and Steinmueller 2000). While attention has been directed to the consequences of persistent economic power in media, information, and communication markets (McChesney and Schiller 2003; Mosco 1996; Schiller 1999), there has been a substantial amount of enthusiasm about the Internet and new opportunities for self-publishing using social networking sites as a means of consumer and citizen empowerment.

If we turn away from these two dominant traditions, towards the work of more critical scholars who are informed by theoretical traditions that are more open to uncertainty and to the messiness of the everyday lives of the producers and users of ICTs, we find many studies of the relationships between ICT networks, information flows, and time–space reconfiguration. This work has proliferated since the mid-1990s (Castells 1996, 1997, 1998, 2001, 2009; Slevin 2000; van Dijk 2006). Undertaken from many different perspectives, there is considerable agreement that it is in the analysis of the interplay between online and offline activities that we are likely to gain purchase on the social, political, and cultural consequences of innovations in ICTs (Orgad 2007). While many new processes and practices are enabled by the spread of ICTs, warranting in some instances the label "revolutionary," there are also continuities with earlier developments. And indeed, we have skeptical accounts that are dubious about whether societies are being radically altered by the spread of ICTs (Garnham 2000; Webster 2006).

Studies in the diffusion of innovations tradition neglect the fact that ICTs are implicated in complex power relationships. But in those traditions that do consider power relations there are bifurcations between macro- and micro-analytical approaches. As Mattelart suggests, in situated accounts emphasizing mediations and interactions there is a tendency to overlook aspects of technology production and a "technoscientific system remains, more than ever, marked by the inequality of exchanges" (Mattelart 1996/2000: 109). Alternatively, studies that privilege analysis of political and economic power sometimes neglect the agency of individuals.

Research in the physical sciences, computer science, and engineering is mainly devoted to promoting innovations in ICTs (Mansell and Collins 2005) and studies of ubiquitous computing, software automation, the Semantic Web, and or "knowledge management" receive substantial financial support as compared to studies informed by the social science disciplines, but the former give little attention to the uncertainties of the innovation process—ethical, social, economic, or political. Some of those concerned to promote more rapid diffusion of ICTs—whether the Internet or the mobile phone—turn to philosophy to defend their assertion that a wholly new way of thinking is called for in order to understand the consequences of ICTs. However, since it is the interpenetration of the old and the new technologies, and the older and newer practices and meanings that matter for the social order, it seems essential to ensure that the place and consequences of these technologies are considered through an analytical lens that deploys concepts of power and the political to make sense of the transformations that are underway. Thus, we need to apply some of our existing ways of thinking in the emerging context of mediated networks.

## ICTs, power and inequality

Applying some of our existing ways of thinking in the emerging context of mediated networks to make sense of ongoing transformations is especially important because the production and consumption of ICTs are marked by inequalities, frequently, though not always, reflecting the inequalities of societies. As a result the technology diffusion process itself needs to be understood in the context of how these technologies enable changing social practices, offer new methods of communication and information sharing, encourage network forms of organization, and give rise to new learning dynamics and commercial practices. We need a basis on which to assess the desirability of encouraging innovations of certain kinds but perhaps not of others because, while feasible, they may be deemed unethical or simply unhelpful. Judgment—individual and collective—is necessary. Perhaps one of the greatest requirements is for a more broadly based acknowledgment of the need for new literacies and communicative resources for expressing cultural identity, fostering new kinds of "community" and mediating experience (Livingstone 2007, 2008; Livingstone et al. 2008).

In essence, we require deeper insights into the embeddedness of ICTs in different contexts to understand how mediation processes are influenced by them, and we need a consideration of their social consequences, but we also require an acknowledgment of power and the political in all mediated relationships. One approach that has been helpful over the past 30 years in this respect is the analysis of the dynamics of "techno-economic" systems. This work, particularly through its focus on the qualitative aspects of the diffusion of ICTs, has shown that changes in technological, organizational, social, cultural, and political systems occur unevenly in time and space. Each system or subsystem of the innovation process encompasses inequality and power dynamics that influence the distribution of resources and the new spaces of opportunity and insurgency available to any given actor. However, helpful as this tradition of research is in drawing attention to the dynamics of such systems, most of those contributing to it do not acknowledge the need for an explicitly theorized account of power. There have been attempts to do so, of course, but the core of this tradition remains tightly focused on the technologies and the economic impacts of their diffusion.

In *Communication by Design: The Politics of Information and Communication Technologies*, we sought to link the "techno-economic" tradition to studies of the social processes, power relations, and agency of technology producers and consumers. As a result, we were able to make observations about ICT innovations such as the following. These technologies "raise profound concerns about the way advanced information and communication technologies influence industry, government policy and our every day lives ... and about their capacity to contribute to a more equitable and democratic society" (Mansell and Silverstone 1996a: 1). We asserted that an understanding of the complexity of the innovation process was not approachable through traditions typified by the dominant diffusion of innovations approach. This complex cluster of technological and social interdependencies required then, as it does now, a multidisciplinary perspective (Mansell and Silverstone 1996b). However, that in itself is insufficient unless power relations are also examined.

We suggested that the ICT innovation process should be treated as a "dialectic in which power is exercised in the production and use of technological artefacts as well as in the institutionalization of behavior" (Mansell and Silverstone 1996a: 7) and that this entails a methodological challenge to undertake work at the individual, household, firm, or organizational levels. What cries out for investigation is the politics of the innovation process itself. A process that involves "a politics engaged in by participants—individuals as well as institutions ... It is, finally, a politics deeply embedded not just within the institutions that design and distribute technologies and services, but within the technology itself ... containing and constraining behavior, and embodying ... both the normative and the seductive" (Silverstone and Mansell 1996: 213). And crucially, "the structuring of our symbolic environments which these technologies uniquely undertake is an activity which still has to be managed: accepted, rejected, transcended, or transformed" (Silverstone and Mansell 1996: 219). The politics of the management of the technological innovation process were what was at stake and, arguably, continue to be so today.

When we consider the strategies of various actors and the social consequences of each new generation of technology, we need to bear in mind that

these [developments] occur locally, but they are interpenetrated by manifestations of the global in complex ways involving power relations which, at least potentially, enable new opportunities for learning and diversity ... Whether these opportunities make a profound difference in people's lives and whether they are understood as helpful are questions that the scholarly community must continue to assess.

(Mansell 2009)

## Practices and strategies in everyday life

Essential to any perspective on ICT innovation that takes power and the political into account is an acknowledgment of the significance of the practices and tactics of "everyday life." Following Certeau, investigations of innovations in ICTs need to examine the new "ways of operating." "They need to show how these constitute the innumerable practices by means of which users reappropriate the space organized by techniques of sociocultural production ... to bring to light the clandestine forms taken by the dispersed, tactical, and make-shift creativity of groups or individuals already caught in the nets of 'discipline'" (Certeau 1984: xiv). In other words, they need to seek out the spaces of indeterminacy and to allow for potentially empowering developments instead of assuming either that these are inevitably present or that they are always absent.

In addition, following Silverstone, studies of ICT diffusion need to contend with the fact that "the world of globally mediated communication offers and to a degree defines the terms of our participation with the other" (Silverstone 2007: 27) and that "access to, and participation in, a global system of mediated communication is a substantive good and a precondition for full membership of society, and that the distribution of such a right must be fair and just" (Silverstone 2007: 147). Taking these considerations into account in assessments of ICT, and acknowledging that we require analytical tools addressed to the meso- and macro-levels of investigation as well as the individual agent, it is essential to conceptualize and analyze the strategic interests of various groups. This can be achieved in many different ways, one of which is to consider how the interests of institutional actors are being articulated through time. We (Mansell and Steinmueller 2000) developed, for example, an analytical framework focusing on the changing relations of incumbency, insurgency, and virtual community status in the context of ICT innovation.

Absent from that analytical framework was a consideration of whether an intensely networked world yields some form of "collective intelligence" (Sunstein 2006; Surowiecki 2004; Tovey 2008) and a new potential for the expression of individual creativity, yielding the kinds of empowerment envisaged by many forecasters. There is no doubt that big changes are underway as the information society extends to support multiple identities, where individuals engage in switching identities, undertaking, using, and dropping roles at various times. These practices are, arguably, as old as human civilization, but the new networked practices of everyday life are creating a changed social environment as a result of the scale and scope of today's global networks (Rab 2006). As a result, there are many signs of a blurring of boundaries between formerly distinct realms of social activity. These changes are visible in the realms of commerce, entertainment, and learning; in the personalization and professionalization of services; in the conflation of private and public life; and in the growing emphasis on the feasibility of using electronic services to support civic participation and democracy.

## Renewal of contestation

There is contestation over the meaning and social significance of these developments and little sign of consensus over whether networked individualism or collective action is the predominant trend, much less as to whether the latter should be privileged over the former. There is even less agreement about whether access to increasingly vast resources of information is consistent with the availability of reliable or useful information, and, indeed, whether the fact that many individuals can become active communicators means that citizens can also be co-decision makers about the parameters within which they live their lives. Still to be investigated more deeply is whether it is important to distinguish between "real" and "unreal" intimacy; whether intensive networking favors heightened sociability or enhances anti-social behaviors; whether the new forms of mediation favor social isolation or inclusion; and whether "always on" networking yields a work–life balance that is empowering or increasingly stressful. The mediated matrices of everyday life are raising issues about trust, privacy, identity management, and safety. At the individual level there are divergent views, from the complacent to the defeatist, in these areas, as suggested by this quotation:

> Me, myself and I: manage online identity more safely. A scrap of information here, a little detail there … the web is safe if you guard what personal and financial information you provide. Or is it? Identity theft and cyber-spying are on the rise, and keeping track of what you reveal is nigh on impossible.
>
> (Cordis 2010: n.p.)

From spammers and phishing to hoaxes, hackers, spyware, adware, and Trojan Horse viruses, we are finding that surfing the web is becoming synonymous with new forms of surveillance and that security and actual and perceived risk are becoming intertwined in new ways. Surveillance brings new threats to civil liberties even though it is widely accepted by some citizens. And despite the fact that the growing use of databases on children, migrants, and "pre-criminals" is contested, it is growing nonetheless in the UK and elsewhere in Europe (Anderson et al. 2007). Improvements in data management are occurring relatively slowly compared with the proliferation of databases containing personal information. Although the political considerations and civil rights issues associated with the use of "social sorting software" are becoming better documented (Braman 2010; Gandy Jr. 2009;

Lyon 2007), the increasingly pervasive use of these techniques is reinforcing social inequalities. This is inconsistent with the simple claims about the relationship between networking and citizen empowerment.

These developments suggest the importance of digital literacies, broadly understood to embrace cognizance of the uses to which networking is being put and a capacity to make judgments about whether or not these practices are justified or should be resisted through individual and collective means. Also needed is a better understanding of the respective roles of family and peers in inculcating these literacies, what educational practices should be encouraged, and whether literacies should be seen as political issues or left to the resolution of the market. There is, of course, also the issue that even if citizens can speak about these kinds of issues in a knowledgeable way, it is not at all clear where they will find the deliberative forums in which their voices might be heard (Dahlberg 2001; Dahlgren 2005; Lievrouw 2010).

## Conclusion

There are many social dilemmas confronting all those who conduct their lives in mediated spaces alongside their material lives. We do not know whether, for example, the predominant trends are towards social contacts widening or narrowing, enhanced democratization or the concentration of power, enhanced individuality and personality multiples, a new form of mass society in which "mass self-communication" comes to predominate (Castells 2009, 2007), or greater user creativity and empowerment or the next generation of the digital divide (Hargittai 2002; Jung 2008; Zhao and Elesh 2006).

Technologically mediated societies have been with us for centuries, but the most recent changes are associated with codifying and manipulating information. In this context, we know that 'assembling the "tools" is only part of the task ... Measures must be taken to assemble the human capabilities and related technologies to make the best use of the new opportunities offered by ICTs' (Mansell and Wehn 1998: 261), yet it appears that there is resistance to learning this lesson, just as there was historically. This is because decisions in this area are highly political and necessarily judgmental. What criteria, for instance, should guide management of our non-place-based identities? What human rights and responsibilities should be associated with the emerging networked order? Once we acknowledge the political, we confront the fact that all choices made concerning the mediated environment of the twenty-first century are political as well.

Yet the headline digital narrative continues to suggest that ICTs mean the end of hierarchy, the ascendance of an open commons for information, and the decline of barriers to information sharing. If this is so, then economic and political power may shift irreversibly to individual information producers and citizens. Acquiring the new literacies may be relatively less problematic for each generational cohort (Williamson et al. 2010), and the active mass media audience may be in the process of becoming the new media content or information producer. There is no doubt that there is a proliferation of blogs, SMSs, email lists, and decentralized networks. As more citizens become information producers there is a tendency to assume that the standard diffusion model will run its course so that all citizens eventually will be embraced by an inclusive information age, and included in ways that are to their advantage.

These universalistic claims appear to take little or no account of the diversity of societies. If diversity is acknowledged and valued then it follows that it is unlikely that there will be a universal information society. Rather there are diverse mediated environments just as there were diverse instantiations of the industrial age (Mansell 2009, 2002). It is a fallacy to think in terms of a "catch-up" process where digital divides are closed, yielding global and local harmony within the information society. Therefore, we need to examine specifically who is being included, on what terms, and, crucially, who is not and what means we have to alter this condition.

A close investigation of whose narratives about ICTs and information societies or knowledge societies are being validated is required. We need investigations of who is participating in decisions about ICT design and deployment with a view to understanding their values so as to reveal the contested political foundations of these developments. Considering these issues in historical and political perspective is very likely to show that, whatever the shifts in power in society accompanying the diffusion of ICTs are, they are partial and temporary. They do, however, provide us with focal points for a material investigation of the contested practices and values that are embedded within our new mediated forms and structures of control.

## Notes

1. Some parts of this chapter were presented in a plenary lecture for SwissGIS 10 Years After—The Future of the Global Information Society, Zurich, November 6, 2009; and in a presentation to the Swedish Presidency High Level Conference "Visby Agenda: Creating Impact for an eUnion," November 9–10, 2009, Visby Sweden, and in Mansell et al. (2007).
2. For historical studies, see Braudel (1984), Castells (1996), Freeman and Soete (1997), Innis (1950, 1951), Marvin (1988), and Mattelart (1996/2000).
3. My purpose is not to set out that body of work in this chapter. The interested reader might start with Castells (2009).
4. For research in this tradition see, for example, Attewell (1992), Brancheau and Wetherbe (1990), Carter et al. (2001), Chin and Marcolin (2001), Deroian (2002), Fichman and Kemerer (1999), Lyytinen and Damsgaard (2001), and Stoneman (2002).
5. See, for example, Lamberton (1971, 2006) on the variety of roles that information plays in the economy, Noam (2001) on the institutional rules governing the development of new markets, and Quah (2003) on the potential of ICTs for creating digital goods such as digital music and novel software algorithms.

# Bibliography

Anderson, R., I. Brown, R. Clayton, T. Dowty, D. Korff, and E. Munro. 2007. "Children's Databases—Safety and Privacy: A Report for the Information Commissioner." Wilmslow, UK: Information Commissioners Office.

Attewell, P. 1992. "Technology Diffusion and Organizational Learning: The Case for Business Computing." *Organization Studies* 3 (1): 1–19.

Bell, Daniel. 1973. *The Coming of Post-Industrial Society: A Venture in Social Forecasting.* New York: Basic Books.

Braman, S. 2010. "Anti-Terrorism and the Harmonization of Media and Communication Policy." In *Handbook on Global Media and Communications Policy*, eds. R. Mansell and M. Raboy. New York: Wiley-Blackwell.

Brancheau, J. C. and J. C. Wetherbe. 1990. "The Adoption of Spreadsheet Software: Testing Innovation Diffusion Theory in the Context of End-User Computing." *Information Systems Research* 1 (2): 115–43.

Braudel, F. 1984. *Civilization and Capitalism 15–18 Century: Vol. III—The Perspective of the World.* London: Collins.

Brousseau, E. and N. Curien (eds.) 2007. *Internet and Digital Economics: Principles, Methods and Applications.* Cambridge: Cambridge University Press.

Carter, F. J. T., V. Jambulingam, K. Gupta, and N. Melone. 2001. "Technological Innovations: A Framework for Communicating Diffusion Effects." *Information & Management* 38 (5): 277–87.

Castells, Manuel. 1996. *The Information Age: Economy, Society and Culture Volume I: The Rise of the Network Society.* Oxford: Blackwell.

Castells, Manuel. 1997. *The Information Age: Economy, Society and Culture Volume II: The Power of Identity.* Oxford: Blackwell.

Castells, Manuel. 1998. *The Information Age: Economy, Society and Culture Volume III: End of Millennium.* Oxford: Blackwell.

Castells, Manuel. 2001. *The Internet Galaxy: Reflections on the Internet, Business and Society.* Oxford: Oxford University Press.

Castells, Manuel. 2007. "Communication, Power and Counter-Power in the Network Society." *International Journal of Communication* 1: 238–66.

Castells, Manuel. 2009. *Communication Power.* Oxford: Oxford University Press.

Certeau, Michel de. 1984. *The Practice of Everyday Life.* Berkeley CA: University of California Press.

Chin, W. W. and B. L. Marcolin. 2001. "The Future of Diffusion Research." *Data Base Advances in Information Systems* 32 (3): 8–12.

Cordis. 2010. *My, Myself and I: Manage Online Identity More Safely.* CORDIS, February 12, 2008 [cited February 20, 2010]. Available from http://cordis.europa.eu/ictresults/index.cfm?section=news&tpl=article&BrowsingType=Features&ID=89504, accessed October 20, 2010.

Dahlberg, Lincoln. 2001. "Democracy via Cyberspace: Mapping the Rhetorics and Practices of Three Prominent Camps." *New Media & Society* 3 (2): 157–77.

Dahlgren, Peter. 2005. "The Internet, Public Spheres, and Political Communication: Dispersion and Deliberation." *Political Communication* 22 (2): 147–62.

Deroian, F. 2002. "Formation of Social Networks and Diffusion of Innovations." *Research Policy* 31 (5): 835–45.

Egan, J. 2008. *Relationship Marketing: Exploring Relational Strategies in Marketing—Third Edition.* New York: Prentice Hall.

Fichman, R. G. and C. F. Kemerer. 1999. "The Illusory Diffusion of Innovation: An Examination of Assimilation Gaps." *Information Systems Research* 10 (3): 255–75.

Freeman, C. and L. Soete. 1997. *The Economics of Industrial Innovation, Third Edition*. London: Pinter A Cassel Imprint.

Gandy Jr., O. H. 2009. *Coming to Terms with Chance: Engaging Rational Discrimination and Cumulative Disadvantage*. Burlington, VT: Ashgate.

Garnham, N. 2000. *Emancipation, the Media and Modernity: Arguments about the Media and Social Theory*. Oxford: Oxford University Press.

Hargittai, E. 2002. "Second-Level Digital Divide: Differences in People's Online Skills." *First Monday* 7 (4): 267–83.

Innis, H. A. 1950. *Empire and Communication*. Toronto: Oxford University Press.

Innis, H. A. 1951. *The Bias of Communication*. Toronto: University of Toronto Press.

Ito, Youichi. 1991. "'Johoka' as a Driving Force of Social Change." *Keio Communication Review* 12: 35–58.

Jung, J. 2008. "Internet Connectedness and Its Social Origins: An Ecological Approach to Postaccess Digital Divides." *Communication Studies* 59 (4): 322–39.

Lamberton, D. M. 2006. "New Media and the Economics of Information." In *The Handbook of New Media*, eds. L. A. Lievrouw and S. Livingstone. London: Sage.

Lamberton, D. M (ed.) 1971. *The Economics of Information and Knowledge: Selected Readings*. Harmondsworth: Penguin.

Le Guyader, H. 2009. "eInclusion Public Policies in Europe." Brussels: European Commission.

Lievrouw, L. A. 2010. *Understanding Alternative and Activist New Media*. Cambridge: Polity Press.

Lyon, D. 2007. "National ID Cards: Crime-Control, Citizenship and Social Sorting." *Policing* 1 (1): 111–8.

Lyytinen, K. and J. Damsgaard. 2001. "What's Wrong with the Diffusion of Innovation Theory? The Case of a Complex and Networked Technology." In *IFIP 8.6 Working Conference*. Banff, Canada.

Machlup, F. B. 1962. *The Production and Distribution of Knowledge in the US Economy*. Princeton NJ: Princeton University Press.

Mansell, R. (ed.) 2002. *Inside the Communication Revolution—New Patterns of Social and Technical Interaction*. Oxford: Oxford University Press.

Mansell, R. (ed.) 2009. *The Information Society—Critical Perspectives in Sociology Volume 1*. London: Routledge.

Mansell, R., C. Avgerou, D. Quah, and R. Silverstone (eds.) 2007. *The Oxford Handbook of Information and Communication Technologies*. Oxford: Oxford University Press.

Mansell, R. and B. S. Collins (eds.) 2005. *Trust and Crime in Information Societies*. Cheltenham: Edward Elgar.

Mansell, R. and R. Silverstone. 1996a. "Introduction." In *Communication by Design: The Politics of Information and Communication Technologies*, eds. R. Mansell and R. Silverstone. Oxford: Oxford University Press.

Mansell, R. and R. Silverstone (eds.) 1996b. *Communication by Design: The Politics of Information and Communication Technologies*. Oxford: Oxford University Press.

Mansell, R. and W. E. Steinmueller. 2000. *Mobilizing the Information Society: Strategies for Growth and Opportunity*. London: Oxford University Press.

Mansell, R. and U. Wehn (eds.) 1998. *Knowledge Societies: Information Technology for Sustainable Development*. Oxford: Published for the United Nations Commission on Science and Technology for Development by Oxford University Press.

Marvin, C. 1988. *When Old Technologies were New: Thinking about Electric Communication in the Late Nineteenth Century*. Oxford: Oxford University Press.

Mattelart, A. 1996/2000. *Networking the World: 1794–2000*, trans. by J. A. Cohen. Minneapolis, MN: University of Minnesota Press.

McChesney, R. and D. Schiller. 2003. "The Political Economy of International Communications: Foundations for the Emerging Global Debate about Media Ownership and Regulation." Geneva: UNRISD Working Paper.

Monge, P. R. and N. S. Contractor. 2003. *Theories of Communication Networks*. Oxford: Oxford University Press.

Mosco, V. 1996. *The Political Economy of Communication*. London: Sage Publications.

Noam, E. 2001. "Two Cheers for the Commodification of Information." New York: Columbia University.

OECD. 2005. "Guide to Measuring the Information Society, Working Party on Indicators for the Information Society." Paris: OECD DSTI/ICCP/IIS/2005/6/final.

Orgad, S. 2007. "The Interrelations Between Online and Offline: Questions, Issues, and Implications." In *The Oxford Handbook of Information and Communication Technologies*, eds. R. Mansell, C. Avgerou, D. Quah, and R. Silverstone. Oxford: Oxford University Press.

Porat, M. U. and M. R. Rubin. 1977. *The Information Economy, Nine Volumes*. 9 vols. Washington, DC: Department of Commerce Government Printing Office.

Quah, D. 2003. "Digital Goods and the New Economy." In *New Economy Handbook*, ed. D. C. Jones. London: Academic Press Elsevier Science.

Rab, A. 2006. "Real Life in Virtual Worlds: Anthropological Analysis of MMO Games." In *Identity in a Networked World, FIDIS Future of Identity in the Information Society*, eds. D.-O. Jaquet-Chiffelle, E. Benoist, and B. Anrig. Berne: Berne University of Applied Sciences and European Commission.

Rogers, Everett M. 1962. *The Diffusion of Innovations*. New York: Free Press.

Rogers, Everett M. 1995. *The Diffusion of Innovations* (fourth edition). New York: The Free Press.

Schiller, D. 1999. *Digital Capitalism: Networking the Global Market System*. Cambridge MA: MIT Press.

Silverstone, R. 2007. *Media and Morality: On the Rise of the Mediapolis*. Oxford: Polity Press.

Silverstone, R. and R. Mansell. 1996. "The Politics of Information and Communication Technologies." In *Communication by Design: The Politics of Information and Communication Technologies*, eds. R. Mansell and R. Silverstone. Oxford: Oxford University Press.

Slevin, J. 2000. *The Internet and Society*. Cambridge: Polity Press.

Stoneman, P. 2002. *The Economics of Technological Diffusion*. Oxford: Blackwell.

Sunstein, C. R. 2006. *Infotopia: How Many Minds Produce Knowledge*. New York: Oxford University Press.

Surowiecki, J. 2004. *The Wisdom of Crowds: Why the Many are Smarter than the Few and How Collective Wisdom Shapes Business, Economies, Societies and Nations*. New York: Doubleday.

Tovey, M. (ed.) 2008. *Collective Intelligence: Creating a Prosperous World at Peace*. New York: Collective Intelligence Network.

UNCTAD. 2009. "The Information Economy Report 2009: Trends and Outlook in Turbulent Times." Geneva: United Nations Conference on Trade and Development.

Van den Bulte, C. and Y. V Joshi. 2007. "New Product Diffusion with Influentials and Imitators." *Marketing Science* 26 (3): 400–21.

van Dijk, J. A. G. M. 2006. *The Network Society: Social Aspects of New Media Second Edition*. London Sage.

Webster, Frank. 2006. *Theories of the Information Society. Third Edition.* London: Routledge.

Williamson, B., D. Lewin, P. Marks, and B. Glennon. 2010. "Demand-Side Measures to Stimulate Internet and Broadband Take-Up: A Report to Vodafone." London: Plum Consulting.

Zhao, S. and D. Elesh. 2006. "The Second Digital Divide: Unequal Access to Social Capital in the Online World." In *Annual Meeting of the American Sociological Association.* Montreal.

# 1.1
# The World Summit Awards Benchmarking Arabic Websites: A Case Study of Governance

*Ilhem Allagui*

The first Arab country connected to the Internet was Tunisia in 1991. Internet usage in other Arab countries did not take off until the mid-1990s, and in general the development of the Internet in the Arab region was slow and timid. The major impediments to Internet adoption among Arab countries have been the lack of infrastructure—primarily because of poverty, high illiteracy rates, the use of non-Latin script leading to the poor quality of Arabic content on the Internet, and the political and religious threats posed by the Internet. Among the strategies and initiatives to promote the development of e-content, the organizing body of the World Summit on the Information Society (WSIS) suggested identifying success stories, stimulating competition and rewarding the best content to be set as benchmarks for the industry. This was done under the framework of the World Summit Awards (WSA),[1] a global competition that aims to identify the best e-content practices and strategies considered as a benchmark for the region. Although it is based on the concept of "governance," this case study reviews the rewarded Arabic websites and questions the outcome of the governance practice in this context.

Governance is a practice (Chhotray and Stoker 2009) that deals with "establishing prescriptions and proscriptions that steer or guide" (Drake and Wilson 2008: 8). In the context of this study governance is described as encompassing activities that promote e-content, as well as identifying what kind of Internet content needs to be advanced in order to achieve a better information society. Inspired by the WSA competition and results, this case study questions the quality of the outcome under the current governance arrangement (Chhotray and Stoker 2009). A content analysis of the ten websites rewarded in this WSA competition[2] shows that they primarily use text and graphics, with less than 10 percent using animation and/or audio-video content. These websites are primarily made up of Pan Arab content, since more than 66 percent of them deal with Pan Arab issues. Another 20 percent deal with local or national content specific to where the website originates from and only 9 percent generate content that has international interest and/or deals with international issues. Four out of ten of these websites are state owned by public or government institutions, and six are privately owned. Furthermore, 30 percent of the websites are cultural, 26 percent are religious and the others deal with social, women's, political or economic issues. Additionally, these websites show a huge deficiency in web properties. First, they lack hypertextuality; for instance they do not offer any hyperlinks that serve as pathways to other websites. When users arrive at these sites, they are "virtual prisoners," as only a small proportion (15 percent) of the 642 coded elements open up to cyberspace. Second, the interactivity of website-to-user or user-to-user is also absent; 90 percent of the elements coded offer no interactivity at all. This is a passive website model, made for users who can only read the content. These findings display a similarity to Arab media, mainly before the growth of satellite television market,[3] most of which are under the control of the state, and broadcast Arab cultural and religious content to the detriment of serious social, community or political issues (UNDR 2003; Roy 2002). This duplication of Arab media systems and the fact that the selected sites are deprived of the core Internet features of interactivity and hypertextuality (Burnett and Marshall 2003) make one wonder why the WSA considered them.

When examining the organization of the competition (Allagui 2007), one can note the involvement of different layers in the selection process and, most importantly, the selective role of the jury at the national or country level. It is striking to see how conventional, unobtrusive and non-innovative most of these Arab websites are. Knowing how the media in Arab countries are also restrained, and the political restrictions applied to media in Arab countries (Hafez 2008), it is no surprise that the jury members had to work within contextual, political and cultural restrictions. Many other innovative websites exist but do not qualify to enter the competition because they are restricted in access, censored or even self-censored. This implies the existence of a greater power of interference of the state, either directly or indirectly, to control the outcome. This was noted by WSIS critics at different levels, for example Gogging (2009: 55), who stated: "I certainly agree with critics that the key models that guided WSIS embodied political positions and significant limitations that proved fatal." For the WSA, this direct or indirect state interference has cast a shadow on governance, whose essence in this context is to liberate creativity, and help to provide a benchmark for a better and richer Arab web content, but has failed to do so.

## Notes

1. The WSA was organized during and within the context of the World Summit on Information Society (Tunis, Nov. 2005), organized by the United Nations. It aimed to discuss the Internet, its policies and regulations in order to achieve a better information society.
2. For a full discussion of methodology and results, see Allagui (2007).

3. The growth of Arab satellite channels introduced some dynamism and changes to the Arab media market on the political level, with Arab governments adopting TV-friendly modes of conduct (Sakr 2007).

## Bibliography

Allagui, I. 2007. "The Awarded Arab Cyberspace: Reproducing or breaking off with Arab media?" *The International Journal of Technology, Knowledge and Society* 3 (6).

Amin, H. 2008. "Arab Media Audience Research." In *Arab Media: Power and Weakness*, ed. K. Hafez. New York: Continuum Publishing.

Burnett, R. and D. Marshall. 2003. *Web Theory: An Introduction*. London: Routledge.

Chhotray, V. and G. Stoker. 2009. *Governance Theory and Practice: A Cross-Disciplinary Approach*. Basingstoke UK: Palgrave Macmillan.

Drake, W. 2008. "Introduction: The Distributed Architecture of Network Global Governance." In *Governing Global Electronic Networks: International Perspectives on Policy and Power*, eds. W. Drake and E. Wilson III. Cambridge: The MIT Press.

Gogging, G. 2009. "The International Turn in Internet Governance: A World of Difference?" In *Internationalizing Internet Studies: Beyond Anglophone Paradigms*, eds. G. Gogging and M. McLelland. New York: Routledge.

Hafez, K. 2008. "The Unknown Desire for "Objectivity: Journalism Ethics in Arab and Western Journalism." In *Arab Media: Power and Weakness*, ed. K. Hafez. New York: Continuum Publishing.

Roy, O. 2002. "L'islam mondialisé." Paris: Éditions du Seuil.

Sakr, N. 2007. "Approaches to Exploring Media-Politics Connections in the Arab World." In *Arab Media and Political Renewal: Community, Legitimacy and Public Life*, ed. N. Sakr. London: I.B. Tauris.

UNDR. 2003. "Arab Human Development Report: Mass Media, Press Freedom and Publishing." Retrieved February 4, 2004. http://www.undp.org/rbas/ahdr.

# 1.2
## Test Driving E-Participation: The Case of Gipuzkoa (Basque Country)

*Alfonso Unceta*

The idea of e-participation is emerging as a possibility for redefining the field of politics. We must look to civil society as a construction zone for participatory political citizenship; in so doing we will perceive new places and forms of expression that are reorienting the relationship between politics and society (Alexander 1998, Warren 2001, Magnette 2003). We can foresee new forms of interaction between the representatives and the represented and yet, at the present time, we scarcely have even partial knowledge of these innovative scenarios.

Representation based on a more direct link between the representatives and the represented has been hampered by the increasing role of political parties, which have taken over the most prominent place in politics. Undoubtedly, this has not only caused a clash of priorities but also altered the way in which the relationship comes about between the representatives and the represented. Furthermore, it is becoming increasingly mediated, uncertain, vague, and distant.

In this context, the application of information and communication technologies (ICTs) is an important element in participatory processes, as these technologies capitalize on many electronic and digital platforms: e-mail, mailing lists, websites, wikis and blogs, discussion forums, and newsletters. Most are user friendly and require no special means of operation (Norris 2001).

It is true that electronic participation is still of very limited scope and far from being a commonplace tool for political participation (Davis 1999). Even so, population centers that are not excessively large are suitable for test-driving participation and e-democracy (Tsagarousianou et al. 1998).

Given all of the above, making political involvement a reality will require getting beyond the talking stage and exercising focused leadership that favours participatory processes. The provincial government of Gipuzkoa[1] is a case in point: it implemented a participatory process to develop a Standard for Citizen Participation, among other initiatives. Besides establishing four working groups (Experts, Political Parties, Citizen Associations, and City

Councils), it rounded out the project by setting up a blog that was active during two of the five months the process lasted (http://blog.gipuzkoaparte hartzen.net/lang/es/).

In a comparison of the five forums, how they evolved and what they contributed to the process, the following should be highlighted:

1. The four working groups conducted 16 meetings with 36 different people in attendance, resulting in a total of 23 hours and 50 minutes of work. These meetings generated 452 feedback comments and resulted in 19 work documents.
2. The blog gipuzkoapartehartzen.net received 1532 visits, with an average of 4 minutes and 31 seconds per visit, and generated 57 comments. Of these visits, 96 percent originated in Spain; of these, 92 percent were direct traffic and 8 percent came through search engines.

In this process, the blog played a role that was more informative than participatory, as it brought a political issue to the public's attention, but it did not become the best tool for participation. In the environment we were analysing, at least, the culture of e-participation was not yet sufficiently widespread.

The study presented here shows that, in this type of process where both initiative and interest are institutional, in-person participation is still more effective than e-participation in the Basque Country. This may be because traditional forms of interaction such as face-to-face meetings, telephone calls, and even e-mails are still far more prevalent in the culture than new forms of communication—blogs, wikis, forums—among citizens born before the digital era. This point is exemplified by Wikipedia, where, although 700 articles are added daily, half of the editing work for the English version is completed by 0.7 percent of the users; for the Spanish version, 8.1 percent of the users do 90 percent of the editing (Sunstein 2006: 152).

The new Web communications tools function when communities of practice exist around them—users who are far apart, geographically, but share a common interest. In these cases, when a user publishes an opinion in a forum or a blog, that user expects immediate responses from the rest of the users. Although they participate through pseudonyms and aliases, users know each other "virtually," and there is a sense of group that is lacking in relationships with various local and national governments. Thus interactions with the government are completely impersonal; perhaps this is why users become skeptical and assume that their opinions will not be heard, that they will fall on deaf ears.

Applying e-participation to political processes requires a change in individual and group behaviour and is still today just a possibility, an idea that has not yet taken hold throughout society. Therefore, e-participation and e-democracy will not be so unidirectional in the future and will provide increased and improved interactivity through innovative practices. In the absence of a common interest, governments must find an element that political representatives and the represented have in common to bolster the relationship between them, especially in the Basque Country where, just as in the rest of Spain, e-participation indicators are still quite low in comparison with northern European Union countries.

## Note

1. Gipuzkoa is a territory located in the Basque Country with a population of 700,000. It has its own government and parliament, which have a wide range of areas of responsibility.

## Bibliography

Alexander, J. (ed.) 1998. *Real Civil Societies*. London: Sage.

Davis, R. 1999. *The Web of Politics: The Internet's Impact on the American Political System*. New York: Oxford University Press.

Forse, M. and C. G. Lafaye. 2008. "Participative Democracy and the Hopes of Citizens." *Archives Europeennes de Sociologie*, 49(2): 173–204.

Magnette, P. 2003. "European Governance and Civic Participation: Beyond Elitist Citizenship?" *Political Studies* 51: 144–60.

Norris, P. 2001. *Digital Divide: Civic Engagement, Information Poverty and the Internet Worldwide*. Cambridge UK: Cambridge University Press.

Norris, P. 2002. *Democratic Phoenix. Reinventing Political Activism*. Cambridge, UK: Cambridge University Press.

Putnam, R. 2002. *Democracies in Flux. The Evolution of Social Capital in Contemporary Societies*. New York: Oxford University Press.

Sunstein, C. R. 2006. *Infotopia: How Many Minds Produce Knowledge*. New York: Oxford University Press.

Tsagarousianou, R., D. Tambini, and C. Bryan (eds.) 1998. *Cyberdemocracy: Technology, Cities and Civic Networks*. London: Routledge.

Warren, M. 2001. *Democracy and Association*. Princeton, NJ: Princeton University Press.

# 1.3
# Visualizing Electricity and Magnetism: The Collaborative Development of a Multimedia Text

*Jennifer A. George-Palilonis and John Belcher*

As digital learning trends change the way we teach and learn, many educators are experimenting with and developing new tools for content delivery. They not only seek new ways to deliver traditional content via digital means, but they also are developing entirely new digital content to better harness the power of multimedia technology. "Visualizing Electricity & Magnetism" (http://web.mit.edu/viz/EM/flash/E&M_Master/E&M.swf) is one such example.

This multimedia text is based on a number of 3-D animations, interactive applets, and videos focused on six units of electromagnetic theory. The module allows students to visualize concepts invisible to the naked eye and better understand relationships among subtopics. The authors' collaboration grew from two separate efforts. The first was the Technology-Enabled Active Learning (TEAL) Visualization Project at MIT, which produced more than one hundred visualizations used to teach electromagnetism in freshman courses at MIT. The second was the Ball State Digital Publishing Project (DPP), intended to address questions related to the design, development, and distribution of multimedia texts, and explore how storytelling changes when educational content takes digital forms.

In spite of their differing backgrounds and fields, the authors have a shared belief in the power of visualization to enhance learning. Through this text, they hope to help transform how the foundations of electricity and magnetism are taught, moving from a traditional, equation-based system to a visualization-based multimedia experience. Their work was driven by two main assumptions: (1) that by providing an engaging storyline, a better sense of coherence and connection would be established; and (2) that by developing a nonlinear presentation for the text, student enjoyment would increase because the user is given more control over content navigation. The authors believe that classical electromagnetism is an ideal subject for such an approach because it is a fundamental underpinning of a technical education, and one of the most difficult subjects for students to master because mathematical complexity quickly overwhelms physical intuition.

The traditional instructional approach uses equations to teach the subject and fails to help students establish more conceptual models. A purely textual or mathematical approach does little to connect the dynamics of electromagnetic fields to students' everyday experiences. However, by adding animations that further illustrate equations, concepts begin to seem natural and intuitive. Both explanations should be provided. The first is quantitative and appeals to students who are analytical thinkers. The second is qualitative, more intuitive, and comprehensible to students of all persuasions, because it can be understood by analogy to concepts they already have. The authors argue that one year after taking a course in electromagnetism, average students will not remember details of the textual or mathematical approach. However, if they have seen the concepts in action, they will have a mental model that will endure.

This collaboration is also a great example of the doors that can open for educators when they step outside the silos that often confine them. It is unlikely that a single person would have expertise in a specific field like physics as well as the technical and design expertise to develop a single multimedia text. Thus collaboration is nearly a necessity. After one face-to-face meeting, this text was developed from a distance. The physicist developed PowerPoint presentations with the storyline, appropriate graphics, and links to visualizations. Then the designer used this as a guide to develop an Adobe Flash-based module with structure and functionality similar to that suggested by the physicist. They then iterated content until they were both satisfied before proceeding to the next topic, and the process converged within a few iterations. "Visualizing Electricity & Magnetism" is currently used in freshman level physics courses at MIT, and the authors are working on a second text for advanced electromagnetism students and a project that will explore how this content might be positioned in a 3-D virtual world.

# Part II
# Defining New Media

# 2
# Emergent Journalism and Mass Media Paradigms in the Digital Society

*Carlos Elias*

"Life is more and more about being glued to the screen or being online." This is how Gilles Lipovetsky and Jean Serroy, authors of *La pantalla global* (The Global Screen), characterize the extent to which the Internet has altered our way of life. "Journalism Wounded but Not Dead," "Death to the Leading Journalists!," "Long Live Citizen Journalism!," and "The Post-Journalism Era in the Digital Society." Titles such as these from recent media studies conferences could make up this entire chapter, as a sign of the tremendous insecurity the new technologies have triggered in this field. No one foresaw this great revolution, not even a few years ago. One of the very few exceptions, perhaps, was Alvin Toffler in his book entitled *The Third Wave* (Toffler 1980), in which, for the first time, there is reference to "several-to-several" communication as well as to the progressive "demassification" of media production and the resulting emergence of personalized media communication.

Mass communication, so enjoyed by journalists for the controlling power it afforded them, would no longer exist, and—the book's most revolutionary concept—newspaper readers and television watchers would be able to have a say in what they were reading and watching. When it was published in the early 1980s, Toffler's book was viewed, by university communications departments, at least, as science fiction akin to the "time traveling" of physicists. In the middle and late 1990s, judging by the bibliography, the book was still garnering credibility more on the basis of its sociological aspects, such as the disappearance of the traditional family, increasing social isolation, and the "culture of childlessness," than its speculations on the evolution of the mass communications media.

At the beginning of this century, when what is known today as Web 1.0 was already fully operational, the belief was that the only way to guarantee the quality, objectivity, truth, and credibility of informative content of a journalistic nature on the Internet was to have journalists write this information. Moreover, protocols from the previous era such as "brand prestige" were stressed, and a brand image such as that of the *New York Times*, for example, was considered crucial to guaranteeing the reliability of the information. If it is good on paper, it will be good on the Internet.

In 2005, however, this began to change. An analysis of free newspapers' web portals showed that they were competing with subscription media by offering personal communication space and that their best front-page articles were written by readers (Álvarez 2005). The article selection process, however, remained in the hands of the newspapers' editorial staff, who decided which article was the most interesting and which blog should be blog of the day. In other words, while journalists still had the power to hierarchize the news, one of the profession's major powers, there were signs of a shift toward citizen-generated information. People began to speak of "citizen journalism." Newsgathering services, Google itself, for example, began to be seen as communications media in their own right. YouTube, which was very heavily used in the 2008 Obama electoral campaign, is seen as a television channel comparable to CNN. Early in 2010, the BBC announced a project being undertaken by leading British networks BBC, ITV, and BT Vision along with Channel 4 to launch a single Internet platform called IPTV (Internet Protocol Television), which represents the future of television.

What happened between 2001 and 2005? Basically, the so-called Web 2.0 philosophy burst onto the scene. Web 2.0 is a term that was coined in 2004 by Tim O'Reilly to describe a second stage of Internet history in which user communities interact via social networks, blogs, and the like. The key feature of Web 2.0 is the option to modify both the content and structure of texts and, in addition to easy access, the opportunity to place material online and have personal as well as collective participation. In other words, Web 2.0 enables citizens to play a leading role in the informative process—a role once reserved for journalists alone. The online Encyclopedia Britannica, which we can all passively consult (Web 1.0), is not the same as Wikipedia, on which we can all collaborate (Web 2.0). Naturally, the entry for Madonna, the singer, is likely to be more extensive than the entry for Aristotle, but in this new era knowledge hierarchies are breaking down.

Citizens, including journalists, of course, as well as the great media companies, can play a direct role in developing Journalism 2.0. New terminology has appeared: we no longer speak of journalistic genre but rather "the transmedia narrative;" we refer to the old news publishing networks as "corporate hybridization;" and the ever-more antiquated concept of the audience, even the concept of public opinion, has been replaced in Web 2.0 by the concept of "collective intelligence" (Lévy 1997) or "intelligent multitudes" (Rheingold 2002). We no longer speak of editing journalistic texts (formerly the job of section chiefs or editors-in-chief) but rather "beta reading," a concept in which, on the Internet, readers themselves can rework or edit a text written by a journalist and claim that they published it.

Sometimes we wonder whether we will even be able to keep using the simplest definitions of the journalism profession, although those are usually the definitions that survive longest. "Journalism is reporting to people what's happening to people," as the great journalist Indro Montanelli rightly pointed out. But, what reality do you choose for reporting what is happening to people? Which reality do people spend the most time in each day: the real world or *Second Life*? The concept of MMORPG (Massively Multiplayer Online Role-Playing Games), that is, games where thousands of people interact through avatars in a fantasy environment, has given way to the more sophisticated concept of "alternative reality," coined by Jane McGonigal. Those alternative realities interact with the "real" one, so, what to report?

In 2006, we journalists were shocked when Reuters, one of the world's major news agencies, opened its own avatar, "office and its journalist," in *Second Life*. Can a journalist report on alternative realities? Can information be exchanged between one reality and another? Is a journalist required to do that? Is that journalism? When Gaspar Llamazares, leader of the Communist Party in Spain, decided in 2007 to hold a political rally in *Second Life*, what was remarkable was not the speech he gave, rather it was the fact that the "real" media gave this rally far more extensive coverage than any other rally happening in the "real" reality. Llamazares was emulating French politicians Nicolas Sarkozy and Ségolène Royal. From my point of view, *Second Life* has not suffered the repercussions that cybernetic theorists, in particular, had predicted. Reuters closed its *Second Life* office in 2008, but it had demonstrated that the world is not what it used to be. Thus, we are going through a period of uncertainty where the old seems to be disappearing but we do not yet know what is to come.

## The convergence culture: A new paradigm

One of the most appealing explanations of what has happened in the communications media with the emergence of the new technologies, Web 2.0, above all, appeared in 2006 when Henry Jenkins, professor of Media Studies at MIT, formulated his so-called convergence culture paradigm. This 'convergence culture' would be the combination of three sub-processes connected with the Web 2.0 philosophy on different levels: media convergence, participatory culture, and collective intelligence.

We will introduce this phenomenon by describing a case study that is being very closely analyzed (Rosenzweig 2003; Jenkins 2006; Lozano 2008). In the fall of 2001, following the September 11 attacks in New York, an ordinary Filipino–American high school student named Dino Ignacio designed a digital collage using Photoshop. In the utter innocence of youth, he merged the image of Bin Laden, whom the communications media considered the intellectual author of the attacks, with the image of the cantankerous Bert on *Sesame Street*, then Dino's favorite children's program, which first aired in 1970 and continues to this day. On his personal website, he posted a series of images entitled "Bert is Evil." In reality, it was only a joke but, as the blog continued to circulate, its search engine positioning steadily rose. Bert and his friend Ernie are icons widely known in the West among those who were born after the mid-1960s, that is, among the majority of Internet users.

This youngster's small success may explain how a publisher in Bangladesh found and chose the collage when he was searching the Internet for images of Bin Laden to print on anti-American posters and T-shirts to be worn by thousands of Pakistani demonstrators opposed to US policy, following the September 11 attacks. Although there is a program modeled on *Sesame Street* that is televised in Pakistan, the Bert and Ernie characters do not appear in it.

The demonstrations were picked up by CNN and broadcast around the world. Representatives of Children's Television Workshop, creator of the series *Sesame Street*, were horrified by the CNN images showing enraged demonstrators carrying posters with the paired images of Bin Laden and Bert. They threatened to take legal action, but CNN argued that it was only broadcasting newsworthy images of reality, and those demonstrations with the paired images of Bert and Bin Laden were reality.

Right away numerous forums of joker fans connecting terrorists with *Sesame Street* characters sprang up, and Dino Ignacio, the joke's involuntary protagonist, became an Internet cult personality. This was news in the traditional media, including some as prestigious as the BBC, and appeared in Wikipedia as a page entitled "Bert is Evil." Publicity swelled to such a point that Ignacio became worried and took down the website, posting the following message: "I am doing this because I feel this has gotten too close to reality." His personal fantasy had become reality. What had begun as a joke on his website became a news item. Later, it turned into an international and business crisis reported in the traditional media and, primarily, the digital media. Since 2003, when the prestigious journal *American Historical Review* (Rosenzweig 2003) published the case as a "scientific study," the communications media have also taken a full-scale, scientific approach in publishing it. This is the world of convergence culture where traditional communication schemes such as the Jakobson sender–message–receiver model are radically distorted. It is a world where mass culture becomes high culture and high culture, scientific results, for example, becomes mass culture.

Is this an isolated case? I think not—such cases are coming to light more and more often. Gaspar Llamazares, the Communist Party leader in Spain mentioned above, made the news again in January 2010 when the FBI used his image (filtered by the Internet) to update the robot portrait of Bin Laden. Some ordinary internauts took notice and, once again, it became worldwide news, even sparking a diplomatic clash between Spain and the US.

What makes this paradigm shift possible? To begin with, one of the elements of the convergence culture, media convergence, would be a factor (Jenkins 2006). The circuits Bert has traveled are the product of not only globalization but also the convergence and synergy of communications media. Convergence has a new meaning, however; it refers not so much to the convergence of media platforms, as defined at the turn of this century, as to a cultural shift, primarily, spearheaded by the Internet, where media consumers search for other information and create new connections between scattered media content and where the end observer–user–receiver becomes the sender. In other words, convergence happens not in the media machinery itself, however technologically sophisticated it may be, but rather in the brains of senders and receivers (Jenkins 2006), in the brain of an FBI agent, for example.

The other new concept essential to understanding the convergence culture paradigm is that of "participatory culture," which is the opposite of the old "passive media spectator" idea. We can no longer speak of media producers and media consumers as separate entities performing different functions, as the audience can be an information producer via websites or blogs, and the source can even become a mass communications medium.

The last term I would like to describe is "collective intelligence" (Lévi 1997). This concept is based on the premise that none of us can know everything, especially in an interconnected world where there is more information than the human brain can handle. Because each of us knows something and knows about something, however, the best option is to exploit Web 2.0's more advanced social networking and work together to build knowledge, contributing our experiences and joining all the pieces to create new, more precise information. These Internet conversations create a "buzz" that is of ever-increasing importance in the media industry and will define its future in the years ahead. In the convergence culture paradigm, "online buzz" is crucial to an understanding of how communications media function in the Web 2.0 era.

## Journalistic sources become communications media

The convergence culture model is going to radically alter the concept of communications media, for a source can now become a communications medium in its own right. For example, in 2004, there was cause for celebration among scientists who were studying Martian geology, climate, and geography: NASA's rovers *Spirit* and *Opportunity* landed on Mars to gather extremely valuable scientific data. Nothing unusual was expected or found. Journalism students experienced quite a shock, however: during the 24 hours *Spirit* had spent on Mars, NASA's website had 225 million hits. (Every time an internaut accessed any NASA web page—text, photos, or any other content—it was counted as a hit.) Over the course of that 90-day mission, the number of visits to the website reached 6530 million, a figure NASA was quick to rectify because it was "more than the earth's total population" of 6300 million that year. One-fifth of the traffic was from outside the US, and it was determined via a survey that one-fourth of all visitors were professors or elementary and high school students.[1]

A study conducted in 2007 showed that in May of that year—a month when there was no spectacular mission—only 3,952,000 unique individuals visited NASA's portal, with an average stay of 12 minutes. The number of unique visits to the science and technology portals of traditional communications media was very much lower, however. CNN had the most, at 502,000 visits, which was almost eight times less. Visits to the science portal at FOX and CBS could not be quantified because the set minimum of 360,000 visits was not reached. Only one website, Space.com, which has nothing to do with the traditional media, came close to NASA's figure, with 1,178,000 unique visits (Hedman 2007). This persuaded such media giants as the Boston Globe and CNN to close their science sections in 2008. And where is this leading us? To a disturbing conclusion: journalism, in large measure, now happens at the source.

What is NASA—a media source or a communications medium in its own right? Perhaps it is both. As a communications medium, it apparently has a larger audience than CNN (the science section), CBS, and FOX. NASA is aware of that, in fact, and keeps two avenues of communication open, one with journalists (the traditional sources, though this has varied, also) and another with society directly, thus bolstering its brand image in both directions.

NASA's communication with journalists has been enhanced in that any journalist anywhere in the world, regardless of their medium, can now subscribe to NASA's news service and receive timely email notification of the latest news about space missions, communiqués on scientific findings, and videos with statements by and press conferences with the space agency's scientists. In other words, NASA is not dependent on the major communications media, such as news agencies and international television networks, because they all have the same information. This works against journalism, however, because it has made it easy for the media to copy, literally, NASA's communiqués, which include feature articles and interviews. In the digital age, it seems that young people on scholarships who know how to copy and paste the NASA-supplied content are all that is needed, and then the old journalists are fired. Here is a case of the source, NASA, via its website, being a mass communications medium in its own right, and it infects the traditional communications media, as well.

What is truly novel about the digital society is that information that was once received, evaluated, and published only by the mass media is now also received directly by the whole of society—without the need of a journalist as intermediary. Any ordinary citizen interested in NASA can access its website and find there practically all the same content a journalist would copy. Besides that, however, the technology allows any citizen to subscribe to NASA's news service under the same terms as a journalist would subscribe. That used to be physically impossible, but not now. Any citizen who yearns to be a reporter, or even to be vindictive, whatever the motive may be, can report NASA news in their blog, copy images, and post a video of a press conference or of the astronauts at the International Space Station (ISS).

People truly interested in information about space have more respect for these bloggers than for journalists who work in traditional media. Consequently, these bloggers have a great deal of power and opportunity to generate "online media buzz," which may, in turn, influence NASA's decisions. The buzz that arose from the dismissal of James E. Hansen, Chief of Service in the Atmospheric Department at NASA's Goddard Institute, is a paradigmatic case. What surfaced in the online buzz was that the Bush administration had forbidden him to hold conferences that would corroborate global warming. This online buzz found its way into the *New York Times* (January 29, 2006). Another very interesting case, which we cannot discuss in depth here, is the Internet buzz being generated to have Pluto declared a planet again. The Internet is capable of conditioning not only political and commercial but also scientific decisions.

There is another angle on this, however. NASA is broadcasting longer and longer video clips of the crew members' daily life at the ISS. Established in 2001, it is without question one of the major milestones of present-day science, indeed, all of human history, because, in its philosophy, there is the underlying concept of a kind of research city where human beings would always be present. In other words, the idea was to have human beings located somewhere apart from planet Earth at all times—nothing more, nothing less. The ISS is very important from both a scientific and cultural standpoint, which is why all journalists have access to the website where video clips of the crew members' life there are broadcast live. Any citizen can watch them, too, however, as if they were a reality show. What distinguishes those live broadcasts of the ISS crew members' life from *Gran Hermano* (*Big Brother*)? Once again, we have a blend of mass culture and high culture.

Professors of journalism, watching in dismay as their mission to train journalists vanishes in the new order, have registered one commentary: that "journalists will always be needed because someone has to write the feature articles, even at NASA." This is true but, first of all, NASA does not recruit journalists; they recruit people with a degree in science who also have communication skills. Something similar is happening at Nature, as well, but this is quite different from the traditional concept of a journalist with a traditional education in journalism. These communicators have become the sources, that is, they write feature articles for NASA's website that are literally copied by media around the world. Now NASA scientists, who were once the sources, have become the journalists, that is, they comment on the news in their blogs, and some of them have a large following. Their intention is to create "media buzz," but that sometimes becomes a primary source for the traditional media. They have the power to unleash a political storm, even in economics. When *Financial Times* blogger Izabella Kaminska labeled Spanish cabinet minister José Blanco as "paranoid," it set off a crisis within the Spanish government: the communications medium is now the source.

NASA takes very good care of the traditional media by giving them an abundance of information. The agency understands perfectly, however, that it is a mass communications medium in its own right and has a brand image, as it creates links on its website for parents, teachers, and children, among others. In other words, this is the source speaking directly to society without intermediaries. This strategy is beginning to bear some very interesting fruit: independent institutions such as universities and scientific associations are adopting NASA's approach and transforming themselves into mass communications media. Unlike sources such as NASA, however, which are subject to criticism for being government-supported, these institutions have a free and more influential voice. The most notable among them are Britain's Royal Society and the American Association for the Advancement of Science, which have become communications media with websites visited by millions of users in search of news or documents on climate change, Darwinism, and other scientific issues. Their leadership is beginning to look for ways to dominate the online media buzz.

It is truly intriguing how becoming mass communications media bolstered these sources' brand image, and their media impact. The traditional media now cite them as sources much more often so that they echo the media buzz more and more every day.

## Traditional journalism in the digital age

What fate is in store for the old journalism in the digital age? At the moment, that is a mystery. There is no reason, however, why a change in technology should wipe out an essential element of democracy such as the journalism profession. A scenario similar to this one has played out before, when the telegraph appeared in 1845. Being able to get information faster led to newspapers placing far more emphasis on reporting events immediately, rather than explaining them calmly. That year, James Gordon Bennett, publisher of the *New York Herald*, came to the conclusion that the telegraph would put many newspapers out of business. But journalism actually flourished more than ever after 1845.

A key facet of journalism in the digital age is that traditional journalists might no longer be needed or might not be so influential, at least, since providing informative content on the Internet will be their only work. The "professionalism" so carefully cultivated all these years is no longer relevant: the professional journalist is disappearing, and the expert economist, scientist, attorney, or political scientist, skilled in communication, is emerging. In recent worldwide media events, from the Iraq war to the Obama campaign, we have witnessed rigid behavior on the part of the traditional media and the hidden, self-interested agendas that characterize them (Kamiya 2009).

Moreover, "professionalism" can be a dubious epithet, as evidenced, for example, in the pathologically close relationships many Washington political journalists have with their government sources. Such "professional" relationships engender bias and false perceptions in these journalists, and they become critical of "amateur" bloggers. It has been shown, however, taking the Iraq war as an example, that the most truthful information and the most interesting and accurate analyses were written by academics specializing in the Middle East. They may never have contacted a White House source, but they had perfect knowledge of the region's language and history, and that is how they came up with the right diagnosis.

Internet journalism guru Jeff Jarvis maintains that, for the most part, specialization will take control of journalism. No longer will we all be doing the same thing, making the news worthwhile. Instead, we will all be going out to make our mark by providing in-depth coverage on a specific segment (Jarvis 2008).

In December 2008, Bree Nordenson described this other approach in one of journalism's leading academic journals, the *Columbia Journalism Review*. As it turns out, explanatory journalism may have a promising future in the market of news. On May 9, in partnership with NPR News, *This American Life* dedicated its hour-long program to explaining the housing crisis. "The Giant Pool of Money" quickly became the most popular episode in the show's 13-year history. *Columbia Journalism Review* praised the piece as "the most comprehensive and insightful look at the system that produced the credit crisis."[2] And Nordenson added: "Rather than simply contributing to the noise of the unending torrent of headlines, sound bites, and snippets, NPR and This American Life took the time to step back, report the issue in depth, and the explain it in a way that illuminated one of the biggest and most complicated stories of the year."

In the journalism of the future, which is already here, with massive amounts of information on the Internet but no business model for costly, in-depth journalism, media studies graduates can only aspire to working with what Nordenson calls "the noise of the unending torrent of headlines, sound bites, and snippets," that is, the very noise busy citizens are trying harder and harder to escape. The experts, such as the specialized journalists who have some credential beyond a journalism degree, will be those qualified for what Nordenson considers the future of journalism: "Took the time to step back, report the issue in depth, and then explain it."

As Richard Lanham explains in *The Economics of Attention*, "Universities have never been simply data-mining and storage operations. They have always taken as their central activity the conversion of data into useful knowledge and into wisdom. They do this by creating attention structures that we call curricula, courses of study" (Lanham 2006). This is the role of communication experts, however, not journalists who have no in-depth knowledge in areas such as economics, law, science, and policy.

In 2009 we saw world-renowned icons of journalism lose readership and advertising. The *New York Times*, for example, registered a 16.9 percent loss of revenues, and its third-quarter profits were 35.6 million dollars less than in the third quarter of 2008. This financial hemorrhaging was very grave news for traditional journalism: a policy of termination with severance package for *Times* journalists (100 in 2009) and a 5 percent reduction in salary for those who remained on staff. Laying off journalists is the worst possible strategy for a newspaper such as the *New York Times* because, obviously, this would undermine the quality that distinguishes it from free newspapers and the Internet. Other exemplary newspapers suffered similar losses: for example, *USA Today* (a 17% loss), the *Los Angeles Times*, and the *Washington Post*. Among the weekly news magazines, *Time* saw losses of 30 percent and *Newsweek* losses of 40 percent in the second quarter of 2009. There is a similar trend in Europe.

Oddly enough, however, there is one sector that is surviving: the specialized newspapers. The *Wall Street Journal* grew by 12,000 copies in 2009 for a total of 2.02 million, and its revenues increased by 10 percent that year. Its leadership attributes that financial success to the new iPod and Blackberry applications. While the majority of weekly news magazines were in decline, *The Economist* doubled its circulation from 600,000 copies in 2001 and 1,200,000 copies in 2009. What is their secret? Specialized information, as they supply trustworthy, specialized information, and yet they are also open to other subject matter once covered by the generalists. The strategy at the *Wall Street Journal* and *The Economist* has been to snatch a portion of the market from the generalists while maintaining their specialist focus. In the second quarter of 2009, specialized magazines such as the Disney publication *Family Fun* grew by 16.7 percent, and the National Rifle Association magazine *American Rifleman* realized a 20.2 percent increase in revenues.

The traditional media's response has been to try to specialize, not by areas but by subject matter. It is too soon, however, to speculate on the outcome. One experience along this line was the agreement the *New York Times* and the *Washington Post* entered into with Google's project "Living Stories" in 2009. For the time being, it is in the experimental section (livingstories.googlelabs. com) where stories are grouped by subject, for example, healthcare reform in the US, or the war in Afghanistan.

One interesting but disturbing observation is that the journalism that survives is the journalism that focuses on very specific disciplines such as economics, politics, science, or armaments, not the journalism of first-class writing that rises to literary excellence. It appears that the good stories will be found on the Internet while the analysis by experts and highly specialized journalists will be found in print. The most striking case is that of the *Atlantic Monthly*, the celebrated periodical where journalists of such renown as Faulkner, Twain, and Hemingway have put their signature on feature articles. In 2007, after a glorious, 150-year history of legendary journalistic reporting, it was forced to discontinue its print edition and move to the Internet. Considering that its website has more visitors than that of its great competitor, the *New Yorker*, this has been a relatively successful move. Such success, however, derives from replacing journalists and journalistic reporting with blogs and editorials by such figures as Joseph Stiglitz, Nobel Prize winner in economics, and by professional politicians who are at odds with each other such as Christopher Hitchens (staunch defender of the Iraq invasion and an avowed atheist) and Andrew Sullivan (homosexual, HIV-positive, Roman Catholic, and economically liberal Republican), for this generates the controversy that brings visitors to the website. All of that, however, is a far cry from those feature articles of days gone by that were written with such literary flair that they had power to change the course of history, for example, helping to bring about the abolition of slavery in the US.

## Conclusion: freedom of choice is the tyranny that saves journalism

Journalism magnate Rupert Murdoch, owner of the *New York Times* and the *Wall Street Journal*, has stated that Internet journalism will survive only if the public is willing to pay for that information. He has also observed, however, that the public is willing to pay for information only if it is highly specialized and very well explained and contextualized—in other words, expert information.

Philip Meyer has calculated that 2043 will be the last year that newspapers are printed in the US. He arrived at this date by extrapolating the decline in readership US newspapers have been seeing for decades until he reached "the last reader" (Meyer 2006). Obviously, newspapers will disappear long before that, however, because they will not last until there is only one reader left.

Yet Meyer is optimistic. While he believes that newspapers definitely will not continue to enjoy the revenues and readership they have had until now, because readers, primarily, but also revenues are forsaking them for the Internet, they do still have a future, and it lies in selecting the information and providing the analytical and explanatory content that keeps the financial and intellectual elite informed. Those elite are very important because they represent newspaper customers, Meyer argues. This is the role *Le Monde Diplomatique* plays in the realm of European politics, for example, which brings us right back to our original premise, for *Le Monde Diplomatique* has hardly any journalists. The majority of its writers are sharp political, financial, and scientific analysts who belong, primarily, to the elite among intellectuals and university professors. To put it another way, in this model of the future, a generalist education in journalism studies is good for nothing.

The alternative would be for the generalist, the traditional communications studies graduate, to focus on breaking news. However, today's readers reject the continually updated website news. As Barry Schwartz correctly points out, the passivity resulting from a lack of control is known in psychology as "learned helplessness." Even though logic would suggest that more news available would mean more control for consumers, what happens is just the opposite: having too many options can get to be cumbersome. Instead of feeling that we are in control, we feel unable to cope (Schwartz 2004). "Freedom of choice eventually becomes a tyranny of choice," Schwartz states. The public loses interest and becomes voluntarily uninformed when it comes to that type of brief, breaking news that has always been top priority in journalism. It was suitable for television in the 1960s, for example, because someone who wanted to watch television at a particular time would be forced to see that news. People have no reason to do that now, however, which is why television journalism has drifted into news-entertainment. That will also happen with web journalism.

Nonetheless, an influential elite does indeed need and value contextualized information, and that information must be created by people with a full, university-level command of the discipline (science, economics, politics, law, and the like) in which they report. This is how a website earns the "quality" label and the "brand name" guarantee that are essential to being chosen from among hundreds of millions of websites on the Internet.

Approaching it from a different standpoint, as described a few months ago on Salon.com, the influential media trends blog, Gary Kamiya reached the same conclusions. He believes that journalism will survive, but what will disappear is the news (Kamiya 2009). According to Kamiya, the Internet gives readers what they want and gives the newspapers what they need. The physical layout of a newspaper means that people read information even if they are not looking for it. A person buys a newspaper to read the science news, for example, and on the front page, there is some international news or a political analysis. He or she was not looking for that news but, in the end, reads it. This does not happen on the Internet because the reader proceeds directly to specific portals. Audiences are split between the Internet and digital television, and that involves specialization, too, but in the opposite sense. Only the intellectual elite are capable of making the effort, financially and mentally, to read content they need but, in principle, have no *a priori* interest in. That new journalism will be in print and intended for these influential elite, however, so it must be written by journalists among the elite who have university-level and even doctoral-level knowledge in the specific discipline they are reporting, as well as enough understanding of journalism to write engaging text.

## Notes

1. Information obtained from a NASA press conference and analyzed on the educational portal http://www.distance-educator.com/ among other media.
2. In "Boiler Room," the essay by Dean Starkman, September–October Issue of *CJR*.

# Bibliography

Álvarez, Franco G. 2005. "A New Way to Make Free Journalism: The Printed Weblog. Quémadrid." *Ámbitos, Revista Internacional de Comunicación*. 13–14: 177–84.

Hedman, E. 2007. "The Fragility and Resilience of NASA." *The Space Revew* (http://www.thespacereview.com/article/924/1, accessed December 4, 2007).

Jarvis, J. 2008. "Three Reflexions about Journalism in the Web," posted in his blog *BuzzMachine* in 2008 (http://www.buzzmachine.com/), accessed July 12, 2008.

Jenkins, H. 2006. *Convergence Culture*. New York: New York University Press.

Kamiya, G. 2009. "The Death of the News." http://www.*Salon.com* (February 17, 2009).

Lanham, R. 2006. *The Economics of Attention*. Chicago: The University of Chicago Press.

Lévi, P. 1997. *Collective Intelligence: Mankind's Emerging World in Cyberspace*. Cambridge: Perseus Books.

Lipovetsky, G. and J. Serroy 2009. *La Pantalla Global*. Barcelona:Anagrama.

Lozano, J. 2008. "Cultura de masas/Cultura alta: ¿Convergencia o transducción?," Conference in *Congreso Septenio Canarias* (recogida en www.septenio.es), accessed January 21, 2008. La Palma.

McGonigal, J. 2008. "Making Alternate Reality the New Business Reality" Op-Ed. *Harvard Business Review*. Special Issue: Top 20 Breakthrough Ideas for 2008. February 2008 (http://www.harvardbusinessonline), accessed February 10, 2008.

Meyer, P. 2006. *The Vanishing Newspaper: Saving Journalism in the Information Age*. Missouri: University of Missouri Press.

Nordenson, B. 2008. "Overload! Journalism Battle for Relevance in an Age of too Much Information." *Columbia Journalism Review* (November–December).

Rheingold, H. 2002. *Smart Mobs: The Next Social Revolution*. Cambridge, MA: Basic Books.

Rosenzweig, R. 2003. "Scarcity or Abundance? Preserving the Past in a Digital Era." *American Historical Review*, 108.

Schwartz, Barry. 2004. *The Paradox of Choice: Why More is Less*. New York: Harper-Collins.

Starkman, Dean. 2008. Boiler Room, *Columbia Journalism Review*, Sept/Oct 2008.

Toffler, A. 1980. *The Third Wave*. New York: William Morrow.

# 2.1
## CitizenShift and Parole Citoyenne: The Democratization of Media

*Reisa Levine*

CitizenShift (CS) and its French language sister site Parole Citoyenne (PC) are two award-winning web platforms exploring today's social issues through films, photos, articles, blogs and podcasts. Born out of the National Film Board of Canada in 2004, and now housed at the Institut du Nouveau Monde, the CS and PC social media networks are community-driven spaces designed for filmmakers, activists and 'ordinary' citizens to share content and voice opinions.

CS and PC were inspired by the famous Challenge for Change program, which ran out of the NFB in the 1960s and 1970s, putting filmmaking tools directly into the hands of ordinary citizens.

> The mandate of Challenge for Change was to provide the voiceless population of Canada with the means to communicate... it used film to promote social change and allowed the people themselves, not the experts, to express the social issues that concerned them.
>
> NFB Our History

Over 30 years later CS and PC, the digital offspring of the Challenge for Change program, picked up on the spirit of community driven media and were at the forefront of the Web 2.0 revolution. CS and PC were online before YouTube and Facebook and were among the earliest pioneering websites to aggregate and produce social issue media.

Since the projects' inception, CS and PC have reached out to diverse audiences across Canada and around the world with a constantly growing collection of media works. But CS and PC are so much more than two websites. Through live events such as screenings, discussions, workshops and film launches, our sites are deeply rooted in communities and we connect with people through a wide variety of networks.

If the twentieth century was the age of broadcasting, the twenty-first century is the era of conversation and the new language is visual media. In order to support media-driven conversations, we offer the basic things people need to take advantage of this emergent digital language. This includes small production grants to filmmakers, the server space to house their media, training on using social networking and outreach techniques and, perhaps most importantly, a context in which to make their voices heard.

One crucial lesson that we have learned is that real engagement and community building is not automatic; it takes work, demonstrated integrity and commitment to the causes. The role of 'community manager' was something that our teams practiced intuitively, long before this job title was recognized as integral to any participatory web project.

CS and PC are constantly breaking new ground as we run the projects using a bottom-up, community-driven approach. We skirted the controversial copyright debate by adopting the Creative Commons licensing model, we use Open Source software and we employ the latest collaborative work tools and processes. As a project of the National Film Board of Canada, a federal government agency, we had to carefully craft our editorial line to be inclusive and respectful; we are a platform to host debate, but the point of view comes from the community not from our staff. This policy has served us well over the years, and has kept us on track through controversies that can arise when discussing politically charged issues.

In the summer of 2009 the NFB decided to cut CitizenShift and Parole Citoyenne because of budgetary constraints. It was also recognized that these two media platforms were now mature enough to 'make it on their own.' And so, the CS and PC projects were transferred to the Institut de Nouveau Monde, a not-for-profit organization with the mandate to encourage citizen debate.

In this new context, we are once again redefining ourselves in order to continue to meet the needs of our community. With camcorders, cell phones, digital cameras, and laptops in the hands of ordinary citizens, people are now telling their own stories more than ever. What we know as the "mainstream media" is actively being challenged by a citizen media revolution: stories told from multiple perspectives by those who are directly affected. CS and PC continue to thrive as important players within this global movement to empower individuals through the language and tools of media.

## Bibliography

CitizenShift Web Platform: http://citizenshift.org, accessed February 24, 2010.

Parole Citoyenne Web Platform: http://parolecitoyenne.org, accessed February 24, 2010.

"Citizen Media". 2007. In Wikipedia, The Free Encyclopedia. Retrieved June 10, 2007, from http://en.wikipedia.org/w/index.php?title=Citizen_media&oldid=129435027.

"Copyleft". 2007. In Wikipedia, The Free Encyclopedia. Retrieved June 11, 2007, from http://en.wikipedia.org/w/index.php?title=Copyleft&oldid=137032969.

Haughey, Matthew. 2007. Some Community Tips for 2007. "Seven Tips on How to Run a Successful Community." Retrieved June 10, 2007, from http://fortuito.us/2007/05/some_community_tips_for_2007.

Institut du Nouveau Monde (INM). Retrieved January 22, 2010 from http://www.inm. qc.ca/english.

Levine, Reisa. 2007. "Technology Integration, Team Process and Citizen Media: Stories from the Inside of a Fast-Paced Web Production Team at the National Film Board of Canada." Department of Education (Educational Technology) (Master's thesis) Concordia University.

Levine, Reisa. 2010. "Comments on CRTC 2009-661—Review of Community Television Policy Framework." Intervention number 2353. Retrieved April 26, 2010 from http:// support.crtc.gc.ca/applicant/docs.aspx?pn_ph_no=2009-661&call_id=127557&lang= E&defaultName=Institut%20du%20Nouveau%20Monde&replyonly=&addtInfo=& addtCmmt=&fnlSub=.

"NFB Our History." National Film Board of Canada website. Retrieved January 22, 2010 from http://nfb.ca/historique/1960-1969.

O'Reilly, Tim. 2005. "What is Web 2.0." Retrieved June 10, 2007 from http://www. oreillynet.com/pub/a/oreilly/tim/news/2005/09/30/what-is-web-20.html.

Tyler, Tom R. 2002. "Is the Internet Changing Social Life? It seems the More Things Change, the More They Stay the Same." *Journal of Social Issues* 58 (1): 195–205. New York: The Society for the Psychological Study of Social Issues.

Waugh, Thomas, Brendan Baker and Ezra Winton (eds.) 2010. *Challenge for Change—Activist Documentary at the National Film Board of Canada*. McGill: Queen's University Press.

Winton, Ezra. 2010. "Beyond the Textbook." *Point-of-View* Magazine 77 (Spring): 14–17. Published by the Documentary Organization of Canada.

# 2.2
## Presidential Rhetoric in 140 Characters or Less

*Nathan Angelo*

As a presidential candidate, Barack Obama was known early on for his tech-savvy campaign strategies, which included maintaining accounts on social networking sites like Twitter and Facebook, but the establishment of President Obama as the "first wired president" (Griggs 2009) was undoubtedly confirmed on January 19, 2010 when he personally submitted an update to the Red Cross Twitter feed—a first for a sitting president. Three months later Twitter again entered public discourse when the Library of Congress announced that it would house all of the so-called tweets posted on the rather expansive website—sparking almost instant public debate (Bliumis and Scoble 2010). Much of the controversy surrounding this decision was based on physical location, as it was the proximity of the archive to important documents such as the Gettysburg Address that caused concern. The critics questioned the need to archive arbitrary announcements such as banal statements of one's sour mood in such a prestigious institution. But one thing that these critics do not address is how to archive new digital forms of rhetoric. The Library of Congress' decision did clear up one relevant question: where the archive of President Obama's collective rhetoric on Twitter would be housed.

Before the Library of Congress' announcement no one seemed to know where the archive would be held. The Government Printing Office, which prints daily and annual compilations of presidential speeches and documents, made no mention of President Obama's twitter account in its Daily Compilation of Presidential Documents. Although one could potentially argue that Twitter itself acts as a kind of archive, it seems essential to maintain a separate public archive for presidential speeches rather than relying on a private corporation for this service. Certainly scholars who are interested in rhetoric would be concerned with a newly emerging digital form of rhetoric that reaches over three million people and represents a direct and unfiltered line of communication between the president's administration and the public. Posts on Barack Obama's Twitter feed, which are increasing in frequency and, at the time of writing, numbered 672—appear multiple times during the day and reflect Obama's well-known rhetorical style. There are multiple accounts associated with Obama. One is maintained by Organizing for America, which represents Obama's agenda, and another is updated by White House staff. While neither is directly updated by him, both reflect the rhetoric of his administration. Like all forms of rhetoric, the Obama administration is using Twitter as another way to speak directly to citizens in an attempt to influence public opinion.

But how exactly is President Obama using Twitter? During his 2010 State of the Union Address, the speech itself appeared on Twitter, but only selected quotes were posted. The shortening of the speech itself was most likely due to the well-known 140-character limit placed on individual posts. The Twitter version of the speech was condensed to exactly 12 quotes and comprises a mere 270 words, much shorter than the full 70-minute speech, which was close to 7500 words. While one can safely assume that Obama hopes most American citizens will watch the more extensive speech, those who cannot watch, do not have time, or simply want the short version can read the condensed Twitter version.

How should scholars treat this new form of condensed digital rhetoric? And what questions does it open up for scholars who study presidential rhetoric? For political strategists, Twitter equates to a new way to communicate directly with the public and, in turn, creates new rhetorical speech-writing methods. These individuals will be faced with decisions such as how to work the tweets that make up the speech, so that it will have the most impact in its condensed form. Scholars interested in technology and communication will hope to understand the way in which media affects the message; in other words, they will hope to answer the question: what will be the impact of Twitter's 140-character limit on presidential rhetoric? Or, perhaps, scholars will study the way in which Twitter is contributing to an increasingly flattened communication system, one where nuanced arguments are becoming replaced with easily digested rhetorical chunks. Whatever the approach, one thing is clear; a new form of presidential rhetoric is emerging that needs new forms of analysis. With the Library of Congress' recent decision to archive Twitter, new and interesting questions will arise for scholars who focus on presidential rhetoric, communication, technology, and political strategy. President Obama's presence on Twitter demonstrates new ways in which presidents can use their rhetorical power; it also demonstrates the relationship that Obama intends to have with his audience of "tech-savvy" followers. However, most importantly it opens up presidential rhetoric to a new medium, which may, one day, fundamentally change the way that US citizens relate to their president.

# Bibliography

Bliumis, Benjamin and Robert Scoble (2010) "Tweets Alongside the Gettysburg Address: So Wrong." Retrieved April 25, 2010 from Businessweek.com: http://www.businessweek.com/debateroom/archives/2010/04/tweets_alongsid.html
Griggs, Brandon (2009). "Obama Poised to be First 'Wired' President." Retrieved April 25, 2010 from CNN.com: http://www.cnn.com/2009/TECH/01/15/obama.internet.president/index.html

# 2.3
# Web 2.0 Technologies and the Museum

*Seth Thompson*

While the definition of a museum is often contested, it appears that the notion of the museum has shifted and expanded from that of a storehouse or temple of objects to that of a visitor-centered educational repository of objects and information (Schweibenz 1998). This paradigm shift is evident in such books as Eilean Hooper-Greenhill's *Museums and the Interpretation of Visual Culture* and Gail Anderson's *Reinventing the Museum: Historical and Contemporary Perspectives on the Paradigm Shift.* Although many of the essays in Anderson's book exhibit the shift from being a collection-driven institution to a visitor centered one, it is Hooper-Greenhill's book that leads to ideas for the future of the museum, especially when it entails the use of computer technology in order for the museum to "play the role of partner, colleague, learner (itself), and service provider" (Hooper-Greenhill 2000).

Using digital media and, more specifically, Web 2.0 technologies, the museum may not only enhance its ability to act as a mediator between object and audience, allowing for visitors or users to learn, question and engage in ways that have not been possible before, but also change the notion of the museum from a material complex to that of an interface in which the museum itself becomes a communicative device. With a museum's digital assets such as images, video, audio and text, museum media design that employs Web 2.0 and social media technologies, while still being invested in the more authoritative Web 1.0 model as "publisher," potentially allow the museum to assume multiple roles as authority, partner and learner in regard to the assembling, dissemination and interpretation of knowledge. For example, the Indianapolis Museum of Art website (www.imamuseum.org) encourages online visitors to interact and share information using such social media applications as Facebook, Flickr and blogging for alternative learning and participation. ArtsConnectEd (www.artsconnected.org), a collaborative effort by the Walker Art Center and the Minneapolis Institute of Art, is an online resource primarily for K-12 teachers and students that enables the institutions to share their collections, archives and educational resources digitally, allowing users to sort and sift through its database electronically to gather content, research and annotate material to create customized thematic PowerPoint-like presentations for educational purposes. Launched in 2009, the "Make History" web project (makehistory.national911memorial.org), an initiative of the National 9/11 Memorial and Museum, is a model case of how individuals may create a collective memory of a time, place or event using Web 2.0 technologies. Users of the site have the ability to upload images, videos and personal stories to the site as well as search for different media through its database. These efforts are laying the groundwork for a more inclusive museum environment.

Nevertheless, at this point in time, a great deal of research and risk-taking is still necessary by museums to answer such questions as: How can museums employ questioning strategies to further engage the visitor using social media applications? How might museums employ video strategies to create more meaningful and engaging pre-visit and stand-alone educational material? And how can museums use internet-based technologies to create more visceral experiences?

Eilean Hooper-Greenhill writes,

> Where the modernist museum was (and is) imagined as a building, the museum in the future may be imagined as a process or an experience. The post-museum will take, and is already beginning to take, many architectural forms. It is, however, not limited to its own walls, but moves as a set of processes into the spaces, the concerns and the ambitions of communities ... [T]he post-museum will negotiate responsiveness, encourage mutually nurturing partnerships, and celebrate diversity.
>
> Hooper-Greenhill 2000: 153

If museums continue to recognize that Internet and web technologies can provide a platform that may build and engage audiences in meaningful ways, they may be able to achieve their educational vision by means and scope that may have never been imagined before.

## Note

Research for this article was supported in part by a Faculty Research Grant from the American University of Sharjah.

## Bibliography

Anderson, Gail (ed.) 2004. *Reinventing the Museum: Historical and Contemporary Perspectives on the Paradigm Shift*. Lanham, Maryland: AltaMira Press.
Besser, Howard. 1997. "The Transformation of the Museum and the Way It's Perceived." In *The Wired Museum: Emerging Technology and Changing Paradigms*, eds. Katherine Jones-Garmil and Maxwell Anderson. Washington, DC: American Association of Museums.

Howe, Deborah Seid. 2007. "Why the Internet Matters: A Museum Educator's Perspective." In *The Digital Museum: A Think Guide*, eds. Herminia Din and Phyllis Hecht. Washington DC: American Association of Museums.

Hooper-Greenhill, Eilean. 2000. *Museums and the Interpretation of Visual Culture*. New York: Routledge.

O'Reilly, Tim. 2005. *What is Web 2.0?* Retrieved from http://oreilly.com/web2/archive/what-is-web-20.html (accessed July 12, 2009).

Roberts, Lisa C. 2004. "Changing Practices of Interpretation." In *Reinventing the Museum: Historical and Contemporary Perspectives on the Paradigm Shift*, ed. Gail Anderson. Lanham, Maryland: AltaMira Press.

Schweibenz, Werner. 1998. "The 'Virtual Museum': New Perspectives for Museums to Present Objects and Information Using the Internet as a Knowledge Base and Communication System." Retrieved from http://is.uni-sb.de/projekte/sonstige/museum/virtual_museum_isi98 (accessed November 6, 2009).

Thompson, Seth. 2009. "Redefining the Notion of the Museum in the Digital Age: Web 2.0 Technologies and Contemporary Museum Theory." In *The International Journal for the Arts in Society*, eds. Mary Kalantzis and Bill Cope. Champaign, IL: Common Ground Publishing LLC.

# Part III
# The Texts of Digital Publishing

# 3
# Academic Publishing at the Crossroads

*John B. Thompson*

The new millennium is proving to be a testing time for academic publishers (Thompson 2005). Whereas the "long decade" from the early 1980s to 2000 was a buoyant period for many presses that were active in the field of academic publishing, including many of the university presses, the period since 2001 has been a rude awakening. Growth rates of university presses have fallen to the lowest levels in many years, returns from booksellers have reached unprecedented heights, and some university presses have been faced with the prospect of imminent closure. Nor has it been plain sailing for the big college-textbook publishers. Accustomed to annual growth rates of 6–8 percent, textbook publishers have suddenly found themselves faced with declining unit sales and surrounded by allegations that they are fleecing students with inflated prices (Lewin 2003). Why do academic publishers find themselves in such difficult circumstances and what, if anything, can they do about it?

To understand the problems of academic publishers today, we have to see that their current predicament is the outcome of a long process of development that stretches back to the 1970s and before. Curiously, and despite the unquestionable importance of books for teaching and the dissemination of knowledge in higher education, there has been no serious academic study of the modern book-publishing industry for more than two decades. Probably the best attempt to provide a detailed analysis of the modern book publishing industry is the now classic study by Coser, Kadushin and Powell (1982), but this is now very dated and it sheds little light on the profound changes that have swept through the industry over the last couple of decades. Of course, there is a long tradition of publishers writing about their own industry, and the recent books by André Schiffrin and Jason Epstein are two outstanding examples of the genre. In *The Business of Books*, Schiffrin (2001) offers an impassioned reflection on what has happened to publishing in the age of conglomerates, while in *Book Business*, Epstein (2001) reflects on the transformation of publishing from a cottage industry to a business dominated by big corporations and retail chains. But their accounts are inextricably entangled with their own experiences and career trajectories. These are not even-handed accounts of an industry in the midst of change, nor do they purport to be: They are personal memoirs with a critical edge, and they only underline the need for a more systematic study of the changes that are transforming the nature and prospects of the book-publishing industry today.

It is with that aim in mind that I set out, in the summer of 2000, to gain a deeper understanding of the industry in the US and Britain. My focus was on academic and higher education publishing. By "academic publishing" I mean the world of scholarly book publishing—the world of the university presses and commercial academic publishers, which despite their differences are united by the fact that they publish high-quality scholarly books (or what are commonly called "monographs"). By "higher education publishing" I mean the world of the college textbook publishers. To the outsider these worlds may look the same, but anyone who studies them in detail will discover very quickly that they are, quite literally, worlds apart.

To be sure, they are both involved in producing books for higher education, but they belong to universes that are structured in very different ways. I use the concept of "field," borrowed from the work of Bourdieu, to capture their different specific features. A field is a structured space of power and resources with its own forms of competition and reward. Markets are an important part of fields, but fields are much more than markets: They are also made up of agents and organizations and the relations among them, of networks and supply chains, of different kinds and quantities of power and resources that are distributed in certain ways, of specific practices and forms of competition, and so on. Each field has a distinctive dynamic that has evolved over time. The logic of the field defines the conditions under which agents and organizations can participate in the field and flourish or falter within it, that is, the conditions under which they can play the game.

So if we want to understand the problems faced by academic publishers today, we have to reconstruct the logics of the fields to which they belong. Only a deeper analysis of that kind will enable us to see that those problems are not temporary and superficial disturbances in an otherwise smoothly operating system, but rather are symptomatic of a profound structural transformation.

The field of scholarly book publishing has been shaped by two powerful dynamics that have trapped academic publishers, and especially American university presses, in a pincer movement. On the one hand, the kind of book that has been the standard fare of scholarly publishers—the monograph—has undergone a process of continuous decline since the 1970s. Experiences vary from publisher to publisher, but the overall pattern is indisputable. In the 1970s scholarly publishers in the US and Britain would commonly print between 2000 and 3000 hardback copies of a monograph and expect to sell a substantial proportion (if not all) of them. Scholarly publishing was a relatively straightforward business. For the most part, presses could take the market for granted and concentrate their energies on deciding which books merited publication. But by the 1990s that comfortable position had been radically transformed. Today most scholarly publishers find that the total sales of hardback-only monographs are often as low as 400–500 copies worldwide. As unit sales have fallen to a quarter or less of what they were in the 1970s, what was once a fairly straightforward and profitable kind of publishing has become extremely difficult in financial terms.

Why have monograph sales declined so dramatically? Is it because readers are turning to other sources of information like the Internet, as many observers have speculated? The main explanation almost certainly lies elsewhere. Research libraries constitute a principal market for scholarly monographs, and in the course of the 1980s and 1990s they were subjected to intense pressures of their own. One of the most significant pressures was the steep rise in the prices of scientific journals. The big scientific journals publishers like Elsevier realized that they had "must-have" content which was regarded by many academics as essential for their research; they had a stranglehold on the market for scientific journals and they exploited their position of dominance by ratcheting up the subscription prices year on year at several times the rate of inflation. This, together with the increasing costs of information technology in libraries, placed downward pressure on the purchase of monographs. Library budgets were limited and something had to give. In the period from 1986 to 1998–9, the number of monographs purchased annually by research libraries in the US declined by more than 25 percent; similar trends were evident in the UK. As academic publishers were also producing more monographs each year, an ever-increasing range of available titles was competing for a dwindling pool of resources. Academic publishers found that they were selling fewer copies of their monographs into university libraries and that the revenue generated per title was declining with time.

At the same time, many American university presses were coming under pressure from another source: their host institutions. In the 1970s and 1980s some began to find themselves faced with growing pressure to reduce their dependence on direct or indirect subsidies and become more autonomous financially—"self-supporting" was the term often used. Universities faced their own fiscal constraints, and university presses, with their somewhat ambiguous status (were they academic units or business units?), were an obvious target for financial scrutiny. How that has played itself out has varied from one institution to another. Some university presses have experienced substantial reductions in the levels of support received from their host institutions, while others continue to receive subsidies comparable to or even higher than earlier years. There is no single pattern here. But what has unquestionably happened is that university presses have been subjected to more intensive financial scrutiny than in the past. They have been forced to re-examine their practices and priorities and to introduce changes, including some radical and controversial changes, to the ways in which they have traditionally operated.

The professional lives of those working for university presses or other academic publishers over the last decade or two have been shaped above all by these two fundamental developments. Many strategies have been pursued, experiments undertaken and changes forced through, all with the aim of trying, in one way or another, to bolster the finances of academic publishers at the very time when the market for scholarly monographs has been collapsing. Perhaps the most significant change has involved the attempt to shift editorial programs away from monographs and toward other kinds of books that offer the possibility of generating more reliable revenue streams. In other words, university presses and other academic publishers have migrated into other fields.

Trade publishing has been a favored destination for many American university presses, as is regional publishing; some have also tried to commission more books likely to be adopted as so-called supplemental textbooks at colleges and universities. That is not so much a move downmarket or a sacrificing of quality on the altar of commerce as some critics like Schiffrin have suggested. Rather, it has been a perfectly sensible response to a logic of development that has transformed the field of academic publishing and forced academic publishers to look for new sources of revenue to sustain a monograph publishing program that is increasingly unsustainable on its own terms.

The paradoxical outcome of that development is that academic publishers can survive today only if they become something other than academic publishers, that is, only if they are able and willing to move beyond the field of academic publishing per se and publish different kinds of books for different kinds of markets. They are obliged to diversify their lists. The mix of books is the key.

Such pressures have been experienced by all academic publishers to varying degrees, but they have been experienced more intensely by American university presses than by their British counterparts, Oxford University Press and Cambridge University Press. That is partly because OUP and CUP are much larger and more international than American university presses, and partly because they had already diversified into other forms of publishing (most notably, English-language teaching, which was a major source of revenue) before the decline in monograph sales became a serious problem. But they are not immune to the changes that are sweeping through the world of academic publishing and they, too, have been forced to adapt.

The need to diversify presents new opportunities for academic publishers, but it also places new burdens on them. Now they must acquire new skills and develop new forms of knowledge and expertise about publishing fields that work in different ways, and about which they may know very little. They also find themselves faced with risks on a scale to which they have previously been unaccustomed. Trade (or "academic trade") publishing initially looked immensely attractive to university presses, but they have sometimes found themselves paying advances that are too high, overprinting books, underpricing them, seeing their margins squeezed by high discounts and finally ending up with a warehouse full of returns and unsold stock. With monographs it is easy to lose money; with trade books you can lose your shirt.

In many ways, moving into textbook publishing would have been a less risky and more sensible strategy for academic publishers faced with declining monograph sales; indeed, some of the British-based commercial academic publishers did exactly that. However, American university presses have been less inclined to move in that direction. Partly that is because editors in university presses have tended to understand their role as one of publishing original scholarly research, not providing pedagogical materials for higher education; the idea of publishing textbooks has seemed like a formulaic activity at odds with what they perceive to be the essence of scholarly publishing. But partly it is also because textbook publishing has been seen to be the province of the big college textbook publishers, like McGraw-Hill and the various subsidiaries of Pearson and Thomson, and it has been felt by many that those big players have pretty well sewn up the American market.

There is some truth in that, but only some. The field of college textbook publishing is a field with its own distinctive structures and dynamics of change. To understand how it works, one has to see that textbook publishing is based fundamentally on the adoption system, which means that textbooks are marketed to professors and sold to students. The professors are the gatekeepers in the marketing chain. But the person who recommends the textbook is not the person who buys it. Hence the considerations that weigh uppermost in the minds of the gatekeepers are not necessarily the considerations that matter most to the students ultimately required to buy the book. The adoption system thus creates a form of non-price competition—competition among publishers on grounds other than price that has shaped the evolution of the textbook publishing business.

In the attempt to persuade professors to adopt their textbook rather than the textbook of a rival company, publishers have invested more and more resources in producing ever more elaborate and comprehensive textbooks and in developing a range of ancillaries, from instructors' manuals and test banks to packages of software and multimedia products—the so-called package wars. But while the struggle for adoptions ratchets up the scale of investment, the only way of generating a return on that investment is through the sale of printed textbooks to students. Most of the electronic and multimedia supplements are given away to professors with the aim of influencing their adoption decisions. Thus the only way to recoup escalating costs has been to concentrate on lower levels of the curriculum, where student numbers are large, and to increase the prices of textbooks. The big textbook publishers have done both. They have concentrated on the first and second years of the college curriculum, and they have commonly increased textbook prices by at least 6–8 percent per year. But the increase in prices has tended to fuel a second development, which has played a crucial role in the field of textbook publishing: the growth of the used-book market.

The practice of buying and selling used textbooks is not new, but in the course of the 1970s and 1980s the used-book market in the US became increasingly national and organized. That was facilitated by the proliferation of used-book jobbers, who bought up used textbooks and sold them to retail outlets, and by the increasing involvement of the retail sector in the used-textbook business. The major college bookstores, especially national chains like Barnes & Noble and Follett, became not just retail outlets for publishers but also used-book brokers as they entered the increasingly sophisticated marketplace for second-hand textbooks.

While the rise of the used-book market has been good business for retailers, it has been disastrous for textbook publishers. It has meant, in effect, that the sales horizon of textbooks has been greatly shortened. Before the used-book market really took off in the 1980s, textbook publishers generally assumed an attrition rate of 10–20 percent; if a textbook sold, say, 20,000 copies in year one, publishers would generally assume that it would sell around 16,000 in year two. But with the rise of the used-book market, the attrition rate has sky-rocketed to 60–70 percent, in some cases to 80 percent or more. So now the textbook that sells 20,000 copies in year one will typically sell only 6000 to 8000 in year two, and by year three it will be dead in its tracks. With the dramatic increase in the attrition rate, the backlist revenue stream has dried up.

Textbook publishers have tried various strategies to counter the debilitating impact of the used-book market, but at the end of the day there has been only one strategy that works: speeding up the cycle of new editions. Some 20 or 30 years ago, textbook publishers would bring out a new edition of a successful textbook every four or five years, which would enable them to keep the book up to date and help ensure that they didn't lose market share to competing volumes. However, as the used-book market took off, publishers began to speed up the cycle of new editions to render earlier editions obsolete. Now the typical cycle for new editions of the most successful textbooks is two or three years.

Textbook publishers invest a great deal of time and effort in maintaining their successful textbooks in what is virtually a state of continuous revision. That is not an option but a necessity, because in the context of a flourishing used book market, the continuous revision of content and frequent repackaging are the only ways of preventing your assets from rapidly declining in value. Moreover, since a textbook now has only two years of effective life, all the costs involved in developing and producing the books, as well as the costs involved in producing a range of ever more elaborate ancillaries, have to be recouped in a very short time. That, in turn, places tremendous pressure on publishers to increase the price of textbooks. That dynamic, which lies at the heart of the textbook publishing business, can be sustained only as long as the end users, the students, continue to buy the textbooks that are adopted by their professors. But there are growing signs that that can no longer be taken for granted.

The student was always the silent partner in the traditional textbook model. Publishers listened carefully to the gatekeepers because they needed their adoptions to survive, but they did not pay much attention to students because they assumed that students would buy what they were told to buy. Now the silent partner is demanding to be heard in the only voice that really matters in this game. They are refusing to buy. They regard prices as too high and are inventing all sorts of ways to avoid doing the one thing they are supposed to do, which is to buy the books. They are borrowing books, sharing books, going online to shop around for the cheapest books they can find, and so on. Enterprising jobbers are importing cheaper foreign editions and undercutting the sales of American editions. Textbook publishers are experiencing increasing rates of return and declining levels of sell-through. They are worried, and the future is unclear.

So what is likely to happen? In the field of college textbook publishing, the struggles for adoptions will undoubtedly intensify, and there are likely to be further casualties, as the remaining big players continue to absorb smaller publishers and take over lists that are shed by other houses. Much ingenuity will be displayed in attempts to stimulate sell-through by offering a range of lower-cost alternative texts and customized editions and by setting up online bookstores to sell directly to students. Further down the road, textbook publishers may be able to reduce their production and distribution costs by disseminating more content online, but so far the experiments with online textbooks have, for the most part, been disappointing.

As for the university presses, many face an uncertain future. At a time when colleges and universities are facing growing pressures on their finances, some institutions may be forced to make tough decisions about what they regard as essential, and it cannot be taken for granted that a subsidized press will always fall on the fortunate side of the line. Without wishing to suggest that these points are particularly novel, let me highlight five measures that university presses can take to increase their chances of survival.

First, university presses need to reform their monograph publishing practices. Of course, they have changed those practices in many respects over the last decade, but they still tend to publish monographs in ways that were designed for earlier market conditions. The books are overproduced and underpriced. Unlike Oxford, Cambridge, and commercial academic publishers in Britain and Europe, American university presses have been reluctant to increase monograph prices.

There are various reasons for that. Partly it is because they believe, rightly or wrongly, that the market for monographs is elastic, and they will sell more copies to individuals if they keep prices low; partly because they tend to take price bearings from other university presses (and from New York trade houses), and no press wants to be seen to adopt a more aggressive pricing policy than its competitors; and partly because they believe that making scholarship available at affordable prices is part of the mission of university presses. But that is beginning to change, and university presses will come under increasing pressure to gear their prices more accurately to the costs they are actually incurring in their monograph publishing programs.

Second, they should be more selective in terms of their list-building activities. Again, some presses have introduced measures to reduce the number of monographs they produce and to redirect their commissioning activities, but they should be prepared to be even more proactive. The growth in monograph output over the last couple of decades has been driven not by an overall growth in demand but by a combination of other factors (including the demand from academics for credentials that can be used in the tenure-and-review process and the short-term need of presses to meet their sales forecasts). Publishing fewer monographs and concentrating only on works of outstanding quality might result in some friction with local faculty members, and some temporary shortfalls in front list revenue, but if it is accompanied by an effective shift of editorial strategy to other kinds of commissioning, it would strengthen the position of the presses in the long run.

Third, university presses will have to look to other sources of revenue to support themselves. It will not suffice simply to improve the way they publish monographs. While the American university presses have tended to look to trade publishing as a way of generating additional revenue, that is, as we have seen, a path strewn with dangers. It would be prudent for the presses to take a more cautious view of the trade potential of their books and to devote more effort to commissioning other kinds of books, including reference works and books that stand a good chance of securing adoptions at colleges and universities. The presses could strengthen their positions considerably by focusing their attention on publishing for the higher education market, especially for those levels of the curriculum, like upper level undergraduate and graduate courses, that have been neglected by the big textbook publishers, which have been forced by the logic of their own field to concentrate on the lower levels of the curriculum. The commissioning of textbooks and supplemental texts would not compromise the commitment of the university presses to publish original works of scholarship, but would be complementary to it and entirely consistent with their overall educational mission.

American university presses could learn some lessons here from the success that some commercial academic publishers in Britain have had in publishing for the higher-education market. Faced with the same spiraling decline of monograph sales, British-based publishers like Blackwell, Palgrave Macmillan, and Routledge have refocused much of their editorial activity on developing textbooks for university courses. That transition has been facilitated by the structure of the higher education market in Britain, which, compared with the US, is more open and less dominated by large publishing corporations. But some British publishers have also succeeded in securing good adoptions for upper level undergraduate and graduate courses in the US, demonstrating that there is a demand not wholly met by large textbook publishers.

Fourth, university presses need to put more effort into managing their relations with their host institutions and with the academic community they serve. Most academics are woefully ignorant of the real financial conditions of scholarly publishing and the changing circumstances that have left university presses in such difficult straits. They depend on the presses to publish their work, to maintain the vitality of their disciplines, and to lubricate the processes of recruitment, tenure, and promotion, and yet they generally know precious little about the forces driving presses to act in ways that are sometimes at odds with the aims and priorities of academics.

If scholarly monographs are valued by academics as a means of disseminating research and advancing knowledge in their disciplines, then the academic community cannot relinquish all responsibility for ensuring that the organizations saddled with the task of making monographs available are able to do so without being driven to the wall. Expecting university presses to continue to publish monographs in the context of a declining market and without active support is willing the ends without the means. The monograph can survive only if the academic community actively supports it with a vigorous defense of, for example, library book-acquisitions budgets, the provision of subsidies for specific publication projects, and, above all, the willingness of host institutions to support their university presses in difficult times.

Fifth and finally, university presses can achieve real benefits, both organizationally and financially, from the intelligent use of new technologies, but for the foreseeable future the main benefits of new technologies are likely to be quite different from those envisaged by many commentators in recent years. There are many who have dreamed of "an electronic solution to the monograph problem," hoping that the digital revolution will provide the opportunity to discard the printed monograph and replace it with the online dissemination of scholarly book content. While various experiments are still underway, and their outcome is unclear, we now know enough to draw one reasonably firm conclusion: just as the digital revolution was not the origin of the problems faced by academic publishers today, so too it is unlikely to be their salvation.

The main reason technology will not rescue academic publishers is not because there are unresolved technical problems; most of the technical problems associated with digitizing book content and making it available online have been solved. The problem is that the market for the content of books delivered in electronic format is, at present, nowhere near as robust as many people once thought it would be. Whether books are sold as individual e-books or embedded in databases of content marketed through subscriptions to libraries and other institutions, most publishers who have experimented with online dissemination have been disappointed with the low levels of revenue generated. Indeed, many have found that, so far, the costs of employing technical staff members, developing software platforms, digitizing content, clearing copyright permissions, dealing with license agreements, and so on, have significantly exceeded any new revenues gained.

That could change in the future, although on the evidence to date it would not be wise to count on it. But regardless of online dissemination, there are real benefits to be gained by using new technologies in the world of academic publishing. The digital revolution is well underway in book publishing, but it is not so much a revolution in product as a revolution in process. The final product may look the same as the old-fashioned book, but how it is produced is being radically transformed. Among the many benefits of that hidden revolution is that it is increasingly enabling publishers to exercise much greater control over the management of their resources and stock, through, for example, digital printing and print on demand.

Academic publishers are going through difficult times, and they are unlikely to find any magic bullets, technological or otherwise, to resolve the problems they face. But by reconstructing the dynamics and history of the fields of which they are a part, we can gain a firmer grasp of what is at stake, and glimpse some of the steps, however modest, that presses can take to ensure they have a future.

## Bibliography

Coser, Lewis A., Charles Kadushin and Walter W. Powell. 1982. *Books: The Culture and Commerce of Publishing*. New York: Basic Books.

Epstein, Jason. 2001. *Book Business: Publishing Past, Present, and Future*. New York: W.W. Norton.

Lewin, Tamar. 2003. "When Books Break the Bank: College Students Venture beyond the Campus Store", *New York Times*, September 16, 2003, section B, 1.

Schiffrin, André. 2001. *The Business of Books: How International Conglomerates Took Over Publishing and Changed the Way We Read*. London: Verso.

Thompson, John B. 2005. *Books in the Digital Age: The Transformation of Academic and Higher Education Publishing in Britain and the United States*. Cambridge: Polity.

# 3.1
## The Open Textbook: From Modules to Mash-Ups

*John W. Warren*

A growing movement by students, parents, and professors protesting at the high price of traditional textbooks in higher education, and denouncing the weight of textbooks in K-12, has given impetus to an increase in digital textbooks. Scholars, publishers, institutions, and policymakers are struggling with trade-offs, real or perceived, which exist between open access and publisher-controlled content, as well as between fair use and the protections offered by copyright. Open source textbooks—many employing Creative Commons licenses that allow others to share and build on the work of content creators, consistent with the rules of copyright (see creativecommons.org)—are a growing trend making an impact on higher education, and to a lesser extent, thus far, on K-12 education.

Flat World Knowledge, a start-up company backed by over $10 million in venture capital, offers free, online, peer-reviewed textbooks on its website (www.flatworldknowledge.com). The business model is to provide content for textbook adoption that is as good as or better than the textbook a professor currently uses, offer content for free, and encourage the purchase of add-on and convenience products in multiple formats. While the online version of a textbook is free, students can buy a PDF download of a book or chapter, purchase a black-and-white print-on-demand (POD) version for about $30, or a color POD version for about $60. The PDF download also includes print-your-own capability (Frank 2009). Flat World Knowledge also provides tools for professors to create custom books, editing, or adding to content at the sentence level, delivering unique books as online, download, and POD versions to students.

So far, Flat World's model of publishing open source college textbooks appears to be working. Adoption has increased from approximately 1000 students in 30 schools in spring 2009 to approximately 40,000 students at 470 schools in fall 2009. Every chapter of every book includes digital study guides such as flashcards, practice quizzes, and audio guides. Approximately 65 percent of students make some kind of purchase, averaging about $30. Authors enjoy the incentives, which include faster time to publication, a greater ease of creating and updating texts, and a 20 percent royalty on any sale. Royalties are more consistent over time; there isn't a steep drop off of sales and royalties as when a traditional textbook hits the used book market (Frank 2009).

Flat World Knowledge plans to integrate more assessment into the texts. Performance data will help professors better teach their courses, while aggregated, anonymous performance data will help authors develop better texts with the understanding of which modules or concepts need refined explanation or additional material. Self-assessment is one of the key factors that can be automated in digital texts and help students, professors, and authors (Young 2009).

Another example is Connexions (cnx.org), developed by Rice University, which offers a collection of free, open-licensed educational materials in fields such as music, electrical engineering, and psychology, with an expressed mission of offering students (and their parents) an alternative to expensive college textbooks. Connexions presents scholarly content in modular, non-linear format, encourages sharing and collaboration, and claims to reduce the time to publication.

The site, while experimenting with different models of peer review, generally relies on market forces for course review, namely that many users will link to interesting and informative courses and few to the not so interesting or informative courses. Material on the site is offered at no charge, with PDFs of texts available for download also at no charge. Authors and the site are supported, however, through the sale of printed textbooks, in most cases using POD and in some cases in a variety of price-points and feature sets. As of February 2010, Connexions offers 15,831 reusable modules in 977 collections (up from 6989 modules in 393 collections in October 2008), and offers content in 27 different languages, including Chinese, Italian, Japanese, Portuguese, Spanish, and Thai (Connexions 2010).

In the K-12 market, the CK-12 Foundation hopes to make digital textbooks affordable for districts and schools, and more adaptable for teachers and classrooms, by providing access to free texts aligned to state standards with developmentally correct content. CK-12 offers tools to create, distribute, and customize high quality educational content in an open-content, web-based, collaborative model termed the "FlexBook." Educators can create customized digital text from existing texts, chapters, and web pages under a Creative Commons Attribution-ShareAlike license. When seven FlexBooks were submitted by CK-12 to the California Learning Resource Network for state textbook adoption in math and science, they met the state's academic content standards by an average of 95 percent; four met 100 percent of the standards, and none scored below 82 percent (Khosla 2009).

The foundation hopes to encourage collaborative learning via a community where authors, teachers, and students create, access, share, rate, recommend, and publish these free texts. The texts are currently provided through a combination of author donations, licensing partnerships, incentives for community-based authorship, and university collaborations.

Open access business models, while attractive from a common-goods standpoint, are still largely unproven for long-term sustainability. "Free texts" depend on success of "bundling," print-on-demand print sales, enhanced products or services, and/or foundation, grant, or endowment support (Jaschik 2009). Interactive, participatory learning spaces using assessments are expensive to produce and maintain, but may ultimately cost less on a variable cost basis than printed equivalents. Eric Frank reports that the average multimedia textbook costs approximately 30 percent more than a straight digital text, while stressing that there is really no "average" (Frank 2009). The ability to repurpose digital content and spread costs over a range of projects allows authors, publishers, and producers to create content for niches and market segments that would not have been feasible with traditional publishing. A not insignificant, but frequently overlooked, factor is the difficulty in changing the mindsets of educators who claim not to have the time to contribute, and it may not only be a technological capability (digital native teacher) but a very real work overload that proscribes the ability to customize and contribute (Khosla 2009). Keeping content contextualized to local, regional requirements as well as global curriculum standards is another significant challenge. Long-term success of the open textbook model will depend on providing clear value to users, minimizing direct costs, and developing diverse revenue sources including subscriptions, licensing to publishers and users, custom services, corporate sponsorships, author fees, endowments, and grants (Maron, Smith and Loy 2009).

Today's new and emerging technologies require a broad rethinking of books for learning, testing, and scholarship. Lines are already becoming blurred between online learning and digital textbooks, and between producers and consumers of content. This emerging world of open textbooks will produce "mash-ups"—combinations of disparate bits of digital video, audio, text, and graphics refashioned into something new—that will change the way we read and publish.

## Bibliography

Connexions Web site. http://cnx.org and http://cnx.org/aboutus (Accessed February 4, 2010).

Frank, Eric, Co-founder, Flat World Knowledge, phone interview, September 9, 2009; also Flat World Knowledge web site. http://www.flatworldknowledge.com (Accessed October 1, 2009).

Jaschik, Scott. 2009. "The World is Open," *Inside Higher Ed*, August 25. http://www.insidehighered.com/news/2009/08/25/bonk (Accessed October 1, 2009).

Khosla, Neeru, Co-Founder and Executive Director of CK-12 Foundation, phone interview, November 18, 2009; also CK-12 Foundation web site http://www.ck12.org/flexr/ (Accessed October 4, 2009).

Maron, Nancy L., K. Kirby Smith and Matthew Loy. 2009. "Sustaining Digital Resources: An On-the-Ground View of Projects Today: Ithaka Case Studies in Sustainability," Ithaka S+R, July 2009. http://www.jisc.ac.uk/contentalliance (Accessed October 1, 2009).

Warren, John W. 2009. "Innovation and the Future of E-Books," *The International Journal of the Book*, 6, (1): 83–94. http://ijb.cgpublisher.com/product/pub.27/ prod.273; also at http://www.rand.org/pubs/reprints/RP1385/ (Accessed October 10, 2009).

Warren, John W. 2010. "The Progression of Digital Publishing: Innovation and the E-volution of E-books," *The International Journal of the Book*, 7, (4), 2010: 37–53. http://ijb.cgpublisher.com/product/pub.27/prod.369; also at http://www.rand.org/ pubs/reprints/RP1411/ (Accessed October 2, 2010).

Young, Jeffrey R. 2009. "New E-Textbooks Do More Than Inform: They'll Even Grade You," *Chronicle of Higher Education*, September 8, 2009. http://chronicle.com/article/ New-E-Textbooks-Do-More-Tha/48324/ (Accessed October 1, 2009).

# 3.2
## Community and Communion: Books as Communal Artifacts in the Digital Age

*Yasmin Ibrahim*

Benedict Anderson in *Imagined Communities* (1983) and Homi Baba in *Nation and Narration* (1990) contend that books capture the collective consciousness of a society and the zeitgeist of a community. In the process they reveal our cultural codes, providing insights into our consumption patterns, preferences, tastes, and social make-up (Bourdieu 1984). In the digital age books have forged new forms of communities and communion. This has happened through the intertextuality between media platforms such as television and the Internet as well as the nature of the electronic environment, which has enabled new forms of communities to flourish.

In the digital age, books can appear as online text-based versions or as spoken-word audio books. The changing political economy of book production in the digital age and the de-materialization of the book have provided new platforms for the emergence of online communities which convene through literary interests. The digital media has created and transformed the marketplace for many commodities and in the process it has also created new forms of interactions, social exchanges, and networks. With new marketplaces such e-Bay and Amazon, and search engines such as Google, new economies as well as new forms of circulating information about books, forums, and discussion groups have emerged (Hillesund 2007).

Although the digital revolution is often seen as causing the demise of physical libraries, many have argued that in the digital age the traditional social role of libraries has not changed. Public libraries are still charged with the responsibility of preserving forms of recorded knowledge, fostering public education, and building the core foundations for a democratic society (Gorman 1998: 22; Lang 1996; McClelland 2003). The advent of the e-book also raises age-old debates about new technologies revolutionizing and obliterating old ones. The book industry today is characterized by the concentration of publishing firms in large groups owned by multinational media conglomerates, the concentration of booksellers in large bookseller chains, and increasing globalization and digitization of production through the introduction of new technologies (Thompson 2005).

A community is not limited by practice, interest, or the like, but should be regarded in the widest sense where multiple communities make up a society and where one person can belong to more than one community (Pang et al. 2006). This means online communities are just as significant as offline groups in fostering a sense of belonging, participation, and identity (see Rheingold 1993; Wellman et al. 1996; Chan et al. 2003). In the digital environment the book can be socially constructed through discussions about its content in book clubs, blogs, and forums. Thus virtual communities can form through discourses about literary content and in the process the electronic platform can create virtual communities which occur through such discussions.

In earlier periods niche literary communities such as book groups functioned to supplement the educational needs of women (Long 2003). In the age of television and the mass media, book clubs and groups have evolved into something altogether different through celebrity endorsements by television and radio personalities and popular genres on different media. The proliferation of websites devoted to literary matters, as well as the increase in book clubs, literary festivals, and reading clubs, have provided new avenues to engage with books and to form new networks and communities. New technologies have also meant the emergence of new audiences on a potentially global platform, thus enabling people to discover what others are reading (Lichtig 2006). The Internet has also provided new platforms for people to become authors, reframing the rigid dichotomy between writers and readers.

In the age of interactivity and convergence, beyond physical communities, books have created new forms of virtual fraternities online where literary discussions can be crucial for the sustainability of a community, its identity, and sense of belonging.

## Bibliography

Anderson, B. (1983). *Imagined Communities: Reflections on the Origin and Spread of Nationalism*. New York, NY: Verso.

Baba, H. (1990). *Nation and Narration*. London: Routledge.

Bourdieu, P. (1984). *Distinction: A Social Critique of the Judgement of Taste* (trans. R. Nice). Cambridge, MA: Harvard University Press.

Chan, H., H. Teo, K. Wei, and Z. Zhang, (2003). "Evaluating Information Accessibility and Community Adaptivity features for Sustaining Learning Communities." *Journal of Human Computer Studies*, 59(5): 671–97.

Gorman, M. (1998). "Living and Dying with 'Information': Comments on the Report 'Buildings, Books and Bytes'". *Australian Public Libraries and Information Services*, 11(1): 22–8.

Hillesund, T. (2007). "Reading Books in the Digital Age Subsequent to Amazon, Google and the Long Tail." *First Monday*, 12 (9). Retrieved 05/11/2007, from http://firstmonday.org/issues/issues12_9/hillesund/index.html.

Lichtig, T. (2006). "Can the Book Survive in an Era of Sound Bites and Texting?" *Battle of Ideas Website*. Retrieved 05/10/2007, from http://www.battleofideas.co.uk/C2B/document_tree/ViewADocument.asp?ID=272&CatID=42

Long, E. (2003). *Book Clubs: Women and Uses of Reading in Everyday Life*. Chicago, IL: University of Chicago Press.

Pang, N., T. Denison, G. Johanson, D. Schauder, and C. Williamson (2006). "Empowering Communities through Memories: the Role of Public Libraries." Paper for the Conference Proceedings of Constructing and Sharing Memory: Community Informatics, Identity and Empowerment, CIRN: Prato October 9–11, 2006.

Rheingold, H. (1993). "Virtual Community: Homesteading on the Electronic Frontier." Cambridge, MA: MIT Press.

Wellman, B., J. Salaff, D. Dimitrova, L. Garton, M. Gulia, and C. Haythornthwaite (1996). "Computer Networks as Social Networks: Collaborative Work, Telework, and Virtual Community." *Annual Review of Sociology*, 22: 213–38.

# 3.3
# Is the 'E-Incunabula' the One Solution for Scientific Communication?

*José Morillo-Velarde*

## Incunabula

In 1452, with the appearance of Johannes Gutenberg's so-called 42-line Bible and the dawn of the incunabular era of the printing press, two seemingly unrelated realms—technology and culture—merged. The result was the new Renaissance society, where easier access to sources of information and knowledge led to the questioning of all assumptions of the medieval world. That wave has reached even unto our day although, in the 1970s, it was engulfed by another wave that was orders of magnitude larger, the one spawned by digital media.

Prior to the printing press, the instrument for establishing and communicating knowledge was the codex. Monks were occupied in creating it, and although it achieved lofty heights of aestheticism and *auctoritas*, access to it was rare, controlled, and difficult.

In this first stage, books were not able to break free of this model and even replicated its flaws. Books emulated the codex typography and text layout as well as its lack of titles and indexes. Printers were mindful of their production capacity but did not anticipate anyone having to locate a particular piece of information within numerous arrays of works. They had discovered movable type but did not optimize its manufacture, so manuscripts were still fairly difficult to read.

Half a century had to pass before the book escaped this bondage and found its own voice. Libraries were acquiring more and more books, and a person with an intellectual endeavor—we cannot yet speak of scientists—needed to have more and more books available and be able to find information in them as quickly as possible.

Over time, this proliferation of works led to the appearance of evaluation methods that would help the reader to choose.

## Periodical publications

Let us skip ahead in time to where we encounter periodical publications—the sphere in which, since the eighteenth century, scientific information has found expression. Prior to the advent of these publications, scientific communication took place in closed circuits. The need to make science accessible led to the emergence of new types of works in print that were periodical in nature. Eventually, other types of works evolved, such as the concept of scientific articles and peer review. In science, it is particularly important that works be evaluated before publication so that scientists do not have to read through a massive number of papers; thus, besides adherence to these basic standards, bibliometric indicators were used to establish rankings for publications in each specialty.

## E-incunabula

Because the state of affairs described above is so long-established and influential, we attempt to reproduce its model in our day. From a technological perspective, however, the concept of periodicity makes little sense since it is possible to publish almost instantaneously. Dispensing with peer review, we could proceed to a trial publication and wait for the scientific community to accept or not accept our assumptions. We could do without the "article" genre because reading is no longer necessarily linear. Search engines lead us directly to the line of text, even in the middle of a document, that contains what we are looking for. We could substitute the number of accesses and other factors related to the social web for the impact factor. In making these proposals, however, what we are questioning is not how scientific information is communicated, but rather the very foundations of science and its method. The opposite scenario is not more promising. If we remain in this era of emulating materials in print, we will be squandering opportunities the new technologies offer us. Even worse, we will be consolidating privileges that are associated with obsolete technological and economic practices.

## A new state of affairs

Food for thought for scientific communication:

1. Revision of evaluation methods:
   - peer review;
   - prestige of repositories;
   - evaluation of published materials and integration of responses into new communications models.

2. Revision of the scientific article model:
   - new forms of writing;
   - intertextuality as a way of structuring texts.

3. Revision of the model of competition among scientists:
   - new selection criteria for fields of research;
   - institutional evaluation of digital publishing.

# Part IV
# The Digital Citizen

# 4
# Digital Citizenship

*Timothy W. Luke*

## Introduction: Space and action

Digital citizenship is an attractive concept, and mapping the properties of those technologies carrying its forms of enactment in contemporary society is imperative. Ranging from open online elections to surreptitious info-warfare attacks, turning from online match-making to text-message kiss-offs, or running with cell-phone video reporting to Twittering poetry slams, contemporary society is becoming more commonly recognized, in part, as interactions of digital beings. As Internet-life enters its fifth decade, the infiltration of bit-driven modes of action, as well as the insertion of bit-built modes of structure into the material practices of strong states, weak states, and non-states, cannot be denied. In the US, the growing numbers of new cyber-assemblies, ranging from this party.org, association.net, and issue.com, to that company.com, agency.gov, and mil.net, are interacting in concert and/or contention against comparable info-collectives operating at other sites in different established states. Often politics on the Internet is a bit more than the sum total of politics off the Internet as blogs, video servers, and archives enliven ongoing face-to-face debates, but at other times closed worlds of online organizing, debating, fund-raising, voting, managing, or ruling erupt in far greater political struggles.

Citizenship traditionally has been rooted to residence in, and service to, a city and state. Common service and shared habitation, then, marked the duties and rights of citizens. How the informatic reconstitution of these relations, by virtue of the properties of digital technology, changes citizenship as well as how the inhabitation of cyberspaces by "cybercitizens" reshape many mappable parameters of common identity, purpose, and meaning are important questions (Dertouzos 1997). Lefebvre asserts the careful analysis of any social order must investigate "spatial practice," because the material processes involved in developing those locales and activities are what "secretes that society's space; it propounds and presupposes it, in a dialectical interaction" (1991: 38). When considering today's web-based modes of political activity, one can argue that cyberspatial practice clearly "embodies a close association, within perceived space, between daily reality (daily routine) and urban reality (the routes and networks which link up the spaces set aside for work, 'private' life and leisure in the mental and material realms of life)" (1991: 38). These machinic materialities are always foundational. Lefebvre, however, asks all to note how perceived spatial practices fully express the "representations of space," which typically express the dominant order of society and production. It is there that one finds "conceptualized space, the space of scientists, planners, urbanists, technocratic subdividers and social engineers ... all of whom identify what is lived and what is perceived with what is conceived" (1991: 38). Finally, Lefebvre also suggests closer attention must be given to investigations of "representational spaces," or "space as directly lived through its associated images and symbols, and hence the space of 'inhabitants' and users ... this is the dominated—and hence passively experienced—space which the imagination seeks to change and appropriate" (1991: 39). Since informatics technology can be regarded as the key limit and main expanse for these spatial practices, one can consider how cybercitizens, or digital citizenship, typify that interplay of practice, thought, and activity "which exists within the triad of the perceived, the conceived, and the lived" (Lefebvre 1991: 39) online in various networks.

Whether as wireless environments or as hard-wired infrastructures, digital domains for cyberspatial agency operate globally under, across, over, and through nation-states in ways that are yet to be fully understood (Beck 2000). One can blog about legislative affairs anywhere, any time, and any way that digital platforms allow it—whether or not one is in-country, getting public goods, doing public service or even a "native" citizen. Thinking about how these realities complicate the mechanics and logistics of citizenship discloses quite fluid textures of networks' underness, acrossness, overness, or throughness in all nation-states, both in lived and conceived space. Here, seeing variability is crucial, and new codes for disclosing how technics mediate social subjectivity and structure as online spatiality is critical. Web-browsing in the US does not necessarily involve the same set of activities as surfing the Internet in the Peoples' Republic of China. Informatic technology, whether it is regarded as techniques, technics, and tools, "is not simply a means of reading or interpreting space: rather it is a means of living in that space, understanding, and producing it" (Lefebvre 1991: 47–8). Instead, individual coexistence in, with, and by the codes of network technologies is caught within the first means of governance over the spaces of social practice, and those means continuously work at positioning social practice in a new habitus, lived; intuitus, perceived; and, intellectus, conceived, spaces. Like the commercial activities of commodification itself, "every space is already in place before the appearance in it of actors; these actors are collective as well as individual subjects inasmuch as the individuals are always already members of groups or classes seeking to appropriate space in question" (Lefebvre 1991: 57). In contemporary modern life, the commodity usually contains and conceals technology, while technology reciprocally conceals and contains commodification. Strangely, digital citizens often militantly rail against big government, but then naively allow their many personal digital devices to control their daily schedules, travel routes, personal communications, work lives, and individual identities far more than any intrusive state bureaucracy.

Nature, which once was held forth as a supreme force in its own organic right, is being superseded or infiltrated by informatic technologies. One is left with these new second natures, processed worlds, or postmodern conditions of contemporary performativity. Within such spatial practices, those who own and control the material and mental means of enforcing order decide what sets of "doing-how" skills will be embedded in the codes of "knowing-how" abilities, and such decisions concretize new inequalities on a global scale. As spatial practice, digitalization fractalizes its effects across many spaces. It is global and local, industrial and agricultural, urban and rural, built and unbuilt. Technē as informationalized living brings together humanity and nonhumanity into complex regimes of big, if also brittle and bloated, embedded practices. A perfect example of such forces in motion and in place is "the grid," "the system," and/or "the network" that support digital citizenship. Such technics, once again, are merged into "the perceived, the conceived, and the lived" (Lefebvre 1991: 39) of existence as the "deep technology" of cyberspace. Cybernetic technology in one sense, then, is the spatial imbrication of many communities imagining, embedding, and engineering their lives only as informatic technoculture when they occupy the same spaces, keep common times, run parallel lives, and experience comparable normalities. Manifolds of MySpace are OurSpaces, but they also are TheirSpaces when subjects meet only in "the shallows" (Carr 2010) of political informatics, digital marketplaces, or online collectives.

## From citizens to netizens?

Foucault's distinctive vision of discursive formations, such as those surrounding political informatics, e-governance, or digital citizenship, indicates that "truth values" are unstable. Their worth often rebounds more fully in the realm of daily operationality rather than in the domains of final veracity per se. That is,

> What rule could it be obeying by both its existence and its disappearance? If it contains a principle of coherence within itself, whence could come the foreign element capable of rebutting it? How can thought melt away before anything other than itself? Generally speaking, what does it mean, no longer to think a certain thought? Or introduce a new thought?
> (Foucault 1994: 120)

Once broken down into bits, both new and old thoughts about subjects and structures can circulate more easily, change quite often, and coordinate constantly on the fly.

It may well be unwanted from above and resisted by below, but digital citizenship arrives and accumulates "in-between," as flows of web-sustained utility. Electronic media increase interconnections for common memorable meanings, and singular shared places that are shaped from pixels, bits, or sound waves stored for continuous recall and reuse. Flowing through every city and town with access, the reifications of exchange behind new digital discursive and cultural formations blink on and off globally and locally as a new world order (Luke 1994). As Virilio notes,

> in fact, there now exists a media nebula whose reality goes well beyond the frontiers of the ghettos, the limits of metropolitan agglomerations. The megalopolis is not Mexico City or Cairo or Calcutta, with their tens of millions of inhabitants, but this sudden temporal convergence that unites actors and televiewers from the remotest regions, the most disparate nations, the moment a significant event occurs here or there.
>
> (Virilio 2000: 69)

Transformational change can strike from above to serve those who are way ahead or far outside. It also can be felt, however, as another side of transnational flows, as those below, inside, and behind converge in the shared space of informationalized global "in-betweenness" (Luke 1995) as system users in web-mediated events. To think about "the Internet and politics" is perhaps now too late in a world rewoven by the web: the Net is always already politics, economy, and culture for all the agents and structures now operating politically as governance objects of, for, or by this system of systems. No web science will be worth conducting until this foundational reality is acknowledged by cybercitizens in all of digital citizenship's theories, practices, and methods.

Once appraised in this fashion, however, the discursive formations of this media nebula also help to materialize crucial forms of technified concourse in which places are found, circulation and exchange are practiced, transformations take hold, and scarcities managed. Running to and fro as concourse implies running together, and informatics concursivities for digital citizens channel their undetected energy, convey the hidden information they need, or contain the dark matter of structures and relations not otherwise unaccounted for within the ordinary accountancy of economic exchange or political authority.

As Lefebvre suggests, the spatial practices of technologies deployed in society do secrete the space their users occupy and utilize. These connections embrace, production and reproduction, and the particular locations and spatial sets characteristic of each social formation. Spatial practice ensures continuity and some degree of cohesion. In terms of social space, and of each member of a given society's relationship to that space, this cohesion implies a guaranteed level of competence and a specific level of performance.

(1991: 33)

Many of the technostructures propounded and presupposed by the secretion of such space will work only if citizens are accustomed to performing rightly or wrongly in them. Amicable compliance—derived from individual competence and collective performance at particular locations with certain spatial settings—cannot be changed without remaking spatial practices, because technologies create, and then conduct, the strong normative expectations of everyday commodification's spatial code. Every engineered system, then, constitutes its own normative order embedded in materiality through the services of its everyday ordinary normality, whether it is leveraged daily as an element for governance either where it sits or when it is deployed to other sites. This fact is also true of all materials, as well as any agents serving as their caretakers, managers, or vendors. Whether or not this normality is accidental, it plainly is materially normative in the production and consumption circuits of everyday urban life.

Near the close of the Cold War, sophisticated computing use was changing rapidly. With this shift in the early 1990s, the spread of PCs and good browsers of WWW content rapidly advanced the options for digital citizenship, as the information-processing environment shifted, first to thousands and then to millions of individual desktop and laptop computers, first using wired and later more wireless access (Luke 2001: 153–74). As the computing and communication networks became more accessible, inexpensive, and ubiquitous, e-governance experiments were tried in many different settings.

Leveraging this institutional transformation, citizens began toying openly and continuously with e-governance test protocols through the pull of their pooled digital resources and the densely clustered points of and for interactivity that have framed a "digital" citizenship. Such domains are, as Crang and Thrift observe, new non-private spaces best apprehended as being a "process and in process" (2000: 3). As more political "content" cascades out of such informational processing, its root machinic network practices must be made into more flexible spatial formations whose "process in process" enables numerous networks of unknown, unacknowledged, and perhaps even unknowable people to share their political agendas, ranging from revolution on the streets of Tehran or Tbilisi to relief for the victims of war in Sudan or the Congo, round the clock and all around various jurisdictions in the real world. No digital discourse in politics is truly just dematerialized digits, but because it is hollowed out, accelerated, compressed dematerialization—following as pixilated images, digitized sounds, or hypertext linked—it changes the practices of public engagement online (Luke 2006: 197–210).

Still, these bigger changes implied by digital citizenship are only the beginning. With desktop, laptop, or netbook computers still serving as the major access portals and/or personal accumulation points of many political texts, the reading and writing, or capturing and watching, economies of digital political life are caught at a historical conjuncture for existing modes of political literacy. Cities are where citizens are expected to be always at fairly fixed points of intersection offline, but cybercitizens can now interact anytime from anywhere in ways that challenge the traditional notions of fixed citizenship. Embedded intelligence, smart objects, ubiquitous computing, and the so-called m-revolution, which mobile information and communication systems pull together in new ICT assemblages, are quickly recontouring life as continuous netizenship. The fixed material links of e-texts to relatively costly, hard to move, and expensive office-proposed computers are being broken by many smaller, cheaper, multipurpose mobile wireless devices, from cell phones to e-readers to tablet computers, which can, in a fashion, approach modes of versatility, simplicity, and mutability for digital interactivity comparable to the paper book. Such new mobile devices alone are expected to number over one billion by 2010.

In addition, these devices' integration into former citizenship-based social systems, such as state-based regimes of effective monetary payment, electronic structures of personal identification, and efficient streams of everyday work management, ramp up new streams of digital command, control, and communication for e-governance. In addition to control over people, the management of things via densely embedded intelligence can relay information from GPS grids, RFID tags, web-enabled appliances, smart power grids, or telematic traffic controls to immediately control people. Once again, these clustered technical transitions underscore how thoroughly very common everyday behaviors are becoming almost invisibly, but also in plain sight, now quite commonly political concerns, even though many lay and expert communities continue to regard them all as exceptional rarities.

Cyberspace enables its many urban denizens to utilize new sites of engagement, but the restless urban landscape of offline cities is nothing compared with the continuous tumult of such netropolitan cyberscapes inhabited by netizens. System outages, constant software updates, platform redesigns, network upgrades, hardware modifications, and connectivity changes make netizenship in the bitstream a rather challenging way of life. This instability keeps most political actions brief, issue-oriented, mutable, and event-driven, while at the same time e-governance often is recurrent, service-oriented, fixed, and function-based. On the one hand, digital politics in the US can, then, pop out as tea-bag populism, pro-choice and pro-family values fanaticism, 1-800 number and www.org campaigning, and 24x7 news cycle reactionary rage;

while, on the other hand, e-governance is leading to more online municipal service mass e-mailings, online drivers license renewals, e-tax filing systems, and GIS-enabled real estate reassessments.

The netropolis, then, is an ambiguous cluster of domains whose scope shifts frequently and broadly: local to global, inside and outside the law, official government action to illegal insurgent resistance, around the block and across the ocean. Who is interacting with whom is not always known with true certainty, but what is said, when, and to whom can be archived forever while being forgotten immediately. Voices among the 'digerati,' like John Perry Barlow, have long espoused the imperatives of unlimited freedom on the Net since online browsing became easy and popular in the mid-1990s; however, this libertarian faith that open access for all on the Net would lead to liberty, justice, and democracy for all has proven misplaced. Where individual freedom threatens authoritarian governments, the regime often can simply pull the plugs into the networks. And, where personal freedom is regarded more positively, openness often becomes oppressive as information overload, flame wars, super factionalized minorities, and over-zealous rivals blog everyone to a standstill in their fractal decentralized 24x7 debates over tactical trivia, strategic disarray, or ephemeral effectiveness.

## Netropolitan community

Digital citizens must answer the challenges of determining their governance both online and offline. Whose governance, for whom, where, when, and how are decisively important questions that center on knowledge as well as power. The politics of networks are shaped, as Lyotard (1984: 9) claims, by their own functionalities and designs, or "who decides what knowledge is and who knows what needs to be decided? In the computer age, the question of knowledge is now more than ever a question of government." While many might dream of slower times and softer places, the key characteristic of the netropolis for digital citizens is error-proof utility. In gaining such fail-safe functionalities, however, digital citizens are forced to realize hard truths: "in matters of social justice and of scientific truth alike, the legitimating of that power is based on its optimizing the system's efficiency. The application of this criterion entails a certain level of terror, whether soft or hard: be operational (that is, commensurable) or disappear" (Lyotard 1984: 24).

The properties of digital technologies always will remain contested, and their entwinement with cybercitizenship qualities only adds other layers of complexity that leave civic behavior caught within proprietary domains for communicating and acting. Conflations of the commercial with the civic are vexing, especially to the degree that financial barriers of entry can limit who will be "a digital citizen." The Net is a layered assemblage of multiple machines under no truly centralized ownership, management or control.

Protocols of performance bind together its functionalities, but it is basic service rather than elaborate control, definite management or fixed ownership that users ultimately require. Hence, it is utility more often than community, identity, or unity that digital citizens demand.

Digital citizenship, and its reliance upon minimal functionality, provides a useful example of how the web is reweaving the culture, economy, and society of today's global, national, and local communities through global computer networks. On its own, it is neither entirely positive nor negative, so this realization is critical if we are to explore more fully how the wired and wireless articulation of discursive power and knowledge express their ambiguous effects at a local, national, or global level in digitally mediated social relations. Certainly, no study of contemporary economies and societies can ignore how more individuals and groups are engaging in collective action via digital discourses and online organizations.

Being concerned with digital citizenship is another indicator of how, as O'Hara and Hall (2008) suggest, "the Web influences the world, and the world influences the Web." As a machinic manifold of linked spaces of, for, and about continuous human interactions staged through linked documents, archived images, stored data, intricate graphics, and audio files, civic uses of the web have proliferated over the past decade as these assets ballooned from less than two billion easily accessed and searchable pages in 2000 to tens of billions of such pages by 2010, along with hundreds of billions of other documents in the "dark," "deep," or "denied" web. Each of these myriad files is to some degree a marker of how human discourse and culture is undergoing rapid digitalization, and then netizen-like civic actions can leverage those resources and opportunities for civil, governmental, and political action.

## Operationality as options and orders

During the past four decades the information and communication technologies that Lyotard (1984) addressed in delivering his "report on knowledge" have changed, but not in their entirety. Tendencies from the dawn of neoliberal networked society, in fact, have been remediated, in fits and starts, as the social interactions of many cultural, economic, and political communities unfold increasingly online through the Internet (Abbate 1999; De Kerckhove 1998; Deibert 1997). As cyberinfrastructure scales up at various locales around the world, many more realms of social structure, working at all scales of aggregation, are being reformatted to become, or be, digital (Negroponte 1995). It creates options, but it also writes orders.

First, computer-mediated communications are reordering the characteristics of the netizen: who can engage with whom, where, when, and how in new spaces; and, second, this increasingly ubiquitous computing regime unfolds hand in hand with neoliberal "fast capitalism" (Agger 1989), which determines those prosperous enough to get access to netizenship. New spaces make possible fresh positioning conditions for individuals and groups, and their politicized interactions often are fluid, flexible, and fast. Converting culture to code and digitizing discourse reformats the individual's experience with built environments, political jurisdictions, economic exchanges, material artifacts, and cultural meanings in ways that often lighten, empty, or minimize their joint significance. For those granted access as well as those denied use, the nature of citizenship changes in being dependent on network access. Instead of arranging meetings with others in one week face to face at some office, one attracts personal contacts 24x7 with all enabled others on Facebook, in MySpace, via Buzz, with Second Life, or through YouTube. Ironically, these digital links between users are believed to define a more advanced sense of freedom, but being freed in this fashion can occur through acceding to the embedded intelligence of inflexible technics driving the Net's digital command, control, and communication (Heim 1998). These shifts are rightly what have to become the subject of "Web Science" (Shadbolt and Berners-Lee 2008), and it might well develop beyond "Social Science" to become a richer epistemic frame for grasping these forms of information-based agency.

Cybersubjectivity in netizen action spins up from individuals and groups linking documents to data to digits to transit networks. Yet this action morphs into much more than the mere measurable effects of tracing such links, hits, downloads, or data flows on the web. Instead, respatializations of agency and structure occur, and human activities shift into new routines, rituals, and roles of "being" on bitstreams. As Hayles has explained, this plane of thought and action can have its own meta-national rules of postmodern embodiment, extraterritorial engagement, and hyperreal enlightenment (Hayles 1999) as digital Iranians, for example, try to use Facebook, MySpace, YouTube to take revolt into Tehran's streets for the whole world to watch. Because so much can change on the web, including some existing face-to-face rules of cultural, economic, and political action, Hayles has touted how those so enabled are becoming new posthuman beings with their own special forms of digital culture. So the netizens of other nations are left watching the posthuman traces of Iranian protesters being beaten, killed, or jailed in scattered scenes of fuzzy cell-phone video. Information moves, but the regime remains. Maybe Hayles is right about posthuman life, but these posthuman beings also remain stuck in Iran, dead on the street, rotting in jail, or pretending their future video files are toppling the Islamic Republic. Although it is true, in part, that a few will develop new empowering identities at the interfaces of bits and bytes with flesh and blood, they must continue to endure existences chained only to cogs and wheels as well as caught by their muscle and bone. Al Qaeda is, in part, a resistance that recruits and thrives in cyberspace, but Western info-warriors are tracking it down there. And, Predator drones wait to rain Hellfire missiles on such netizen-like radicals when they go outside for a cell phone call, appear in satellite surveillance, or log into a chat room.

The web, and the ICTs supporting it, is also the block of a communicative, complementary, collaborative, and concurrent engine. Even as its machinic qualities occlude many of its associative properties, ICTs as "discursive formations" entail their own "concursive formations" in the production, consumption, reproduction, circulation, and accumulation of bits. This concursivity tracks what unfolds spatially together in the machinic assemblages, common traditions, linked networks, cooperative institutions, or other combined effects of digitalization. To concur is to run together, to meet, to converge, to coincide. When combining in action, sometimes by accident, sometimes by design, sometimes by habit, the concursivity of digital agency is apparent. Hence, the industrial "system of objects" (Baudrillard 1996) increasingly must reset, relaunch, and reformat itself with the digitized "objects of systems." Discursivity is one concursivity for network environments as the channel and code, signal and sign, carrier and content interpolate as unified hybrid assemblages (Luke 2001).

Drawing first the Latin verb meaning "to do" or "to act" as the Oxford English Dictionary suggests, agency is essentially the implied faculty of either being an agent or some modus operandi for agents acting. Agency also can be an actively working formation or specific operation for action. The Enlightenment notion of rational autonomous human agency leverages of these notions about embodied agency for the bourgeoisie and proletariat after the eclipse of aristocratic barons and churchly priests. They are held to be more true today for actions taken by the web's digital demos, because this cybernetic construct fulfills the rich myths of modernity first propagated by bourgeois revolutions billowing up out of global markets. With many voices speaking across the web, the civic activity of netropolitan living reaffirms this concursive sense of agency as digital citizenship. In their celebrations of being digital, even though very few people have the knowledge, freedom, or resources to be working persistently at most operations as this sort of digital agent, this opportunity constitutes certain freedoms on the electronic frontier. In turn, such concursive constructions of collective and individual agency are the ones most often coupled with machinic democratization in open discussions of digital agents creating structures for cybernetic culture (Luke 2001).

Such coaligned technologies of production for self and society fuse in the networks sustaining virtual environments; they do make the web, and the web then does remake them. Although it is typically cast as a space of freedom, the web also is promoting capitalist exchange's ideal outcome: "the ultimate realization of the private individual as a productive force. The system of needs must wring liberty and pleasure from him as so many functional elements of the reproduction of the system of production and the relations of power that sanction it" (Baudrillard 1981: 85). Paying for access to speak online commercializes netizenship activity in new ways, but it also advances their sense of agency. The liberatory mythos of digitalization, then, occludes the disciplinary realities of web-based living: all action is trackable, measurable, and forever available as fungible reified information. Face-to-face behaviors of bodies with other bodies can occur; but now in addition one's existence on the networks of networks that function beside, behind, or beneath human behaviors is constantly conforming to the dictates of elaborate "e-structures" for "e-haviors" as the web evolves concursively with its own sui generis meta-national quiddity of, for, and by agents. Plainly, there is a very different "landscape of events" emerging out of these conditions (Virilio 2000).

In digital discourse, each online agent in the network serves as "a post through which various kinds of messages pass," and, "no one, not even the least privileged among us, is ever entirely powerless over the messages that traverse and position him [sic] at the post of sender, addressee, or referent" (Lyotard 1984: 15). Bitstreams build political possibilities for new ethics and tactics out of such language games, and these moves sustain larger systems of web-based digital economic and social relations. Certainly, as Lyotard suggests, these interactions are not the entirety of social relations, but they foster a concursive and discursive basis "to socialize" new opportunities to combat collective entropy, create novel associations, increase overall performance, and exemplify the promise of connectivity.

Emancipation today for people is counter-intuitively embraced as very determinate forms of machine-mediated reification, digitally displayed direction, and code-carried control. Fictions of social origins are displaced by new myths of continuous connectivity. Creating the roles and the scripts for language games of an online political culture in which those with access to ICTs already are the referents, senders, and addressees of more complicated political strategies offline is complicated. And, it moves politics far beyond those rules of action found in simply face-to-face engagement.

Web-centered functionalities, then, spin up new material modes of being immaterially sociable for contemporary society (Luke 1996). From Second Life to Facebook to MySpace, to YouTube, communicative applications translate multiple modes of electronic messaging into "society" itself, as their new clustered affinity and agonistic gaming generate fresh conventional understandings of other and self through digital acts and artifacts (Kelly 1994). All of this can become a politics as elementary components of communicative denotation and connotation become moves and countermoves in digitally driven performative, evaluative, prescriptive, and directive interactions. The social struggles of such YouTubed, MySpaced, Facebooked, or Second Lived communities are not without rules, as Lyotard suggests, "but the rules allow and encourage the greatest possible flexibility of utterance" (1984: 17). Nonetheless, tracking these traces of digital discursivity down in today's concursive circuits of cybernetic culture is the end and the means needed for a web science of digital citizenship (Berners-Lee et al. 2006).

Therefore, narratives of politics emerge where netiquette, cyberculture, and connectivity coalign in the netropolitan concursive circuits of information machines. Along with the discursive details for rules of use, software licensure limits, and ordinary sociolinguistic rules that code jurisdictive boundaries of web-centered communities (Luke 2001), fast tactics, flexible mobilization, and fluid strategies test the political will of friends and foes. Increasingly, digital discursivity, social informatics, and web science do privilege "certain classes of statements (sometimes only one) whose predominance character-izes the discourse of the particular institution: there are things that should be said, and there are ways of saying them" (Lyotard 1984: 17).

On one level, Lyotard's abject predictions about "the computerization of society" appear to be coming true in this new informatic environment inas-much as the web—in its iterative 1.0, 2.0, 3.0, or 4.0 versions—is colonized by business-to-business (B2B), business-to-consumer (B2C), and consumer-to-consumer (C2C) modes of exchange. In this manner, the Net is "the 'dream' instrument for controlling and regulating the market system, extended to include knowledge itself and governed exclusively by the performativity prin-ciple" (Lyotard 1984: 67), and these lean and mean values deeply affect the nature of politics. Still, every such attempt to exert crude disciplinary control sparks sophisticated contra-disciplinary resistances. On another level, then, the web increasingly enables ICT users in "discussing metaprescriptives by supplying them with information they usually lack for making knowledge-able decisions" (Lyotard 1984: 67). Once more and more knowledge-based decisions are made, such power's meta-prescriptive reach grows in strength and scope to order digital citizens' conduct of their conduct.

Of course, new political mobilizations, and maybe even revolutions, are being made globally, nationally, and locally more possible on ICT networks, like the web, as Beck maintains, "under the cloak of normality" (1992: 186) because of the daily working of informatic global assemblages in discursive and concursive environments made possible by polymorphous mediations of HTML, Linix, Microsoft, or IBM. "In contemporary discussions," as Beck suggests, "the 'alternative society' is no longer expected to come from parlia-mentary debates on new laws, but rather from the application of microelec-tronics, genetic technology, and information media" (1992: 223). However, "the alternative society" itself has also become reified in the ever-changing styles of cyberculture for digital citizens. Code morphs into one product and serves as a key producer of such revolutions, which now arrive more often, as new machinic versions of digital discourse rather than fresh organic vari-ations of embodied agency.

With no definitive hegemonic statal force at work in Empire's world society, this ceaseless sweep of politics online appears to be the essence of today's political conditions in more borderless, interlinked, and strategic action by individuals and small groups (Hardt and Negri 2000). As Lyotard claims, each and every relentless pursuit of capitalist restructuring "continues to take place without leading to the realization of any of these dreams of emancipation" (1984: 39). With less trust in any narratives of truth, enlightenment, or progress, he argues the social forces of science, technology, and big business compel most individuals and groups to accept and enter more transnational flows of information by embracing fluid operational values as they assume what can be taken as the roles of "digital citizens." Cybercitizenship plainly can unfold in the forms of digital discourse and cybernetic concourse as a marvelously liberatory space; yet, with the properties of these technologies, the emergent spaces one maps are also always already an essentially "a polymorphous disciplinary mechanism" (Foucault 1980: 106) ironically embraced as a liberating electronic frontier.

## Bibliography

Abbate, Janet. 1999. *Inventing the Internet*. Cambridge, MA: MIT Press.

Agger, Ben. 1989. *Fast Capitalism*. Urbana: University of Illinois Press.

Baudrillard, Jean. 1996. *The System of Objects*. London: Verso.

Baudrillard, Jean. 1981. *For a Critique of the Political Economy of the Sign*. St. Louis: Telos Press.

Beck, Ulrich. 2000. *What is Globalization?* Oxford: Blackwell.

Beck, Ulrich. 1992. *The Risk Society*. London: Sage.

Berners-Lee, Tim, Wendy Hall, James Hendler, Nigel Shadbolt, and Daniel V. Weitzner. 2006. 'Creating a Science of the Web,' *Science* 313 (11) (August): 769–71.

Carr, Nicholas. 2010. *What the Internet is doing to Our Brains: The Shallows*. New York: W. W. Norton.

Crang, Michael and Nigel Thrift. 2000. *Thinking Space*. London: Routledge.

De Kerckhove, Derrick. 1998. *Connected Intelligence: The Arrival of the Web Society*. Toronto: Somerville.

Deibert, Ronald, J. 1997. *Parchment, Printing, and Hypermedia: Communication in World Order Transformation*. New York: Columbia University Press.

Dertouzos, Michael. 1997. *What Will Be: How the New World of Information Will Change Our Lives*. New York: Harper Collins.

Foucault, Michel. 1994. *The Order of Things: An Archaeology of the Human Sciences*. New York: Random House.

Foucault, Michel. 1980. *The History of Sexuality, Vol. I*. New York: Vintage.

Hardt, Michael and Tony Negri. 2000. *Empire*. Cambridge, MA: Harvard University Press.

Hayles, Katherine N. 1999. *How We Became Posthuman: Virtual Bodies in Cybernetics, Literature, and Informatics*. Chicago: University of Chicago Press.

Heim, Michael. 1998. *Virtual Realism*. Oxford: Oxford University Press.

Kelly, Kevin. 1994. *Out of Control: The Rise of Neo-Biological Civilization*. Reading, MA: Addison-Wesley.

Lefebvre, Henri. 1991. *The Production of Space*. Oxford: Blackwell.

Lefebvre, Henri. 1984. *Everyday Life in the Modern World*. New Brunswick, NJ: Transaction.

Luke, Timothy W. 2006. "The Politics and Philosophy of E-Text: Use Value, Sign Value, and Exchange Value in the Transition from Print to Digital Media," *Libr@ries: Changing Information Space and Practice*, eds. Cushla Kapitzke and Bertram C. Bruce. Mahwah, NJ: Lawrence Erlbaum, 197–210.

Luke, Timothy W. 2001. "Real Interdependence: Discursivity and Concursivity in Global Politics," *Language, Agency and Politics in a Constructed World*, ed. Francois Debrix. Armonk, NY: M.E. Sharpe.

Luke, Timothy W. 1996. "Identity, Meaning and Globalization: Space-Time Compression and the Political Economy of Everyday Life," *Detraditionalization: Critical Reflections on Authority and Identity*, eds. Scott Lash, Paul Heelas and Paul Morris. Oxford: Blackwell, 109–33.

Luke, Timothy W. 1995. "New World Order or Neo-World Orders: Power, Politics and Ideology in Informationalizing Glocalities," *Global Modernities*, eds. Mike Featherstone, Scott Lash, and Roland Robertson. London: Sage Publications, 91–107.

Luke, Timothy W. 1994. "Placing Powers, Siting Spaces: The Politics of Global and Local in the New World Order," *Environment and Planning A: Society and Space*, 12: 613–28.

Lyotard, Jean-Francois. 1984. *The Postmodern Condition: A Report on Knowledge*. Minneapolis: University of Minnesota Press.

Negroponte, Nicholas. 1995. *Being Digital*. New York: Knopf.

O'Hara, Kieron and Wendy Hall. 2008. "Web Science," *Alt-N*, 12 (May). Retrieved from http://eprints.ecs.soton.ac.uk/15682/1/OHara-Hall-ALT-N-Web-Science.pdf (accessed October 5, 2010).

Shadbolt, Nigel and Tim Berners-Lee. 2008. "Web Science Emerges," *Scientific American* (October): 32–7.

Virilio, Paul. 2000. *A Landscape of Events*. Cambridge: MA: MIT Press.

# 4.1
## Facebook in Egypt: April 6 and the Perception of a New Political Sphere

*Christopher Wilson*

Suppression of political opposition in Egypt is reinforced through mainstream media control and a 29-year-old state of emergency, which allows for arbitrary and indefinite detention, as well as prohibiting public gatherings of more than five people. Since the turn of the millennium, however, Facebook has been producing important arenas for political information and organization (Khaled 2009).

One of the most widely commented examples is the April 6movement of 2008. A Facebook group created to support striking workers in an outlying industrial town saw its membership explode from 16 to over 60,000 in the course of two weeks. Most members were young and well educated, but the group was remarkable for the bridges it built between laborers, Cairo elites, Islamic reformists, and college students. Support for workers in Mahalla al-Kobra soon morphed into a nationwide strike for democratic reform. Leaflets and graffiti appeared in the streets to inform those without Internet access, and endorsements were issued from established social institutions.

When April 6 came, isolated riots and arrests fueled a national debate on the strike's merits and success (Abdul-Raouf 2008). Internationally, a small frenzy of commentary flamed up and then extinguished itself when subsequent strikes flopped and the group appeared to lose its coherence (Faris 2009). The initial action's efficacy, however, especially in mustering citizens otherwise removed by geographical, socioeconomical, and even political divides, suggests a more profound development in the way politics is organized in Egypt.

The state has certainly taken notice. Arrests and attempts to infiltrate the group were prompt, not to mention allegations that the group's administrator was tortured in an attempt to obtain his password. (He posted pictures of his injuries on Facebook.) Today it is quite impossible to tell how many members of the April 6 group are government agents (Shapiro 2009).

If there was something distinct about the April 6 organization, it is clearly a function of technology rather than substance (Faris 2008). The social network structure of Facebook is distinct from conventional other web-based political

97

activity because it avoids the heavy handed censorship that may befall exclusively political websites; political information is presented effortlessly by a social algorithm that bolsters its relevance; and the button-clicking ease of creating and joining Facebook groups dramatically diminishes the costs of social organization.

Together, these three factors may constitute the conditions for socio-political activity unthinkable in traditional Egyptian fora. The dominance of young people is especially compelling as they are generally regarded as politically disengaged (see Ahram Center study in Shapiro 2009), and although the dangers of dissidence may persist online (group walls have a cloak and dagger feel to them and showing up at a demonstration is not without risk), this has not prevented a continuous stream of new political Facebook groups.

To date, the most prominent of these to follow April 6 is the one on which Mohamed Elbaradai announced his potential presidential candidacy, and saw the group's membership swarm to over 200,000 in two weeks. This group and others (including April 6) then coordinated their Facebook appeals to organize a strike on April 6, 2010 for constitutional reform. Popular expectations were matched by the severity of the state's response, and only a few dozen protesters managed to gather before mass arrests and beatings began.

At the time of writing (the evening of the 2010 strike), it is not yet possible to guess at how or if this will affect online strategizing, but there is little cause to think that state violence will manage to stem the popular perception that Facebook is a place for reform to burgeon, a new political sphere from which to await political change. As one young protester put it, "I am involved in no parties, never, … I just go to Facebook events, wherever they are. I'm in the Facebook party" (Shapiro 2009).

## Bibliography

Abdul-Raouf, Du'aa. 2008. "Was the April 6th Strike Successful?" IkhwanWeb. April 10. http://www.ikhwanweb.com/Article.asp?ID=16634&SectionID=86, accessed April 28, 2010.

BBC News. 2010. "ElBaradei in Facebook Plea for Reform in Egypt." http://news.bbc.co.uk/2/hi/middle_east/8593540.stm (March 29), accessed April 28, 2010.

Faris, David M. 2008. "Revolutions Without Revolutionaries? Network Theory, Facebook, and the Egyptian Blogosphere." *Arab Media and Society* 6, (Fall) retrieved from http://www.arabmediasociety.com/?article=694, accessed April 28, 2010.

Faris, David M. 2009. "The End of the Beginning: The Failure of April 6th and the Future of Electronic Activism in Egypt." *Arab Media and Society* 9 (Fall) retrieved from http://www.arabmediasociety.com, accessed April 28, 2010.

Khaled, Mohamed. 2009. "An Egyptian Blogger Crosses Red Lines." http://www.cpj.org/blog/2009/10/an-egyptian-blogger-crosses-red-lines.php, accessed April 28, 2010.

Shapiro, Samantha M. 2009. "Revolution, Facebook-Style." *New York Times* (January 22). http://www.nytimes.com/2009/01/25/magazine/25bloggers-t.html?_r=1&pagewanted=all, accessed April 28, 2010.

# 4.2
# Grassroots Politics in Popular Online Spaces: Balancing Alliances

*Julie Uldam*

In a context where the grassroots and non-profit sector is argued to have been relegated to the remote margins of the Internet (McChesney 2007), popular online spaces potentially provide social movement organizations (SMOs) with possibilities for reaching wider publics, rather than merely connecting likeminded users and failing to challenge presumptions (Dahlgren 2005).

This case introduces the World Development Movement (WDM), a UK-based SMO, which aims to promote an agenda centered on "a political project of structural change to inherently unjust systems of power" (O'Nions, interview July 2009), advocating alternative approaches to the dominant neoliberal model of globalization. Drawing on interviews with directors and campaign and outreach officers from WDM, the case explores the ways in which WDM uses popular online spaces such as social networking site Facebook and file sharing site YouTube for self-publicity and coalition building, while simultaneously struggling with a sense of disillusionment about the potential of such media for fostering political engagement at an SMO level.

Responding to the conditions of possibility brought about by online fragmentation potentially countered by the popularization of profit-driven social networking and file sharing sites, SMOs are observed to be abandoning alternative or radical online media (Atton 2004; Downing 2001; Rodríguez 2001), which were initially vested with hopes for providing platforms for the proliferation of undistorted counter-discourses that the market principles of "mainstream" online and mass media systems marginalize (Fenton 2008; Couldry and Curran 2003). This perception is shared by WDM staff, in this example from WDM's outreach officer: "We do need to use mainstream tools to get our messages out as widely as we can" (Talbot, interview April 2009).

While the overarching narratives expressed by respondents from WDM are characterized by a cautious take on the potentialities of YouTube in sustaining political engagement, respondents remain hopeful about reaching new publics and gaining visibility for single issues through the promotion of campaign-based coalitions. A perceived tension between organizational positioning and coalition building—between obtaining visibility and commitment to the organization on the one hand and visibility for the cause on the other—is seen as potentially enhanced in an online setting, because new possibilities for self-representations also facilitate the promotion of coherent frames at the coalition level. This perception is exemplified by Katharine Talbot in her comment on WDM's role in the Stop Climate Chaos coalition, which campaigns on climate change issues from the UK:

> Stop Climate Chaos has millions of members and people know it as a brand more so than they know WDM. So it is of value to us to be associated. But also we obviously don't want our brand to get hidden underneath all these other coalition brands.
>
> (Talbot, interview April 2009)

Unifying political identities is seen as critical to political engagement at the organizational level. At a movement level, orchestrating combined opposition to the hegemony of a neo-liberal articulation of globalization processes is also seen as crucial. At the SMO level, the promotion of coalitions is relayed as posing potential tensions, as the possibilities for bypassing mass media filters provided by online spaces is seen as facilitating cross movement visibility rather than sustaining collective identities among WDM supporters. This relates to WDM staff's view of popular online spaces as privileging visibility over engagement, and constitutes a cause for tension, because the latter is essential to WDM to ensure membership contributions, and thus survival independently of corporate funding.

Such alliances blur the particular agendas of coalition SMOs and contribute to what I argue can be seen as a narcissism of small differences, as WDM actors construe their roles in relation to coalition partners, simultaneously conjuring up similarities and constructing each other as significantly distinct, in ardent attempts to engage supporters politically—a process that WDM staff relay as inherently rooted in offline, unmediated practices.

## Bibliography

Atton, C. 2004. *An Alternative Internet*. Edinburgh: Edinburgh University Press.

Couldry, Nick, and James Curran. 2003. *Contesting Media Power: Alternative Media in a Networked World*. Lanham, MD: Rowman & Littlefield.

Dahlgren, Peter. 2005. "The Internet, Public Spheres, and Political Communication: Dispersion and Deliberation." *Political Communication* 22 (2): 147–62.

Downing, John D. H. 2001. *Radical Media, Rebellious Communication and Social Movements*. London: Sage Publications Ltd.

Fenton, Natalie. 2008. "Mediating Solidarity". *Global Media and Communication* 4, (1): 37–57.

McChesney, R. W. 2007. *Communication Revolution: Critical Junctures and the Future of Media*. New York: New Press.

O'Nions, J. 2009. Personal communication. London, July 2009.

Rodríguez, Clemencia. 2001. *Fissures in the Mediascape: An International Study of Citizens' Media*. Cresskill, NY: Hampton Press.

Talbot, K. 2009. Personal communication. London, April 2009.

# 4.3
# From Disability to Functional Diversity: ICT and Amartya Sen's Approach

*Mario Toboso*

In Spain, Palacios and Romañach (2006, 2007) are currently proposing that a new perspective is needed: a new model of disability for society to consider, which is based on a definitive acceptance of human diversity.

To bring about this shift, it is crucial to overthrow the old concept of ability and look for a new concept in which persons with disability can find an identity that is not perceived as negative. The proposal of Spain's independent living movement embraces the concept of "functional diversity" as a much more meaningful, up-to-date, and positive concept than "disability," with its negative connotations (Romañach and Lobato 2005).

This designation is accompanied by proposing a new, non-negative vision that refers to persons who perform some of their functions in a way that is different from the average person (Patston 2007). In this new approach, the point of departure is a basic reality that human beings are diverse in their functioning – whether physical, mental, sensory, or another aspect – and all societies should view this diversity as a source of enrichment.

Attention to human diversity, both the individual's characteristics and external circumstances, constitutes an important aspect of Amartya Sen's "capabilities and functionings" approach (Sen 1984, 1985). His reply to the fundamental question "equality of what?" is equality of personal freedom, understood as the capability of achieving the functionings one considers necessary to have a good life, over equality of income or wealth, for example. Sen points out that, since we are so profoundly diverse, equality in one realm (income) often leads to inequalities in other realms (wealth). Thus, the force of the question "equality of what?" stems largely from the empirical fact of our diversity (Sen 1979).

As elements that alter human capacities and make new actions possible, ICT transforms a person's capability set, which Sen defines as the set of all possible functionings a person can achieve (Sen 1993). Many persons with a disability utilize technological resources, often those associated with accomplishing everyday functionings, in preserving their own quality of life and their opportunities for participating in society (Placencia-Porrero and

Ballabio 1998). This may be the case for a blind person who uses a localization and orientation device to guide his or her movements when away from home.

In this regard, the concept of "functional diversity" may be interpreted as "diversity in functionings," describing the reality of persons who have the potential to access the same functionings as other people but in a different way, often through the use of technical components and technological resources (Hoenig et al. 2003). This perspective suggests that, in an evaluation of wellbeing and quality of life, the capability set that is taken into account should include the whole range of possibilities for functionings in space, in keeping with the fact that different members of the society have different ways of accessing them (Smith 2006). In other words, the functioning "go outside home," for example, should take into consideration the various ways of doing that and not be limited to the legitimate and most common ways used by the majority.

Recognition of all the different expressions of functioning that are possible leads to awareness of and respect for functional diversity. By incorporating the concept of functional diversity into the functionings space in Sen's approach, a capability set with a broader spectrum is generated that, as a basis for analyzing wellbeing and quality of life, serves an equally broader spectrum of society because it is more than just an abstract consideration of the "standard" person's array of possible functionings (Terzi 2005).

If the proposal is approached in this way, consideration of a capability set that incorporates these diverse ways of achieving possible functionings should go hand in hand with consideration of an agency environment that has the capacity to respect that diversity and does not limit this capability set for persons with disability because there are limiting factors that may invalidate functioning alternatives for these persons (Swain et al. 2004).

Failure to respect functional diversity, as a key factor in the design and implementation processes for new technologies, will inevitably lead to problems with social participation, as various groups of people suffer discrimination when they access these technologies and several information society environments (Dobransky and Hargittai 2006; G3ict 2007).

## Bibliography

Dobransky, K. and Hargittai, E. 2006. "The Disability Divide in Internet Access and Use," *Information, Communication & Society*, 9 (3): 313–34. Available at: http://www.eszter.com/research/pubs/dobransky-hargittai-disabilitydivide.pdf, accessed October 4, 2010.

G3ict 2007. *The Accessibility Imperative: Implications of the Convention on the Rights of Persons with Disabilities for Information and Communication Technologies.* Available at: http://g3ict.org/resource_center/g3ict_book-the_accessibility_imperative, accessed October 4, 2010.

Hoenig, H., D. H. Taylor, Jr., and F. A. Sloan. 2003. "Does Assistive Technology Substitute for Personal Assistance among the Disabled Elderly?" *American Journal of Public Health*, 93 (2): 330–37.

Palacios, A. and J. Romañach. 2006. *El modelo de la diversidad. La Bioética y los Derechos Humanos como herramientas para alcanzar la plena dignidad en la diversidad funcional. Madrid: Diversitas.* Available at: http://www.diversocracia.org/docs/Modelo_diversidad. pdf, accessed October 4, 2010.

Palacios, A. and J. Romañach. 2007. *Diversity Approach: A New View on Bioethics from the Functionally Diverse (Disabled) People.* Available at: http://www.diversocracia.org/ docs/Diversity_approach_overview.doc, accessed October 4, 2010.

Patston, P. 2007. "Constructive Functional Diversity: A New Paradigm Beyond Disability and Impairment," *Disability & Rehabilitation*, 29: 20.

Placencia-Porrero, I. and E. Ballabio (eds.) 1998. *Improving the Quality of Life for the European Citizen: Technology for Inclusive Design and Equality.* Amsterdam: IOS Press.

Romañach, J. and M. Lobato. 2005. *Functional Diversity: A New Term in the Struggle for Dignity in the Diversity of the Human Being.* Available at: http://www.leeds.ac.uk/ disability-studies/archiveuk/zavier/Functional%20Diversity%20%20%20%20%20% 20%20%20%20_fv%20Roma%F1ach.pdf, accessed October 4, 2010.

Sen, A. 1979. *Equality of What?* The Tanner Lectures on Human Values. Stanford University, May 22. Available at: http://www.tannerlectures.utah.edu/lectures/ documents/sen80.pdf, accessed October 4, 2010.

Sen, A. 1984. *Resources, Values and Development.* Oxford: Basil Blackwell.

Sen, A. 1985. *Commodities and Capabilities.* Amsterdam: Elsevier Science.

Sen, A. 1993. "Capability and Well-Being." In *The Quality of Life*, eds. M. C. Nussbaum and A. Sen. New York: Oxford University Press.

Smith, J. 2006. *Functional Diversity – A Fundamental Characteristic of Ageing: Implications for Social Policy.* Berlin: Max Planck Institute for Human Development.

Swain, J., S. French, C. Barnes, and C. Thomas, 2004. *Disabling Barriers – Enabling Environments.* London: Sage.

Terzi, L. 2005. "Beyond the Dilemma of Difference: The Capability Approach on Disability and Special Educational Needs." *Journal of Philosophy of Education*, 39 (3): 443–59.

# Part V
# Power, Knowledge, Surveillance

# 5
# Surveillance, Power and Everyday Life

*David Lyon*

## Introduction

Surveillance grows constantly, especially in the countries of the global north. Although as a set of practices it is as old as history itself, systematic surveillance became a routine and inescapable part of everyday life in modern times and is now, more often than not, dependent on information and communication technologies (ICTs). Indeed, it now makes some sense to talk of "surveillance societies," so pervasive is organizational monitoring of many kinds. Fast developing technologies combined with new governmental and commercial strategies have led to the proliferation of new modes of surveillance, making surveillance expansion hard to follow, let alone analyze or regulate. In the past three decades traffic in personal data has expanded explosively, touching numerous points of everyday life and leading some to proclaim the "end of privacy." But although questions of privacy are interesting and important, others that relate to the ways in which data are used for "social sorting," discriminating between groups that are classified differently, also need urgently to be examined. Who has the power to make such discriminatory judgments, and how this becomes embedded in automated systems, is a matter of public interest. Such questions are likely to be with us for some time, because of what might be called the "rise of the safety state," which requires more and more surveillance, and also because the politics of personal information is becoming increasingly prominent.

Literally, surveillance means to "watch over," an everyday practice in which human beings engage routinely, often unthinkingly. Parents watch over children, employers watch over workers, police watch over neighborhoods, guards watch over prisoners, and so on. In most instances, however, surveillance has a more specific usage, referring to some focused and purposive attention to objects, data, or persons. Agricultural experts may carry out aerial surveillance of crops, public health officials may conduct medical surveillance of populations, or intelligence officers may put suspects under observation.

Such activities have several things in common, including that in today's world some kind of technical augmentation or assistance of surveillance processes is often assumed. ICTs are utilized to increase the power, reach and capacity of surveillance systems.

The specific kind of surveillance discussed here is perhaps the fastest growing and almost certainly the most controversial, namely the processing of personal data for the purposes of care or control, to influence or manage persons and populations. In this and every other respect, power relations are intrinsic to surveillance processes. This being so, it immediately becomes apparent that actual "watching over" is not really the main issue, or at least not literally. While camera surveillance certainly does have a watching element, other kinds of ICT-enabled surveillance include the processing of all kinds of data, images and information. Those of which we are most aware include the multiple checks that we go through at an airport, from the initial ticketing information and passport check through to baggage screening and the ID and ticket check at the gate. In this example, both public (governmental; customs and immigration) and private (commercial; airlines and frequent flyer clubs) data are sought. Others of which we may be less consciously aware include "loyalty cards" at supermarkets and other stores, which offer customers discounts and member privileges, but are simultaneously the means of garnering consumer data from shoppers.

All these count as surveillance of one kind or another, in which we are (usually) individuated—distinguished from others and identified according to the criteria of the organization in question—and then some sort of analysis of our transaction, communication, behavior, or activity is set in train. Thus some kinds of surveillance knowledge are produced that are then used to mark the individual, to locate him or her in a particular niche or category of risk proneness, and to assign social places or opportunities to the person according to the ruling criteria of the organization. It is not merely that some kinds of surveillance may seem invasive or intrusive, but rather that social relations and social power are organized in part through surveillance strategies. One can argue that the "surveillance societies" of today are a by-product of the so-called information society.

Surveillance today is often viewed in terms of the Panopticon, introduced by Bentham and discussed by Michel Foucault. Yet several writers have pointed to other features of surveillance that are difficult to squeeze into that frame. Gilles Deleuze, for example, suggested in a brief statement on "societies of control" that we all now live in situations where "audio-visual protocols"—such as cameras, PINs (personal identification numbers), barcodes and RFIDs (radio frequency identifications)—help to determine which opportunities are open, and which closed, to us in daily life (Deleuze 1992). Deleuze's (and Felix Guattari's) idea of the "assemblage" of surveillance activities has also been taken up by a number of sociological authors (such as Ericson and Haggerty 2000).

The notion of assemblage in this context points to the increasing convergence of once discrete systems of surveillance (administration, employment, health, insurance, credit and so on), such that (in this case) digital data derived from human bodies flows within networks. At particular points the state, or totalizing institutions such as prisons, may focus or fix the flows to enable control or direction of the actions of persons or groups. But in this view surveillance becomes more socially leveled out, non-hierarchical, and inclusive of others who might once have felt themselves impervious to the gaze. At the same time, it is suggested, surveillance itself will not be slowed merely by resisting a particular technology or institution.

Others, sometimes indirectly, have also proposed fresh ways of examining surveillance beyond those classic foci on the "state" or total institutions as its perpetrators. Nikolas Rose, for instance, argues that surveillance be seen as part of contemporary governmentality, the way that governance actually happens, rather than thinking of it as an aspect of institutional state activities. He suggests that modern systems of rule depend on a complex set of relationships between state and non-state authorities, infrastructural powers, authorities that have no 'established' power, and networks of power (Rose 1996: 15). Surveillance, which pays close attention to personal details, especially those that are digitally retrievable, contributes to such governmentality. Indeed, governments and institutions may, paradoxically, use "freedom" (conventionally considered in opposition to state power) to further their ends. Consumer "freedom" and surveillance is a case in point.

The ways that contemporary surveillance works frequently leads to new forms of exclusion (rather than control through inclusion that was characteristic of Foucault's understanding of the Benthamite Panopticon). This is clear from empirical studies (such as Norris 2003 on camera surveillance), Bauman (2000) on super-max prisons, and the theoretical work of Giorgio Agamben (which criticizes Foucault for never demonstrating how "sovereign power produces biopolitical bodies" (Agamben 1998). Such exclusionary power has come more clearly into focus since 9/11, not only in the attempts to identify "terrorists" and to prevent them from violent action, but also in the more general sorting of foreign workers, immigrants and asylum seekers into "desirable" and "undesirable" categories. As Bigo and Guild (2005: 3) say, while Foucault thought of surveillance as something that affects citizens equally, in fact "the social practices of surveillance and control sort out, filter and serialize who needs to be controlled and who is free of that control." Such sorting is becoming increasingly evident not only in Europe but in North America and elsewhere too. And it is facilitated by new surveillance measures such as biometric passports and electronic ID cards, currently being established in the UK and the US (Lyon 2009).

The notion of a surveillance society is also given credence by the fact that in ordinary everyday life people are not only constantly being watched, but also willing, it seems, to use technical devices to watch others. Plenty of domestic technologies are on the market, for providing video camera "protection" to homes; cameras are commonplace in schools and on school buses (Monahan and Torres 2009); and many schools are adopting automated identification systems; spouses may use surreptitious means to check on each other; and there is a burgeoning trade in gadgets with which parents may "watch" their children. Day care cams permit parents to see what their toddlers are up to, nanny cams monitor for suspected abuse, and cell phones are often given to children so that their parents may "know where they are." Those technologies that originated in military and police use and later migrated to large organizations and government departments may now be used for mundane, civilian, local and familial purposes. At the same time, the broader frames for understanding surveillance, such as governmentality, which acknowledge its ambiguity as well as its ubiquity, permit consideration of how new technologies may also empower the watched. While global imperial power is undoubtedly stretched by surveillance, and social exclusion is automated by the same means, Internet blogs, cell-phone cameras and other recent innovations may be used for democratic and even counter-surveillance ends. While such activities have none of the routine and systematic character, let alone the infrastructural resources, of most institutional surveillance, they may nevertheless contribute to alternative perspectives and to the organizational capacities of counter-hegemonic social movements.

## Surveillance technologies

The very term "surveillance technologies" is somewhat misleading. If one visits the "spy stores" that seem to spring up in every city, the term seems clear enough. You can purchase disguised video cameras, audio surveillance and telephone tapping equipment, GPS (global positioning satellite) enabled tracking devices, and of course counter-surveillance tools as well. But each of these is intended for very small-scale use—usually one surveiller, one person under surveillance, and they are often people already known to each other—and are decidedly covert. In policing and other investigative activities, such specifically targeted and individually triggered surveillance may be called for, but the kinds of surveillance discussed here are different in almost every respect. Regarding power relations, individual surveillance is one thing, institutional surveillance quite another.

Surveillance that has developed as an aspect of bureaucratic administration in the modern world (see Dandeker 1990) is large-scale, systematic and now increasingly automated and dependent on networked computer power. It depends above all on searchable databases (Lessig 1999) to retrieve and process the relevant data. Although some systems depend on images or film, even these possess far greater surveillance power when yoked with searchable databases. And in most cases surveillance is not covert. It is often known about, at least in a general way, by those whose data are extracted, stored, manipulated, concatenated, traded and processed in many other ways. Those buying houses are aware that checks will be made on them; patients know that health care agencies keep detailed records; video surveillance cameras are visible on the street; Internet surfers know their activities are traced; and so on.

Surveillance technologies enable surveillance to occur routinely and automatically, but only in some cases is the surveillance aspect primary. Clearly, the point of public CCTV is to "keep an eye" on the street or train station (although even here the larger goal may be public order or maximizing consumption). In the UK there are more than four million cameras in public places (Norris and McCahill 2004). Police and intelligence services also use technologies such as fingerprinting devices, wiretaps, CCTV and so on for surveillance purposes and all these depend (or are coming to depend) on searchable databases. For this reason, among others, they contribute to qualitatively different situations, sometimes amounting to a challenge to traditional conceptions of criminal justice (Marx 1988, 1998).

In many cases, however, surveillance is the by-product, accompaniment, or even unintended consequence of other processes and practices. It is sometimes not until some system is installed for another purpose that its surveillance potential becomes apparent. Marketers claim that they "want to know and serve their customers better" and this entails finding out as much as possible about tastes, preferences and past purchases, which has now developed into a multi-billion dollar industry using customer relationship marketing (CRM; see 6: 2005). Retailers may install ceiling mounted cameras in stores to combat shoplifting only to discover that this is also a really good way of monitoring employees as well. In the "privacy" field this latter process is often referred to using Langdon Winner's phrase "function creep" (Winner 1977).

Winner, like David Thomas almost 30 years later, warned that once a digitized national ID number has been assigned – say, to combat terrorism – its use is likely to be expanded to cover many cognate areas. Whatever the specific characteristics of surveillance technologies, they also have to be located culturally in certain discourses of technology. Especially in the western world and above all in the US, technology holds a special place in popular imagination and public policy. Technical "solutions" to an array of perceived social, economic and political questions are all too quickly advanced and adopted, particularly in the aftermath of some crisis or catastrophe. This is not the start of an anti-technology argument; it is simply to say that technical responses have become commonplace, taken for granted.

In the mid-twentieth century Jacques Ellul famously insisted that in the technological society, "la technique," or the "one best way of doing things" had become a kind of holy grail, especially in the US. In a world where from the late nineteenth century progress, associated with undeniable technological advancement (at least in some domains), had been proclaimed, to fall back on technical solutions was both understandable, straightforwardly manageable, and of course lucrative for the companies concerned. By the end of the twentieth century Robert Wuthnow, a sociologist of religion, could argue that technology remains one of the few beliefs that unites Americans (1998). And if it was not clear before the twenty-first century, the challenge of terrorism certainly made it clear that technical responses were highly profitable. Share prices in security and surveillance companies surged after the attacks of 9/11 and also after the Madrid (2003) and London (2005) bombings.

The steady and often subtle adoption of new technologies, including surveillance devices and systems, into everyday life is highly significant from a sociological point of view. If it was ever appropriate to think of social situations in a technological vacuum, those days are definitely over. Because, for example, machines such as cell phones and computers have become essential for so many everyday communications, analyses of networks of social relations cannot but include reference to them. This is the "technoculture." Frequently, however, the focus is on how fresh forms of relationship are enabled by the new technologies rather than on how power may also be involved in ways that limit or channel social activities and processes. In a post 9/11 environment, the key questions are about civil liberties, following the hasty deployment of supposedly risk-reducing technologies in the name of national security. But equally, the mundane activities of shopping using credit and loyalty cards may also contribute to profoundly significant processes of automated social sorting into newer spatially based social class categories that modify older formations of class and status. Sociology itself is obliged to readjust to such shifts (see Burrows and Gane 2006).

## The explosion of personal data

It is difficult to exaggerate the massive surge in traffic in personal data from the 1970s to today. And the quantitative changes have qualitative consequences. It is not merely that more and more data circulate in numerous administrative and commercial systems, but that ways of organizing daily life are changing as people interact with surveillance systems. One of the biggest reasons for this is hinted at in the word just used to describe it— "traffic." There is constant growth in the volume of personal data that flows locally, nationally and internationally through electronic networks. But one cause of this is "traffic" in another, economic, sense, in which personal data are sought, stored and traded as valuable commodities.

Long before notions of the "surveillant assemblage" came to the fore, Australian computer scientist Roger Clarke had proposed another term to capture the idea of "surveillance-by-data"—"dataveillance" (Clarke 1988). A surge in surveillance could be traced, he argued, to the convergence of new technologies—computers and telecommunications that rendered Orwell's ubiquitous two-way television unnecessary. The novel combinations made possible by ICTs permitted quite unprecedented flows of data, illustrated by Clarke in the case of electronic funds transfer (EFT).

It is hard for those who now assume the constant networks of flows (a term appropriated by Manuel Castells) to recall how revolutionary EFT seemed at the time. It enabled supermarket shoppers, for instance, to have their accounts conveniently debited at the point of sale, thus bypassing several stages of financial transaction that would previously have had to occur. Such transfers are not only now commonplace, they also occur across a range of agencies and institutions that once had only indirect and complex connections. Clarke's point about *Nineteen-Eighty-Four* was a critical one, pointing to the potentially negative surveillance capacities of dataveillance. Without minimizing that point, however, it is crucial to note that the major difference between the two is that EFT and its descendants are not centralized. Indeed, on the contrary, they are diffuse, shifting, ebbing and flowing – and yet, as we shall see, not without discernible patterns of their own.

Even when Clarke was writing about dataveillance, a further innovation had yet to become a household word. What is often referred to as the Internet (meaning a range of items, usually including email systems and the World Wide Web) was only coming into being as a publicly accessible tool in the early 1990s. The debate over its threatened commercialization was hot; until then it was the preserve of the military, academics and computer enthusiasts, many of whom saw it as an intrinsically open medium. Its eventual role as a global purveyor of information, ideas, images and data, under the sign of consumerism, signals a major augmentation of surveillance.

Not only were computers and communications systems enabling new data-flows of many kinds, now consumers could participate directly in the process. Online-purchasing of goods and services from groceries to airline tickets to banking meant that personal data was moving on a massive scale. Who had access to these data, and how they could be secured and protected became a central question as quite new categories of crime appeared, such as "identity theft," and as corporations fell over themselves to gain access to increasingly valuable personal data. Knowing people's preferences and purchasing habits was to revolutionize marketing industries, right down to targeting children (Steeves 2005).

A third phase of dataveillance only began to take off at the turn of the twenty-first century. It involves a device that had been in the analytical shadow of the internet during much of the 1990s but which, some argue, may be at least if not more profound in its social implications. The cell phone (or mobile phone) is the single most important item in what might be termed "mobiveillance." If dataveillance started in the world of places, such as supermarkets, police stations and offices, then the use of networked technologies such as the Internet virtualized it, producing what might be called "cyberveillance." Surfing data became significant within the virtual travels of the Internet user. The advent of mobile or "m-commerce," in which the actual location of consumers becomes an important value-added aspect of personal data, using RFID, automated road tolling, or other technologies, as well as cell-phones, brings the activity that characterized "surfing" back into the world of place, only now it can be any place in which signals are accessible (Andrejevic 2004; Lyon 2006).

The result is that personal data now circulate constantly, not only within but also between organizations and even countries. Personal data flow internationally for many reasons, in relation, for example, to police data-sharing arrangements (such as the Schengen Agreement in Europe), especially with the rise of perceived threats of terrorism, or to "outsourcing,"—the set of processes whereby banks, credit card companies and other corporations use call centers in distant countries for dealing with customer transaction data. While for much of the time the public in countries affected by such increased data flows seem to assume that their data are secure and that they are used only for the purposes for which they were released, notorious cases of fraud and sheer error do seem to proliferate with the result that some consumers and citizens are more cautious about how they permit their data to travel.

### The end of privacy?

From the late twentieth century a common response to the massive growth of surveillance systems in the global north has been to ask whether we are witnessing the "end of privacy." What is meant by this? On the one hand, as many socially critical authors assert, there are fewer and fewer "places to hide" (see for example O'Harrow 2005) in the sense that some surveillance systems record, monitor, or trace so many of our daily activities and behaviors that, it seems, nothing we do is exempt from observation. On the other, a different set of authors see the "end of privacy" as something to celebrate, or at least not to lament. In the face of growing e-commerce and the consequent mass of personal data circulating, Scott McNealy, of Sun Microsystems, most famously declared: "Privacy is dead. Get over it!" Privacy is a highly mutable concept, both historically and culturally relative. If privacy is dead, then it is a form of privacy—legal, relating to personal property, and particularly to the person as property—that is a relatively recent historical invention in the Western world. At the same time, this western notion of privacy is simply not encountered in some south-east Asian and eastern countries. The Chinese have little sense of personal space as Westerners understand it, and the Japanese have no word for privacy in their language (the one they use is imported from the west).

The best-known writer on privacy in a computer era is Alan Westin, whose classic book *Privacy and Freedom* (Westin 1967) has inspired and informed numerous analysts and policy makers around the world. For him, privacy means that "individuals, groups or institutions have the right to control, edit, manage and delete information about themselves and to decide when, how and to what extent that information is communicated to others." However, although this definition seems to refer to more than the individual, the onus of responsibility to "do something" about the inappropriate use of personal (and other) data is on data-subjects. That is, rather than focusing on the responsibilities of those who collect data in the first place, it is those who may have grievances who have rights to have those addressed.

However, Priscilla Regan (1995) adds, importantly, that privacy has intrinsic common, public and social value, and that that therefore not only may individuals have a right to seek protection from the effects of misused personal data, but also organizations that use such data have to give account. The huge increase in surveillance technologies, for instance in the workplace and in policing, underscores this point. Today, data are not only collected and retrieved, but analyzed, searched, mined, recombined and traded, within and between organizations, in ways that make simple notions of privacy plainly inadequate. Valerie Steeves maintains that while Westin began, in the 1960s, with a broader definition of privacy, the overwhelmingly individualistic context of American business and government interests, in conjunction with pressure to adopt new technology "solutions," has served to pare down privacy to its present narrow conception (Steeves 2005).

## Surveillance as social sorting

To argue that privacy may not have the power to confront contemporary surveillance in all its manifestations is one thing. To propose an alternative approach is another. For, as in the case of the Orwellian and the panoptic imagery for capturing what surveillance is about, the language of privacy has popular cachet. It is difficult to explain why "privacy" is not the only problem that surveillance poses (Stalder 2002) when this is so widely assumed by lawyers, politicians, mass media and western publics. The best way of deflecting attention from a singular focus on privacy, in my view, is to consider surveillance as "social sorting."

One might say that "to classify is human" but in modern times classification became a major industry. From medicine to the military, classification is crucial. As Geoffery Bowker and Susan Star show, the quest for meaningful content produces a desire for classification, or "sorting things out" (Bowker and Star 1999). Human judgments attend all classifications and, from our perspective, these are critical. Classification allows one to segregate undesirable elements (such as those susceptible to certain kinds of disease) but it is easy for this to spill over into negatively discriminatory behaviors. South Africa under apartheid had a strong population classification system but it served to exclude, on "racial" criteria, black people from any meaningful access to opportunity structures. Classification may be innocent and humanly beneficial but it can also be the basis of injustice and inequity. The modern urge to classify found its ideal instrument in the computer.

One way of thinking about surveillance as social sorting is to recall that today's surveillance relies heavily on ICTs. Both security measures and marketing techniques exploit the interactivity of ICTs to identify and isolate groups and individuals of interest to the organizations concerned. By gathering data about people and their activities and movements and analyzing secondary data by "mining" other databases, obtained through networked technologies, marketers can plan and target their advertising and soliciting campaigns with increasingly great accuracy. Equally, security personnel use similar strategies to surveil or monitor "suspects" who have been previously identified or who fit a particular profile in the hope of building a fuller picture of such persons, keeping tabs on their movements, and forestalling acts of violence or terror.

These actuarial plans for opportunity maximization (marketing strategies for widening the range of target groups for products and services) and for risk management (such as security strategies for widening the net of suspect populations) represent a new development in surveillance. Though they have a long history, they contrast with more conventional reactive methods of marketing or security delivery. They are future rather than past oriented, and are based on simulating and modeling situations that have yet to occur. They cannot operate without networked, searchable databases and their newness may be seen in the fact that unsuspecting persons who fit, say, an age profile, may be sent email messages promoting devices guaranteeing enhanced sexual performance and others, much less amusingly, who simply fit an ethnic or religious profile, may be watched, detained without explanation or, worse, by security forces.

The "surveillant assemblage" works by social sorting. Abstract data of all kinds—video images, text files, biometric measures, genetic information and so on—are manipulated to produce profiles and risk categories within a fluid network. Planning, prediction and pre-emption, permitting all these and more goals, are in mind as the assemblage is accessed and drawn upon. Social sorting is in a sense an ancient and perhaps inevitable human activity but today it has become routine, systematic and above all technically assisted or automated, and in some sense driven. The more new technologies are implicated, however, the more the criteria of sorting become opaque to the public. Who knows by what standards a credit application was unexpectedly turned down or an innocent terrorist suspect was apprehended? Of course, the sorting may be innocent and above question—surveillance, after all, is always ambiguous—but it is also the case that social sorting has a direct effect, for good or ill, on life chances (see Lace 2005: 28–32 for consumer examples).

The main fears associated with automated social sorting, then, are that through relatively unaccountable means, large organizations make judgments that directly affect the lives of those whose data are processed by them. In the commercial sphere, such decisions are made in an actuarial fashion, based on calculations of risk, of which insurance assessments provide the best examples. Thus people may find themselves classified according to residential and socio-demographic criteria and paying premiums that bear little relation to other salient factors. Equally, customers are increasingly sorted into categories of worth to the corporation, according to which they can obtain benefits or are effectively excluded from participation in the marketplace (Gandy 2010). In law enforcement contexts, the actuarial approach is replicated; indeed, Feely and Simon warned in the mid-1990s that forms of "actuarial justice" were becoming evident. The "new penology," they argue, "is concerned with techniques for identifying, managing and classifying groups sorted by levels of dangerousness" (Feely and Simon 1994: 180). Rather than using evidence of criminal behavior, newer approaches intervene on the basis of risk assessment, a trend that has become even more marked after 9/11 (Monahan 2010).

Little has been said about how so-called data subjects of contemporary surveillance engage with and respond to having their data collected and used by organizations. Much depends on the purposes for which those data are collected. Righteous indignation at being shut out of a flight may be the response of a passenger with a "suspicious" name, even though that same passenger may be delighted with the "rewards" from his frequent flyer program with which he "bought" the ticket. In each case, extensive personal data is used to determine the outcome, whether the privileged category of an "elite" passenger or the excluded category of a name on the no-fly list. Consumers appear most willing to provide their personal data when they believe that some benefit awaits them; employees and citizens are much more likely to exercise caution or express complaint at the over-zealous quest of organizations for their details.

## Conclusion

Questions of surveillance and privacy have become more important as so-called information societies have developed since the 1970s. Thus ICTs are centrally implicated in these developments because their establishment may be prompted by these technologies, which may be harnessed to add power to surveillance systems. At the same time, surveillance grows because of certain economic and political priorities and because of the emergence of cultural contexts in which self-disclosure is not merely acceptable but sometimes positively valued and sought. Surveillance has also been expanding rapidly since 9/11.

Calls for greater privacy, once the standard response to increased surveillance, continue to be made, with varying results. Yet regulative bodies, especially those based on legislative regimes, have a very hard time keeping up with the changes occurring. At the same time, the onus of law has tended to be on the individual who feels (assuming she even knows) that she has been violated or invaded, and not necessarily on the organizations that process the data in the first place. Data protection regimes have more to offer here, dependent as they are on registering their activities, and more recent laws— for instance the Personal Information Protection and Electronic Documents Act 2001 (PIPEDA) in Canada—do require organizations, in this case including commercially based ones, to attend to the stipulations of the law.

But large and urgent questions about social sorting remain, even after privacy and data protection policies and laws have done their work. It is quite possible for negative discrimination to be carried out, automatically and systematically, against ethnic minorities (such as categories relating to the likelihood of terrorist involvement) or social-economic minorities (such as those living in low-income districts of cities), despite having such policies and laws in place. The codes by which persons and groups are categorized are seldom under public scrutiny, and if they relate to "national security," they may well be veiled in official secrecy, and yet they have huge potential and actual consequences for the life chances and choices of ordinary citizens. Thus both in terms of accurate analysis and informed political action, much remains to be done in the emerging realm of database-enabled surveillance. It seems unlikely that the issues will be tackled in ways appropriate to the present challenge while the mass media encourage complacency about self-disclosure; high technology companies persuade governments and corporations that they have surveillance "solutions" to their problems; actuarial practices deriving from insurance and risk management dominate the discourse that support surveillance; and legal regimes are couched in the language of supposed rights to individual privacy. The politics of information in the twenty-first century will increasingly be about how to increase the accountability of those who have responsibility for processing personal data.

## Note

This is a revised version of an article that appeared earlier in Mansell, R., Chrisanthi Avgerou, Danny Quah, and Roger Silverstone. 2007. *The Oxford Handbook of Information and Communication Technologies*. Oxford: Oxford University Press.

## Bibliography

Agamben, G. 1998. *Homo Sacer: Sovereign Power and Bare Life*. CA: Stanford University Press.
Andrejevic, M. 2004. *Reality TV: The Work of Being Watched*. New York: Rowman and Littlefield.
Ball, K. and F. Webster (eds.) 2004. *The Intensification of Surveillance: Crime, Terrorism and Warfare in the Information Age*. London: Pluto Press.

Bauman, Z. 2000. "Social Issues of Law and Order." *British Journal of Criminology* 40: 205–21.

Bigo, D. and E. Guild (eds.) 2005. *Controlling Frontiers: Free Movement Into and Within Europe*. Aldershot, UK: Ashgate.

Bowker, G. and Susan Star. 1999. *Sorting Things Out: Classification and Its Consequences*. Cambridge MA: MIT Press.

Burrows, R. and N. Gane. 2006. "Geodemographics, Software and Class," *Sociology* 40 (5): 793–812.

Clarke, R. 1988. "Information Technology and Dataveillance." *Communications of the ACM* 31(5): 498–512.CRM 6, P. 2005. "The Personal Information Economy: Trends and Prospects for Consumers." In *The Glass Consumer: Living in a Surveillance Society*, S. Lace ed. Bristol UK: Policy Press, 17–43.

Dandeker, C. 1990. *Surveillance, Power and Modernity*. Cambridge: Polity Press.

Deleuze, G. 1992. "Postscript on the Societies of Control", *October* 59. Cambridge, MA: MIT Press, 3–7.

Ericson, R. and K. Haggerty. 2000. "The Surveillant Assemblage," *British Journal of Sociology* 51(4): 605–22.

Ericson, R. and K. Haggerty. 1997. *Policing the Risk Society*. Toronto: University of Toronto Press.

Feely, M. and J. Simon. 1994. "Actuarial Justice: The Emerging New Criminal Law." In *The Futures of Criminology*, ed. D. Nelken. London: Sage, 173–201.

Foucault, M. 1979. *Discipline and Punish*. New York: Vintage.

Gandy Jr., O. 2009. "Coming to Terms with Chance: Engaging Rational Discrimination and Cumulative Disadvantage." Farnham, UK: Ashgate.

Gandy Jr., O. and A. Deanna. 2002. "All that Glitters is not Gold: Digging Beneath the Surface of Data Mining." *Journal of Business Ethics* 40: 373–86.

Gandy Jr., O. 1993. *The Panoptic Sort: A Political Economy of Personal Information*. Boulder, CO: Westview Press.

Genosko, G. 2006. "Tense Theory." In *Theorizing Surveillance: The Panopticon and Beyond*, ed. D. Lyon. Cullompton, UK: Willan Publishing.

Hardt, R. and A. Negri. 2000. *Empire*. Cambridge, MA: Harvard University Press.

Lace, S. 2005. *The Glass Consumer: Life in a Surveillance Society*. Bristol UK: The Policy Press.

Lash, S. 2002. *Critique of Information*. London: Sage.

Lessig, L. 1999. *Code and Other laws of Cyberspace*. New York: Basic Books.

Lyon, D. 2009. *Identifying Citizens: ID Cards as Surveillance*. Cambridge: Polity Press.

Lyon, D. 2007. *Surveillance Studies: An Overview*. Cambridge: Polity Press.

Lyon, D. 2006, "Why Where You Are Matters: Mundane Mobilities, Transparent Technologies and Digital Discrimination." In *Surveillance and Security: Technological Politics and Power in Everyday Life*, ed. T. Monahan. New York and London: Routledge.

Lyon, D. 2004. "Surveillance Technology and Surveillance Society." In *Modernity and Technology*, eds. T. Misa, P. Brey and A. Feenberg. Cambridge, MA: MIT Press, 161–84.

Lyon, D. (ed.) 2003. *Surveillance as Social Sorting: Privacy, Risk, and Digital Discrimination*. London and New York: Routledge.

Lyon, D. 2003. *Surveillance after September 11*. Cambridge: Polity Press.

Lyon, D. 2001. *Surveillance Society: Monitoring Everyday Life*. Oxford: Open University Press.

Marx, G. T. 1998. "Ethics for the New Surveillance," *The Information Society* 14: 171–85.

Marx, G. T. 1988. *Undercover: Police Surveillance in America*. Berkeley CA: University of California Press.

Monahan, T. 2010. *Surveillance in a Time of Insecurity*. New Brunswick: Rutgers.

Monahan, T. and R. Torres. 2009. *Schools Under Surveillance: Cultures of Control in Public Education*. New Brunswick: Rutgers University Press.

Norris, C. 2003. "From Personal to Digital: CCTV, the Panopticon and the Technological Mediation of Suspicion and Social Control." In *Surveillance as Social Sorting: Privacy, Risk, and Digital Discrimination*, ed. D. Lyon. London and New York: Routledge, 249–81.

Norris, C. and M. McCahill. 2004. "CCTV in London, Berlin: Urban Eye." www.urbaneye.net/results/ue_wp6.pdf, accessed March 18, 2008.

O'Harrow, R. 2005. *No Place to Hide*. New York: Free Press.

Raab, C. D. 2005. "Governing the Safety State," inaugural lecture at the University of Edinburgh, Scotland (June 7).

Regan, P. 1995. *Legislating Privacy*. Chapel Hill: University of North Carolina.

Rose, N. 1996. *Powers of Freedom*. Cambridge UK: Cambridge University Press.

Stalder, F. 2002. "Privacy is Not the Antidote to Surveillance," *Surveillance and Society* 1(1): 120–4. Available: http://www.surveillance-and- society.org/articles1/opinion.pdf, accessed November 23, 2007.

Steeves, V. 2005. "It's not Child's Play: The Online Invasion of Children's Privacy." *University of Ottawa Law and Technology Journal*, 2 (2).

Westin, A. 1967. *Privacy and Freedom*. New York: Athenaeum.

Winner, L. 1977. *Autonomous Technology: Technics Out of Control as a Theme in Human Thought*. Cambridge MA: MIT Press.

Wuthnow, R. 1998. *The Restructuring of American Religion*. Princeton, NJ: Princeton University Press.

# 5.1
## Full Spectrum Surveillance: NYPD, Panopticism and the Public Disciplinary Complex

*Brian Jefferson*

Recent controversy surrounding the "Lower Manhattan Security Initiative" has yet again breathed new life into the notion of the Panopticon. This initiative marks the latest development in an ultramodern policing strategy that saturates public space with cutting-edge surveillance devices, all linked into the New York Police Department's (NYPD's) central database (Buckley 2007). Indeed, the NYPD's expanding surveillance apparatus sheds considerable light on the symbiotic evolution state power shares with technological innovation, drastically altering the scope in which governmental machinery monitors and regulates political subjects.

A critical study of the security initiative requires consideration of broader technological renovations in the NYPD; particularly in its implementation of Comstat and the "Ring of Steel." The former, short for "comparative statistics," refers to a weekly procedure in which the commissioner and ranking personnel analyze elaborate crime data at NYPD headquarters (Silverman 1999). Data analysis is followed by the exploration of customized, area-specific crime-fighting tactics. This information-centered approach constitutes a computerized mode of policing wherein the city itself serves as a digital map that informs the intensity and style of law enforcement according to precinct. In short, the NYPD has used computer technology to render public space a transparent field of criminal activity, effectively virtualizing the art of crime-fighting.

The ominously titled "Ring of Steel" pertains to thousands of "crime cameras" in select neighborhoods, strategically deployed according to the aforesaid crime pattern mapping. Over the past five years, however, we have witnessed the Ring of Steel evolve into a much more hi-tech and versatile apparatus. Indeed, it gained critical attention with the enunciation of the Lower Manhattan Security Initiative, a project involving the radical extension of what was already a dense web of thousands of closed circuit televisions, license plate readers and counterterrorism mechanisms connected to Comstat (Lisberg 2007). The scope of this augmented surveillance was punctuated by the introduction of helicopters equipped with state-of-the-art cameras capable of facial and movement recognition of two and twelve miles, respectively. Much like the Panopticon, this policing method relies on anonymous mechanisms of surveillance, albeit it with some crucial modifications.

In the twenty-first century the panoptic model is vested in not only institutions (the prison, the military, the school), but also public space itself. Buildings, traffic signs, license plates and airspace now function as conduits to the NYPD's burgeoning surveillance machine. To fully grasp the implications of this development requires a case study of what I term "full spectrum policing": a policing apparatus sharing close affinity with the prerogatives found in the Pentagon's "Joint Vision 2010" and 2020. Much like NYPD strategy, "Joint Vision" doctrine is predicated on a novel conception of technologies' roles in conflicts (Office of the Joint Chiefs of Staff 1996, 2000). Moreover, it is also premised on an exploitation of the state's "digital edge," approaching battlespace as a virtual landscape subject to an overwhelming combination of "non-kinetic" weaponry. This futuristic approach aimed to achieve what the Department of Defense deemed "Full Spectrum Dominance," a doctrine paralleling contemporary NYPD strategy in many aspects.

Full spectrum policing therefore characterizes an ultramodern apparatus, which by dint of hi-tech equipment transforms civic space into what is ultimately a "public disciplinary complex." This concept shares much with the concept of a "new securocracy"—a state-led "emphasis on security generalized throughout the public sector" (Oswick, Harney and Hanlon 2008). However, the concept of the "public disciplinary complex" emphasizes the novel functions of the policing technologies in reorienting processes of political subjectification. Thus public space is altered to create a surveillance apparatus for monitoring and regulating the political character of all who are situated within it. Such a profound mutation is of great significance for the state's role in orienting the political subjectivity of citizens, and thus requires us to revisit and revamp Foucault's apposite theories of panopticism.

## Bibliography

Buckley, Cara. 2007. "New York Plans Surveillance Veil for Downtown." *The New York Times*, 9 July.

Lisberg, Adam. 2007. "NYPD's 'Ring of Steel' Camera Rolls, Monitors License Plates." *New York Daily News*, 1 October.

Office of the Joint Chiefs of Staff. 1996. *Joint Vision 2010*. July.

Office of the Joint Chiefs of Staff. 2000. *Joint Vision 2020. America's Military: Preparing for Tomorrow*. May.

Oswick, Cliff, Harney, Stefano and Hanlon, Gerard. 2008. 'The New Securocracy and the 'Police Concept' of Public Worker Identity'. *International Journal of Public Administration* 31.

Silverman, Eli. 1999. *NYPD Battles Crime: Innovative Strategies in Policing*. Boston: Northeastern University Press.

# 5.2
# The Wired Body and Event Construction: Mobile Technologies and the Technological Gaze

*Yasmin Ibrahim*

Our recent recognition and celebration of the body as a site of consumption means that we have allowed into our lives an array of technological and mobile artifacts designed to lure our cognitive senses away from the communal and into the personal. The personal space is a coveted commodity where new technologies, innovative designs and convergence occur and coalesce. The inbuilt surveillance mechanisms within mobile technologies and the constant circulation of bodies nevertheless constitute new forms of gaze, consumption and surveillance, which have wider implications for postmodern societies. This counter-gaze of the technologically connected bodies presents the potential for empowerment and connection with wider society, yet it inadvertently raises new conundrums where the politics of gazing present new ethical and moral dilemmas for humanity.

Traditional media like radio and television in their heydays heralded the ability for families to come together and experience consumption as one entity. The decentering of technologies of entertainment and consumption from domestic sites into the physical site of the body has meant new forms of configuration in design and convergence. More importantly, many mobile technologies are designed to "speak" to other new media technologies and constitute part of a new media ecology where individuals can customize their content and have the agency to transgress the boundaries between producer and consumer. In these new production and political economies, private content can be linked to a wider economy of information production and dissemination.

The rise of soft news and infotainment (Patterson 2000) in the mainstream media, the integration of personal narratives and the trivialization of politics have meant that since the 1980s there has been an increasing reliance on "eyewitness" accounts (Livingston and Bennett 2003: 370) to construct media reports. Event-driven news (Lawrence 2000), unlike pseudo events (Boorstin 1977) orchestrated by the media, include spontaneous events like natural disasters and terrorists attacks. Bennett and Livingston (2003) have argued that event-driven news is overtaking institutionally based news, particularly in the technologically charged environment of cable television international affairs news. New discursive spheres, whether originating from mobile interactive technologies or the Internet, will become integrated in event creation and memory construction in mediated societies. This signifies new media rituals in conveying global events, where the telling of the story through the media gaze alone is incomplete without the inclusion of the participatory gaze of the civilian through mobile technologies. Technology in this sense increases the occurrence of traumatic acts and access to them (Zelizer 2002: 697).

Beyond bearing witness and facilitating citizen journalism, mobile technologies as devices of connectivity contain the potential for empowerment and for initiating collective action. While re-negotiating the politics and aesthetics of gazing in traditional societies, mobile telephony has also enabled citizens to circumnavigate entrenched and dominant patterns of information dissemination, political communication and expression. In a society where there is ubiquitous watching, this omnipresent gaze is accommodated within a wider paradigm of constant vigilance. This individual gaze facilitated through mobile technologies co-exists within institutionally entrenched CCTV cultures, which codify movements and circulation of bodies as data, creating a post-surveillance society that implicitly accepts the degree of monitoring and surveillance (practiced top-down and bottom-up) while embedding it in daily lives in new and innovative ways.

This post-surveillance society is part of the accelerated modernity in contemporary societies where speed and movement of people and activities has to be reconciled with the ability to record and freeze-frame images through the narcissistic gaze of the individual and the centralized gaze of institutions. The participatory counter gaze of moving bodies is decentralized and disparate but contains the potential to challenge the entrenched and centralized gaze of institutional authority, while raising new and ethical dilemmas about bearing witness and capturing images and circulating them within the wider information economy and data infrastructure in postmodern societies. In this environment of pervasive gaze, the possibilities for re-contextualizing data and images are also endless where "it can become the portal for voyeurism and unauthorized images of unsuspecting prey whereby everyone can become anonymous actors in someone's film" (Hjorth 2006).

## Bibliography

Livingston, S. and Bennet, W. L. 2003. "Gatekeeping, Indexing and Live-Event News: Is Technology Altering the Construction of News?" *Political Communication*, 20: 368–80.
Boorstin, D. 1977. *The Image: A Guide to Pseudo-Events in America*. New York: Atheneum.
Hjorth, L. 2006. "Being Mobile: In Between the Real and Reel." Paper Presented at Cultural Space and Public Sphere in Asia, Seoul, March 15–16, 2006. Retrieved

November 5, 2007 from http://www.cct.go.kr/data/acf2006/mobile/mobile_0101_Larissa%20Hjorth.pdf.

Lawrence, J. (2000). *The Politics of Force: Media Construction of Police Brutality.* Berkeley: University of California Press.

Patterson, T. 2000. "Doing Well and Doing Good: How Soft News and Critical Journalism are Shrinking the News Audience and Weakening Democracy – and What News Outlets Can Do about It." Retrieved November 5, 2007 from http://www.hks.harvard.edu/presspol/publications/reports/soft_news_and_critical_journalism_2000.pdf.

Zelizer, B. 2002. "Finding the Aids to the Past: Bearing Personal Witness to Traumatic Public Events." *Media, Culture and Society*, 24: 697–714.

# 5.3
## Configuring the Face as a Technology of Citizenship: Biometrics, Surveillance and the Facialization of Institutional Identity

*Joseph Ferenbok*

Images of face have been used as legal representation of individuals since the Renaissance when oils on canvas allowed for unprecedented detail and realism (Snyder 1985). However, it was not until well after the invention of photography that faces became integrated into identification documents. Although police departments in the UK began taking mug shots as early as the 1850s (Norris and Armstrong 1999), it was the Geary Act (1892), an extension of the Chinese Exclusion Act (1882), which became the first legislation to require photographs of faces to accompany identification documents for re-entry into the US. The nineteenth century proponents of photography argued that photographs established a concrete indexical link between the document and the individual. However, in practice, immigration inspectors found that "photographic truth" (Tagg 1988) could be co-opted to construct "paper families" (Pegler-Gordon 2009). Despite the experiences of immigration inspectors, in 1914 the US State Department called for photographs to be added to passports (Lloyd 2003), and since then facial images have become a dominant component of globally recognized identification documents

In 1992 Japan became the first nation to print images of faces digitally onto their passports (Lloyd 2003); and in 2003, the International Civil Aviation Organization (ICAO), a body of the United Nations, selected the face as the primary biometric feature for global air travel (ICAO 2003). Computerization and digitization have brought with them new affordances for face-based identification practices. Computer-assisted authentication represents a fundamental paradigm shift from analogue images. Digital faces, on passports, driver's licenses and other institutionally authenticated identity documents have become material tokens of underlying "surveillance assemblages" (Haggerty and Ericson 2000).

Where analogue images of faces are generally localized, either in an archive or at a point of interaction, digital faces may be stored on chips incorporated into the documents and linked to searchable databases. When an individual gets a new driver's license, for example, an image of their face is taken. The digital image is stored in a database and face-recognition algorithms are used to match that image to previous tokens of the same individual to deter identity theft, and against all the faces in the database to search for aliases. This virtual line-up uses algorithmic surveillance (Introna and Wood 2004) to inspect everyone in the database each time a new face is enrolled; individuals are under "virtual" suspicion until they are verified by the system.

The trouble is that biometric systems are measures of probability, the probability that the new input is the same person as the enrolled image, and therefore these systems can never be absolutely accurate. The face, when compared with fingerprints and iris patterns, is a relatively dynamic physiological trait. Faces change considerably with lighting, viewing angles and position, and with time, emotional states, surgery and trauma. Therefore, face recognition technologies confront a number of technological challenges when comparing faces. Systems must not only be able to recognize a face within a particular image, but they must also extrapolate the information in order to be able to account for the dynamic nature of the input (Friedmann and Nissenbaum 1996). Even when presented with images taken in controlled situations, the individual may not be recognized as the individual he or she claims to be.

As faces, and the people behind them, are becoming more readable by the surveillance authorities, the technologies and overall socio-technical assemblage supporting the surveillance practices are becoming more sophisticated, complex and opaque. Image sequences of facial expressions or walking gait can be processed to assess the affective state of an individual and assess intentionality. Face recognition algorithms are already being used to "recognize" a real-world entity and link it to an online profile. The iPhone face-recognition application, Recognizr, allows users to find other users by automatically searching photographs taken on a mobile device. It is certain that the digitization of faces and the introduction of face-recognition technologies will further heighten civil liberties issues and social controversies that have long accompanied the spread of institutional monitoring.

## Bibliography

Friedmann, B. and H. Nissenbaum 1996. "Bias in Computer Systems." *ACM Transactions on Information Systems* 14(3): 330–47.

Haggerty, K. D. and R. V. Ericson 2000. "The Surveillant Assemblage." *British Journal of Sociology* 51(4): 605–22.

ICAO 2003. *Biometric Identification to Provide Enhanced Security and Speedier Border Clearance for Traveling Public*. Montreal: International Civil Aviation Organisation.

Introna, L. D. and D. Wood 2004. "Picturing Algorithmic Surveillance: The Politics of Facial Recognition Systems." *Surveillance & Society* 2(2/3): 177–98.

Lloyd, M. 2003. *The Passport: The History of Man's Most Travelled Document.* Phoenix: Sutton Publishing.

Norris, C. and G. Armstrong 1999. *The Maximum Surveillance Society: The Rise of CCTV.* Oxford: Berg.

Pegler-Gordon, A. 2009. *In Sight of America: Photography and the Development of U.S. Immigration Policy.* Berkeley: University of California Press.

Snyder, J. 1985. *Northern Renaissance Art: Painting, Sculpture, the Graphic Arts from 1350 to 1575.* Upper Saddle River, NJ: Prentice Hall.

Tagg, J. 1988. *The Burden of Representation: Essays on Photographies and Histories.* Massachusetts: University of Massachusetts.

# Part VI
# Digital Property

# 6
# Whose Property? Mapping Intellectual Property Rights, Contextualizing Digital Technology and Framing Social Justice

*Phillip Kalantzis-Cope*

## Introduction

Over the last 30 years we have witnessed a number of consequential shifts in the various domains of intellectual property, such as copyright, patents and trademarks. One of the most important of these has been the use of intellectual property rights to articulate and formulate a new global economic regulatory order. Through international agreements such as the World Trade Organization's (WTO's) Agreement on Trade-Related Aspects of Intellectual Property Rights agreement (TRIPS), there has been enormous pressure for nation-states to revise national frameworks to align with a singular definition of intellectual property as a private property right. This definition extends the logic of "real property" within a commodity-driven, market mediated economic system, into the domain of knowledge and information production. Importantly, this definition imposes the restrictions of economic scarcity onto a domain of infinite productive capacity.

This move has raised intensely controversial questions. For example: what are the effects of patented seeds on food security in the developing world, or food production in general? What effect does copyright have on the commodification of information and access to shared cultural knowledge? What effects do pharmaceutical patents have on access to medicine and global public health in general? How does this logic of ownership displace traditional knowledge systems? What are the substantive and semiotic effects of the language of "piracy" and "theft"? In sum, to what extent does the globalization of a one-size-fits-all, standardized intellectual property regime bring into focus emerging dynamics of (mal)distribution, (mis)recognition and (non-)participation in global social life?

At the same time, within the digital domain, there has been a proliferation of new forms of bottom-up, networked community approaches to intellectual property. The transition from commercial, print-based encyclopedias to

Wikipedia is a paradigmatic case in point. Such global "counter-movements" are founded on the ways in which digital knowledge communities bring together local and transnational communities of practice, bypassing the market, the state and international regulatory mechanisms. These emerging knowledge communities create self-instituting relations to property rights around the idea of the commons in a gift or reputational economy, formulating their own ecologies of order based on communal relations to property. The shared base of these paradigms centers on the free software, "copyleft," open access and open content movements, whose proponents argue that "free" content is not only critical to the freedom of the digital domain and its networks, but also a definitional feature of the technologies themselves.

Moreover, owing to its particular technological affordances, it is argued that the domain of the digital offers new modalities for participatory social production based on collaborative, commons-located and peer-to-peer networks, bypassing traditional hierarchies of knowledge and cultural production. While the arguments of the supporters of the commons focus on the digital as important in and of itself, they also use the digital to speak to the non-digital. For example, how can the lessons of commons-based digital production be transferred to the development of generic pharmaceuticals or seed hybridization? The social visions articulated through the use of intellectual property rights within these models raise a number of key questions. For example: how does the digital divide shape participation in digital networks? Are these self-instituting property relations sustainable, and how do they map onto a material world determined by "real" property? How do we address new modalities of "digital" hierarchical power? To what extent can these arguments be extended beyond the digital, for instance to the pharmaceutical industry, agribusiness or biotechnology? Thus the digital provides new pathways and presents new dilemmas in understanding (mal)distribution, (mis)recognition, and (non)participation, within contemporary global society.

Contestations over the right to intellectual property are increasingly emblematic of the struggles and visions for social justice in times that have been variously defined as "late capitalism," "post-industrial society," or the "information age." The analytical point of departure of this chapter will be to map and define five paradigms for interpreting intellectual property: "information protectionism," "information exceptionalism," "information terra nullius," "network distributionism," and "information materialism." This mapping will be based on iconic expressions of each paradigm. The interpretative spine of the chapter will address the ways in which each paradigm interprets, and advocates visions for, the dimensions of social justice. The final section of the chapter will position the five paradigms within Nancy Fraser's framing of social justice around the integrated logics of economic distribution, cultural recognition and social participation. By asking the question "whose property?," this chapter not only seeks to make intellectual property a focal point for conceptual analysis. It also sets out to explore the ways these debates may illuminate the particular demands of an emancipatory politics within our contemporary globalized epoch.

## Mapping intellectual property rights

The aim of the mapping process in this chapter is to outline the terms of the debate. Each paradigm prompts a set of challenges for conceptualizing social justice. Adding an important early qualification, to categorize the debates as five approaches, is not to imply that each is neatly bounded. The categories are also presented here as a preliminary sketch, pointed towards a research agenda rather than a conceptual prescription.

### Information protectionists

The first paradigm to be mapped is information protectionism. The iconic expression of this paradigm is found in the Trade-Related Aspects of Property Rights agreement (TRIPS). By way of background, the TRIPS framework was a foundational component in the birth of the World Trade Organization (WTO), and is enmeshed within the neo-liberal free trade agenda at its core. With dispute mechanisms and regulations that reach far into domestic regulatory domains, TRIPS has created one of the deepest international legal regimes the world has seen. In fact, it could be argued that within this domain of property law, we have the basic elements of a rights-based global constitution. As Christopher May writes, "for the first time a multilateral trade agreement has required not merely changes in the manner in which imports and exports are regulated at national borders but has also required significant undertakings as regards to national legislation for non-internationally traded products" (May 2007).

A definitional component of information protectionism as an approach to intellectual property is what this "rule of law" enforces. As it stands, TRIPS establishes a single scheme of regulation that covers patents, copyright, trademarks, geographical indications, industrial design and integrated circuit layout design. We engage with these forms of property every day as painting, literature, music, machines, software, genetic material, seeds, brand names, logos, the name of one's region and so on, as manifestations of the cultural products that we as a species create. So TRIPS simultaneously imposes a transnational legal (un)boundedness between people and states, and a (un)boundedness within the knowledge systems and cultural production of the world's peoples, creating a new "territoriality" of the mind.

"Protectionism" is deployed as a framing category for two symbiotic reasons. First, the concept forces into focus a move that makes intellectual property a private good, positioning the work and play of the mind within a discourse that is profit-orientated and market mediated. This applies not simply to protecting the research and development (R&D) and culture-intensive industries that are the base of "post-industrial society," a foundational motivation of TRIPS. It also reflects a capitalist social order that supports knowledge innovation and cultural production motivated by private profits. In a so-called knowledge economy or knowledge society, protection of private intellectual property supports significant economic sectors. These sectors are regarded as critical to the current wave of capitalist development. At its broadest, the idea of the market is being protected, or there is what Christopher May calls the "metaphorical continuity" between "real" and "intangible" property (May and Sell 2006: 18–19). Second, the concept of protection brings into focus a contradiction in the idea of the "free" market and "free" trade. Patents, copyright and other forms of private intellectual property create market-restricting monopolies for defined periods of time. It is not the "free" market for goods that is being protected, but the exclusive and temporary anti-market rights of knowledge owners.

Information protectionism poses a distinct set of challenges in thinking about social justice. Foundationally, it presents a challenge for the scope of rights. In building a universalizing approach to the regulation of intellectual property, who are the members of this rights regime? Who dominates TRIPS? How are intellectual property producers affected by TRIPS' regulatory authority? What democratic feedback loops are there at national and international levels? The processes around global regulation of intellectual property are not focused on fair and equitable material distribution, but on the consolidation of property ownership in the domain of the work and play of the mind. So, what are the material effects of this privatization and commodification of knowledge? Moreover, in the institutionalization of these rights, there has been little or no recognition of alternative non-commercial models of knowledge creation, nor the identities that are defined by these modes of knowledge production.

The next four paradigms emerge as responses to the expanding regime of information protectionism.

## Information exceptionalists

The second paradigmatic approach to intellectual property is "information exceptionalism." The iconic expression of this category is found in the work of Lawrence Lessig and Yochai Benkler. For these thinkers, the vision and possibility of social justice embedded in the property question begins with the proliferation of digital technologies. As Benkler writes, the declining relative costs of digital hardware, such as computers, digital cameras, data storage devices and so on, has "placed the material means of information and cultural production in the hands of a significant fraction of the world's population" (Benkler 2007: 3). The proliferation of digital technologies, Lessig argues, enables "an extraordinarily range of people to become a part of the creative process" (Lessig 2002: 10).

Information exceptionalists such as Lessig and Benkler argue that peer-to-peer production is a defining characteristic of the digital networks of the information age. These foster a new "digital" public sphere where culture and knowledge, rather than being developed top-down, emerge through a participatory culture, ordered by "collaborative filtering" and "accreditation." In turn, these networks provide an opening for a "newfound autonomy" to engage in "personal and collective expression" (Benkler 2007: 466). Integral to this participatory culture and the freedom of production is the ability to secure intellectual property, not as a rivalrous good as articulated by information protectionism, but as a non-rivalrous good. Lessig uses Thomas Jefferson's famous statement to explain a non-rivalrous good. Jefferson argues that "no one seriously disputes that property is a good idea, but it is bizarre to suggest that *ideas* should be property. Nature clearly wants ideas to be free," because, "no matter how many people share it, the idea is not diminished. When I hear your idea, I gain knowledge without diminishing anything of yours. In the same way, if you use your candle to light mine, I get light without darkening you. Like fire, ideas can encompass the globe without lessening their density" (Lipscome 1903: 193). The transformative potentiality of commons-based intellectual property is intimately tied to this understanding of information as inherently non-rivalrous.

The institutional guarantor of this approach is the "information commons," or to be more specific, the "liberal" commons. The defining character of the liberal commons is the ability to opt in and opt out. As Hanoch Dagan and Michael Heller argue, a "participatory commons regime" allows "members the freedom to come and go ... ensuring autonomy to individual members who retain a secure right to exit" (Dagan and Heller 2001: 554). An important qualification, as Lessig makes clear, is that commons is not incompatible with capitalism: "in advocating the commons, I have not argued for a world with only a commons. Not all resources can or should be organized as a commons" (Lessig 2002: 93). In other words the "commons" logic, in this model, does not seek to obliterate profit, capitalism or intellectual property rights. Rather the supporters of this version of the commons argue that non-market spheres of knowledge and cultural production feed back into the vibrancy of the market by creating "systems of sustainable value creation" (Weber 2004: 1). This argument is based on a view that not only will innovation be promoted, but also one's freedom to act in the world will be extended. The aim, as Yochai Benkler writes, is to balance "freedom" and "justice" without sacrificing "productivity" (Benkler 2007: 464).

This informational exceptionalist position is based on what its proponents posit to be the uniquely non-rivalrous nature of intellectual property. This exceptional sphere needs to be secured and institutionalized in the form of an information commons. Information exceptionalism, as a paradigmatic approach to intellectual property, focuses on the importance of the constitutive movement: the freedom to decide what to do with one's intellectual labor. This notion of freedom sits at the core of the information exceptionalist approach to social justice.

Information exceptionalism poses a number of key challenges in thinking about social justice. While arguing that the means of production has been "redistributed," as Goldsmith and Wu detail, there is a real "geography of internet hardware" (Goldsmith and Wu 2006: 55). So to whom, and to what degree, has the means of production been redistributed? Furthermore, can this logic of "redistribution" be extended to all domains of intellectual property? For example, can it apply to genetic engineering, pharmaceuticals, trade secrets and other non-digital domains of knowledge production—domains where proprietary model of intellectual property are strong? Moreover, while it is argued that the "liberal commons" opens cultural production to multiple modalities of identity formation, whose culture is the base? Rather than intellectual property being regarded as something defined publicly by governments in the form of copyright and patent laws, does the commons shift intellectual property to an individualized contact approach to rights as found in commons licenses? What are the implications of this fundamental shift in regulatory approach? Finally, within informational exceptionalism the question of participation becomes one of self-regulation, with a mutually agreed commons being the institutional guarantor. What are the implications when these spaces are constructed through voluntary associations built on contact law? Who has the ability to participate in these commons? In other words who has the liberty and the privilege to give away the fruits of their intellectual labor? In sum, to whom, and on what basis, can justice be afforded, within the information exceptionalism category?

## Information terra nullius

A third paradigmatic approach to intellectual property is "information *terra nullius." "Terra nullius"* or "empty land" is a Latin concept that was used in the justification of European colonization. Even though indigenous peoples were pre-existing occupiers of colonized lands, these lands were regarded as empty by the legal fiction of *terra nullius*. While not using the definition "information *terra nullius"* themselves, within the intellectual property debate this idea of "empty lands" has been applied, specifically, by anti-imperialist and eco-feminist responses to the globalization of intellectual property law. One of the most prominent of these is Vandana Shiva.

Shiva argues that a shared ideological worldview underwrites both the globalization of intellectual property rights, as represented by what this chapter has called information protectionism, and processes of colonialism. She argues that just as Papal decrees and land titles were issued by European monarchs to "explorers" to create legitimacy for the conquest of foreign lands and wealth, intellectual property rights are becoming the legal authority, secular decrees if you will, issued to corporations that pave the way for their ownership of life forms and knowledge systems. For instance, whereas indigenous peoples base their common knowledge in deep understandings of the natural world of plants, multinational corporations make colonial claims over genetic materials. What ideologically underpins this historical move is an outright denial of any prior rights of other peoples and cultures. For Shiva, the assumption of "empty lands" that justified one form of colonialism has been reproduced and expanded to "empty life, seeds and medicinal plants" (Shiva 1997: 4). In turn, this has ideologically paved the way for the colonial dynamic to be "extended to the interior spaces, the genetic codes, of life forms from microbes and plants to animals and humans" (Shiva 1997: 3). This colonization is structured around two interrelated themes. One theme is that of western science, its scientific method and the relationships it establishes between people and knowledge. Another theme is related to law and relationships to property. Embedded in both aspects is a "conflict of world views" between peoples of the developed world on the one hand, and the third world and indigenous peoples on the other (Shiva 1997: 120).

Distinctively, Shiva adds an eco-feminist perspective to her criticism of the colonization of science, analyzing the way in which the colonization of "scientific method" is premised on the privileging of the "new" over regeneration, as well as by the way in which it regards the earth as passive and malleable. The implications, for Shiva, are serious. She argues that the "devaluation" of regeneration sits at the core of both the "ecological crisis and the crisis of sustainability" (Shiva 1997: 43). This predicament stems from the way in which modern western cultures have separated out "production" and "creation" into distinctly different, bounded spaces. Shiva argues that this occurred because "creativity became the monopoly of men, who where considered to be engaged in production" whereas women were perceived to be engaged "in mere reproduction or recreation, which rather than being treated as renewable production, was looked upon as non productive" (Shiva 1997: 44). The production boundary, she says, "is a political construct that excludes regenerative, renewable production cycles from the domain of production" (Shiva 1997: 61). The creation boundary then "does to knowledge what the production boundary does to work: it excludes the creative contributions of women ... and views them as being engaged in unthinking, repetitive biological processes." Shiva argues that this same logic also is applied to "Third World peasants and tribes people" (Shiva 1997: 62). They become entwined with a view of the passivity of the earth and the natural world. She writes that "biotechnology reconstitutes the seed as passive, and locates activity and creativity in the engineering mind" (Shiva 1997: 46). She also argues that "the transformation of nature from a living, nurturing mother to something inert, dead and manipulable was eminently suited to the exploitation imperative of growing capitalism" (Shiva 1997: 47). In sum, Shiva argues that "the moment we ignore the useful and necessary and concentrate on only the profitable, we are destroying the social conditions for the creation of intellectual diversity" (Shiva 1997: 17).

Within the envelope of information *terra nullius* there are also more subtle manifestations of this logic of displacement. The debate over cultural iconography, community and personhood not only revolves around bio-technology but other indigenous cultural icons. In one example, a particularly US phenomenon, "tribes do not want their names used as mascots for high school, college and professional sports teams because the use evokes disgusting characters by frenzied fans" (Greaves 1995: 203). This shows the diverse ways cultural heritage can be colonized and appropriated through intellectual property rights.

Information *terra nullius* poses a number of key challenges in thinking about social justice. The dominant modern redistribution discourse emerges out of materialist understandings of economic security. However, a fundamental question about the nature of economic redistribution is posed by this approach. At the heart of this challenge is how models of economic distribution define the exclusive nature of "value" and "productivity" in relation to biological and social life. These definitions, and the decisions that accompany them, at once have cultural, material, and ecological ramifications, so information *terra nullius* forces us to ask what the conditions of economic, cultural and ecologically sustainable knowledge production are. In addressing these questions, from whose standpoint do we begin?

While information exceptionalism tries to carve out new spheres of the commons, information *terra nullius* brings into focus the displacement of traditional world views based on unique relations to the material world around them. So, how do we think about recognition of these world views, and who holds what in common? How can legal orders incorporate multiple modes of regulation, especially within the domain of property? Shiva, as with many within this mode of thinking, equates globalization with imperialism, but does this have to be the case? Are global regulatory regimes inherently a colonizing force? If so, what are the alternatives for thinking about social justice in a global context?

### Network distributionists

The fourth paradigmatic approach to intellectual property as defined in this chapter is that of the "network distributionists." The iconic expression of this paradigm is to be found in the work of McKenzie Wark and Alexander Galloway. This work focuses on a critique of what had been variously called "information," "cognitive," or "network" capitalism. Digital transformations are analyzed in reference to a third phase of capitalism, what Alexander Galloway calls "the third machine age" (Galloway 2004: 11) and McKenzie Wark calls "third nature." The concerns of this period, the digital age, center on decentralized control as a symptom of the postmodern condition (Galloway 2004: 3).

Foundational to this approach is the view that the dominant narrative of power and control—the state, state sovereignty and hierarchal power as the categories of critical analysis—are not sufficient to assess the new modalities of power and control based on the "institutional ecologies" of the digital era. Galloway uses the concepts of "protocol" and "distributed network" to illuminate emerging contours of power. A distributed network, the central network logic of the Internet and digital media generally, is "a structural form without a center." Protocols, in the case of digital technologies, "refer specifically to standards governing the implementation of specific technologies" (Galloway 2004: 7). These network protocols are "vetted out between negotiated parties … then materialized in the real world" (Galloway 2004: 7). In the real world they determine and shape social practice. In this way governance is ubiquitous, a patchwork of networks and protocols. Power self-orders in a diffused manner, at times under the guise of an open, free space. As Galloway states, in the dialectical nature of digital technologies "one machine radically distributes control into autonomous locals, the other machines focus control into rigidly defined hierarchies" (Galloway 2004: 8). Nevertheless, for Galloway, networks are not completely autonomous, as even distributed networks must speak to each other, or they do not associate. Political struggles can, and do, coexist in multiple networks, and it is within these hierarchies that the political struggles over the digital occur.

In a similar vein, McKenzie Wark's work illuminates a new cartography of power within the digital era. Intellectual property rights, he argues, are re-defining the contours of a global class struggle, changing the relations between the creators of information and the owners of information, the "hackers" or information workers and the "vectoralists," the "emergent ruling class of our time" (Wark 2004: 20). The vectoralist class, Wark writes, is so named because its agents seek to "control the vectors along which information is abstracted" (Wark 2004: 29). They either already own, or seek to own, the communication channels along which the work and play of the mind travels. They seek to control not only the vectors but also the information content that moves along such vectors. For the vectoralist class, "politics is about absolute control over intellectual property by means of warlike strategies of communication, control and command" (Wark 2004: 21). In other words, the power of the vetoralists lies in the monopolization of forms of intellectual property as private property and "the means of reproducing their value," the communication systems (Wark 2004: 32). What is at stake for Wark is not only the ownership of work and the play of the mind, but the horizon of possibility in imagining other abstract and material worlds.

For Wark, the "gift economy" represents the promotion and development of a counter-logic, based in the belief that "within the gift relation, nature appears as endlessly productive in its differences, in its qualitative, not its quantitative aspect" (Wark 2004: 43). In other words the gift economy, based in distinct and particular relationships to ownership, is the source of new vectors that sit outside of commodification. They are "a first step towards accelerating the surplus of expression rather than the scarcity of representation." This "surplus of expression" is fertile ground for the imagination of new modalities of networked social, economic and political orders. In this way, the significance of the "hack" as a critical tool is that it challenges not only the material order of society but also the abstraction (private property) on which society is based.

Network distributionism poses a number of key challenges in thinking about global justice. In some ways, this category runs into the same problems as the information exceptionalist category. For example, who can participate in the gift economy and modes of reciprocal exchange? Who is the hacker? What is their economic status, level of technological access, identity, race, and gender—in other words could this itself at times be an exclusionary category? Might this approach fall into technological determinism? Is it too dependent on its location in the domain of the digital?

Nevertheless network distributionism does highlight the importance of intellectual property, as the category and immaterial imagination at the base of its definition. Through the analysis of vectors and protocols it provides a productive site to critique the emergence of new political struggles, antagonisms, and modes of agency. Moreover, through the gift relation, this position gives us the capacity to think through the concept of the commons beyond its liberal manifestation as espoused by information exceptionalists.

## Information materialists

The final paradigm to be mapped in this chapter is "information materialism." This category finds its roots in a Marxist critique—the historical materialist approach—of the trajectory of intellectual property within the broader lens of capitalist techno-scientific development. Network distributionism and information materialism share a similar critique of exploitation within contemporary capitalism and account similarly for the rise of new modalities of class relations based on relations to knowledge and cultural production. Contrary to the visions of a post-worker, automated labor future for post-industrial society, network distributionists and information materialists share the view that "the traditional locus of exploitation between capital and labor in the workplace has not been transcended, but expanded" (Dyer-Witheford 2002: 8). In this way, digital networks become both factory and distribution channels of information capitalism. Free software models, as advocated by information exceptionalists, can and do fall victim to exploitation as corporations use the fruits of unpaid labor and as service workers sell their labor connected with free intellectual property. Whether it is commons-initiated or private digital properties, the appropriation of digital "social labor" is facilitated by digital networks. The post-industrial knowledge worker becomes a "value subject," with intellectual property law assigning value to their mind-work.

Information materialism distinctively emphasizes the need for an integrated analysis of the production of tangible as well as intangible goods. Instead of separating intellectual labor from the material social world, this approach emphasizes the digital network, and the material infrastructure that the networks are built on, into the focus of critique. One part of this argument is that knowledge has never been "free." Rather, a significant proportion of public knowledge has been subsidized by the state in the form of universities, libraries, public cultural institutions and so on. In this sense, the state has played a direct role in the determining the "material" order that contextualizes "immaterial" production. The private appropriation of scientific and cultural knowledge is thus contextualized, refocusing questions of the scope of the "public domain and the conditions of its adequate resourcing."

Another distinctive characteristic of orthodox historical materialist approaches has been crisis analysis. Crisis logic conjectures that the inherent contractions of capitalist development lead inevitably to the transformation of social, economic and political life and the emergence of a new redistributive order. Critics of this approach cast doubt over the inevitability of this crisis within post-industrial capitalism, based in the belief that capitalism evolves to obscure or temporarily resolve its own contradictions. Information materialism takes a more universalistic, institutional approach than network distributionism in thinking about the motivational logic to aid, or replace, the crisis logic in historical materialist approaches. Nick Dyer-Witheford gestures towards such an approach. He replaces crisis theory with the motivational logic of what he calls "commonism." The collective ownership at the base on the idea of "commonism" is not simply based on an abstract vision of property, but rather on a vision of a global commonwealth, and the emergence of what he calls a "planetary New Deal" (Dyer-Witheford 2010). The motivational logic of such a move is tied to an application of Marx's "species being" as the bedrock of knowledge production. Humans are simultaneously materially grounded and deliberative intellectual and cultural creatures. These are universal features of our species being, and dual bases of human emancipation. Information materialism, as defined here, takes a more institutional approach, based on the interlinked character of material and immaterial production in our social lives.

Information materialism poses a number of key challenges in thinking about social justice. In serving both as a critique of the liberal commons model in the information exceptionalist approach, and also the global regulative power representative of the information protectionist paradigm, it forces one to think about the boundaries of the liberal commons, and the depth of global regulatory power. If the commons is the source of the public domain, information materialism positions social justice within an expanded view of the commons in a reconfigured public sector. The question of social justice then takes a path in the direction of global social democracy.

## Contextualizing digital technology and framing social justice

The mapping presented in this chapter is pointed towards a matrix for investigation rather than a conceptual prescription. The aim of mapping and defining these paradigmatic approaches is not to select one perspective at the expense of others, nor to claim that these are exhaustive of the range of possible ways of seeing intellectual property. Rather, there are two underlying motives for the mapping.

First, all too often these paradigms speak only to their own intellectual and social worlds. For example, information protectionism speaks to international lawyers and international relations scholars; information exceptionalism to computer programmers and legal scholars; information *terra nullius* to activists and eco-feminists; network distributionists to media and communication theorists; and information materialism to advocates of social democracy. At times, the social movements and politics sustaining these debates are disassociated from each other, or come into polarizing conflict. Nevertheless, TRIPS may force our hand when thinking about intellectual property. In reconfiguring the global scale and substantive scope of intellectual property, the social worlds connected with the various paradigms may be brought into necessary association. Political claims of social justice in the realm of intellectual property require each paradigmatic approach need to be seen in relation to the others.

Second, within the spectrum of visions for social justice in the postindustrial, information age or networked society, the focus is usually on digital communications. Digital technology holds promise for multiple, and at times contradictory, visions of economic, political and social life. For the information protectionist, the promise of freedom is in market growth and intangible commodity production. For the information exceptionalist, freedom is attributed to the social production of digital content and autonomous self-creation. For the network distributionists, digital networks become sites at which new structures and relationships of power are articulated and instantiated. For the information materialists, the digital is a site in which to define social life using a frame of reference that moves beyond the commodity and capitalist production. However, we cannot simply consider the destiny of intellectual property and the underlying contests for social justice to be linked to the digital. As the proponents of the information *terra nullius* position point out, we also need to pay attention to other, more immediately material domains of cultural appropriation such as seed genetics and pharmaceuticals.

A suggested approach for navigating the generative tensions between these paradigms is through addressing the question "whose property"? This line of argumentation builds on the analysis of social justice developed by Nancy Fraser.

On one level the question of "whose property" demands that we ask what the spatial frame of justice is. As Fraser herself asks, "what is the penitent frame for determining the requirement of justice in a globalizing world?" (Fraser 2004: 18). For example, in addressing the current demands of social justice how can we aim to transform the global regulatory order that is increasingly shaping our global social lives? Can localized commons-based models become sites at which to address claims of justice? How do we restrain the privatization and commodification of knowledge and cultural production, and the displacing effects these have, not only on peoples of the developing world, but also those in the developed world? This chapter suggests, as does Fraser, that the frame must move beyond a Westphalian or nation-state view of social life to one that incorporates the dynamics of interaction in the international system. The realities of intellectual property today demand such an approach.

On another level, issues of cultural recognition and material distribution are drawn together through intellectual property rights. The question of "whose property" focuses attention on the complex dynamic of cultural and material claims within global social life. At a meta level, looking across the five paradigms, the definition of what intellectual property is represents both a cultural order and a distributive vision. Rather than thinking about social justice through an "either/or" lens of recognition or redistribution, we need to build a "comprehensive framework that encompasses both so as to challenge injustice on both fronts" (Fraser and Honneth 2003: 93). As intellectual property represents the ownership of cultural production, a multifaceted starting point is essential.

Finally, participation is at the core of the ideologies of the post-industrial, networked, information age. The question of "whose property" requires one to critically assess the dimensions of participation. For example, on what level is there participation in the global regimes that order intellectual property? Who can participate in peer-to-peer digital networks? What voice do indigenous or third world peoples have in the context of private or collective ownership? Who can participate in the digital gift relation? Fraser locates "parity of participation" as the core actualizing principle in her theory of justice. She argues that the principle of the parity of participation is founded on the "equal autonomy and moral worth of human beings" and "the real freedom to participate on a par with others in social life" (Fraser and Honneth 2003: 231). These two elements provide a productive logic on which to develop a "radical democratic interpretation of equal autonomy" (Fraser and Honneth 2003: 231). Intellectual property, taken at its broadest, demands that we assess these claims to participation and their symptomatic variations across the five paradigms. In so doing we may be able to discern what a radical democratic interpretation of equal autonomy might look like for this historical epoch. At the very least, framing intellectual property through these criteria might provide a perspective which contributes to an emancipatory theory of social justice for post-industrial society.

## Bibliography

Benkler, Yochai. 2007. *The Wealth of Networks: How Social Production Transforms Markets and Freedom*. New Haven: Yale University Press.

Dagan, Hanoch, and Michael Heller. 2001. "The Liberal Commons." *The Yale Law Journal* 110 (549).

Dyer-Witheford, N. 2002. "Global Body, Global Brain / Global Factory, Global War: Revolt of the Value Subjects." *The Commoner* 3.

Dyer-Witheford, N. 2010. *Commonism* 2010 [cited 4/1 2010]. Available from http://turbulence.org.uk/turbulence-1/commonism/, accessed January 4, 2010.

Fraser, Nancy and Axel Honneth, 2003. *Redistribution or Recognition? A Political-Philosophical Exchange*. New York: Verso.

Fraser, N. 2004. *Reframing Justice*. Amsterdam: University of Amsterdam.

Galloway, Alexander. 2004. *Protocol: How Control Exists After Decentralization*. Cambridge, MA: MIT Press.

Goldsmith, J. and T. Wu. 2006. *Who Controls the Internet: Illusions of a Borderless World*. Oxford: Oxford University Press.

Greaves, T. 1995. "The Intellectual Property of Soverign Tribes." *Science Communication* 17 (201).

Lessig, Laurence. 2002. *The Future of Ideas: The Fate of the Commons in a Connected World*. New York: Vintage.

Lipscome, Andrew and Albert Bergh (eds). 1903. *The Writings of Thomas Jefferson*. Washington, DC: The Thomas Jefferson Memorial Asscoation.

May, Christopher. 2007. *The World Intellectual Property Agenda: Resurgence and the Development Agend*. New York: Routledge.

May, Christopher and Susan Sell. 2006. *Intellectual Proeprty Rights: A Critical History*, ed. by R. Marlin-Bennett. London: Lynne Rienner Publishers.

Shiva, Vandana. 1997. *Biopiracy: The Plunder of Nature and Knowledge*. Cambridge: South End Press.

Wark, Mckenzie. 2004. *A Hacker Manifesto*. Cambridge: Harvard University Press.

Weber, Steven. 2004. *The Sucess of Open Source*. Cambridge, MA: Harvard University Press.

# 6.1
## Ethical Concerns about Digital Property: The Case of FLOSS Licenses

*Roberto Feltrero*

Free software programmers have applied some of the values and principles that belong within the ideal of a pluralistic, transparent and collaborative knowledge society to their technological designs (Feltrero 2007). These principles are implemented by means of the licenses that cover their computer programs and the associated documentation. Such licenses, legally rooted on, and protected by, copyright laws, give the users the right to freely use, copy, study and redistribute the modifications of their software developments, provided that they keep those modifications free. In this way, free software supporters have transformed the usual meaning of copyright laws from the motto "Copyright, all rights reserved" to "Copyleft, all rights reversed."

The Open Source Movement, which adopts an entrepreneurial viewpoint, has captured the more practical benefits of this way of producing software. It leaves users and amateur software developers the freedom to study, test and modify the code in order to improve computer programs, but establishes some additional restrictions to protect their business model. From the expression Free Libre Open Source Software, we get the acronym FLOSS, which encompasses both alternatives, whose common principle is that they ensure, at the least, the openness of the software code.

This FLOSS movement, going beyond the field of technological production and development, has initiated a very deep collective reflection about the nature of the 'property' of digital works. Thus, the basic ideas of their legal model have been transferred to other kinds of cultural works in digital format, developing a myriad of free and Open Licenses for books, music, films, paintings and so on. The Creative Commons licenses are the most popular set of Open Licenses that give users, as a minimum, the freedom to get and share copies of any kind of digital works on the Internet.

Some of the achievements of this movement, like the widespread use of software with reverse engineering systems, or the popularity of peer to peer networks, which allow users to share not only free digital content but also copyrighted content, are being interpreted as evidence that free software and the related hacking and sharing technologies are bad for society. But in the context of this fight between the FLOSS movement and the entertainment industries, the basic ethical proposals of Open Licenses on computing and information are being systematically concealed and silenced. Those principles can be unveiled by means of a philosophical renewal of the main targets of computer ethics and information ethics in the twenty-first century.

Computer ethics began in the 1980s with ad hoc reflections on the main issues related to the, at that time, new emerging technologies of personal computers (Johnson 1994; Mason 1986). Nowadays, computation is not just present in personal computers for office tasks. It is also a fundamental technology for communication, culture production and transmission, education and similar areas. Computers are becoming technologies of expression (Lessig 2004).

Since computers are mediating technologies for most human activities nowadays, the main topic of computer ethics should be the orientation of technological designs to ensure the deployment of information architectures that would provide users, at least, with the same freedom of action that they would posses if their activities were not mediated by such technologies (Feltrero 2006). Computer ethics should, therefore, help us to understand the ethical drawbacks of the technologies of control that can be implemented by means of opaque designs (Castells 2001: 197). The transparency of the code that regulates basic computing technologies is an ethical problem, since code is the effective law in cyberspace (Lessig 1999). FLOSS licenses for software allow the free copy and circulation of computer programs and their facilities for everybody. But, beyond this practical benefit, they also ensure the transparency of the code and the deployment of collaborative strategies for technological evaluation and development.

Finally, the philosophy of FLOSS licenses should also be interpreted in terms of so-called information ethics. Since information, in a broad sense, has become a central value for most human activities in the information society, we need an ethical reflection on access to that information and on its preservation, in order to enrich the new infosphere (Floridi 2002). Open access to scientific information, cultural content and so on has been defended on the basis of a number of moral arguments (Willinsky 2006), closely related to those at the core of the FLOSS inspired licenses. These Open Licenses for cultural works ensure their free copy and circulation, and guarantee equal access to culture and information. A more plural and open participation for everybody in the production and evaluation of knowledge is necessary to promote fairness and equality in a knowledge based society.

FLOSS licenses have proved to be an adequate legal interpretation of intellectual property to promote these ethical desiderata. Maybe it is time to re-think old economic arguments on intellectual property rights, still based on old technologies, in order to understand from a new perspective the "property" of knowledge and information in the digital era.

# Bibliography

Castells, M. 2001. *La Galaxia Internet*. Barcelona: Plaza & Janés.

Feltrero, R. 2006. "Ética de la Computación: Principios de Funcionalidad y Diseño." *Isegoría* 34: 79–109.

Feltrero, R. 2007. *El Software Libre y la construcción ética de la Sociedad del Conocimiento*. Barcelona: Icaria.

Floridi, L. 2002. "On the Intrinsic Value of Information Objects and the Infosphere." *Ethics and Information Technology* 4: 287–304.

Johnson, D. G. 1994. *Computer Ethics*. Englewood Cliffs, NJ: Prentice Hall.

Lessig, L. 1999. *Code and Other Laws of Cyberspace*. New York: Basic Books.

Lessig, L. 2004. *Free Culture: How Big Media uses Technology and the Law to Lock Down Culture and Control Creativity*. New York: The Penguin Press.

Mason, R. 1986. "Four Ethical Issues of the Information Age." *MIS Quaterly* 10(1): 480–98.

Willinsky, J. 2006. *The Access Principle: The Case for Open Access to Research and Scholarship*. Cambridge, MA: MIT.

# 6.2
## On/Off the Agenda: Intellectual Property Rights, the UN and the Global Politics of the Internet

*Mikkel Flyverbom*

There was a time when finding a book involved a trip to the local library, reading it involved a search for a quiet corner with a comfortable chair, and a photocopier was necessary if one wanted to distribute it. The emergence of digital technologies has collapsed the boundaries between these three activities—accessing, using and copying—which are so central to our dealings with original and copyright-protected works, such as books. This has consequences for users, who comfortably but precariously can download, enjoy and forward such products. For producers and rights holders it means that while distribution is simple, control over content is difficult. And for those seeking to set the rules and regulations for such products, this collapse means that existing treaties and other forms of regulation no longer suffice. This case study sheds light on how the issue of intellectual property rights (IPR) has been constructed, handled and negotiated as a key concern in the ongoing process of addressing the global politics of the Internet, under the auspices of the United Nations.

The global governance of intellectual property rights is handled by a limited number of organizations, such as the World Trade Organization (WTO) and World Intellectual Property Organization (WIPO), and primarily reflects US and EU positions stressing the need to protect rights holders (Braithwaite and Drahos 2000). Achieving a balance between protecting copyright holders and allowing for the free flow of information is central to IPR and constitutes the main concern in recent conflicts in this area. As pointed out by Siochrú and Girard (2002: 10), "extending the rights of private intellectual property owners is always at the expense of the release of information into the public domain." For this reason, and because the present IPR regime favors rights holders (Lessig 2004), the governance of IPR has emerged as a key concern in ongoing discussions about the global politics of the Internet. With the World Summit on the Information Society (WSIS), held in 2003 and 2005, the UN has become one important forum for such discussions. Organized as multi-stakeholder processes, these discussions have brought together a wide range of social worlds, including civil society groups, governments, technical associations and international organizations, and contributed to the reconfiguration of the Internet as an object of global governance (Flyverbom forthcoming).

This case study highlights how IPR has sparked conflicts in discussions about the global governance of the Internet. IPR has played a central role in attempts to broaden the WSIS agenda from a concern with technical aspects, such as domain names, to a much wider conception of Internet governance, including access, security, privacy, development and human rights. During the consultations held by the Working Group on Internet Governance (WGIG), which was given the task of defining Internet governance and the public policy issues to be addressed in the second part of WSIS, hundreds of participants voiced their concerns. Some, like representatives of the US and Brazilian governments and the film industry, simply wanted IPR off the agenda. According to the US existing IPR treaties should be preserved because "IPR fosters innovation." The Brazilian government argued that UN discussions should not focus on IPR because the matter is already addressed by established organizations like WIPO, which are "accepted by all." As the process unfolded, it became clear that the existing IPR regime was far from being accepted by all. WGIG members prepared short reports on the different public policy issues that could be included in the definition of Internet governance, and the paper on IPR led to one of the most heated conflicts in the group and in later consultations. Much to the dismay of the authors and IPR-critical groups, the secretariat decided not to publish the paper, but post a short summary. But even this spurred a number of reactions, for instance from the International Federation of Film Producers Associations and the International Video Federation, both of which – in online contributions – called the summary "misleading" and asked for it to be revised so as not to do "damage to the credibility and neutrality of the WSIS Internet Governance process." Even governments with concerns about existing IPR governance, such as India, suggested that IPR should not be on the WGIG agenda, saying for instance that there is "no need to worry too much about" IPR, since this could be discussed at WIPO and "will not be overlooked in WSIS." In an attempt to move discussions forward, the WGIG chairman stated that there was "a certain amount of consensus, I sense that there are a whole class of policy matters which are salient to Internet, but where the real locus of action is elsewhere, somewhere else in the international system," such as IPR. And as "we are not going to be able to solve the IPR issue in the Internet governance system," which "is a much larger problem," this issue should not be addressed by WGIG. However, civil society participants did not agree that any "consensus on taking IPR off the agenda" had emerged during discussions. From the perspective of civil society, WGIG was the appropriate, even the only, place where discussions about the "intersection between IPR and Internet governance" could take place, because WGIG is the "only agency that can take a holistic view, and is not encumbered by bureaucratic special interests." This statement also tied in with an earlier point made by civil society participants that "all relevant issues important to stakeholders groups should be addressed by WGIG. Unresolved controversies should be documented in papers and statements and not used as the basis to omit particular issues."

While the final WGIG report proposes a very inclusive definition of Internet governance including IPR, and stresses the need to balance the rights of holders and the rights of users, it is also very vague. It simply states that "there are different views on the precise nature of the balance that will be most beneficial to all stakeholders, and whether the current IPR system is adequate to address the new issues posed by cyberspace" (WGIG 2005: 7). So, while IPR was not taken off the agenda in the UN processes to address the global politics of the Internet, the issue was addressed in very vague terms. At the time of writing there is no obvious forum where the balance between the concerns of users and owners of copyrighted materials and intellectual properties can be negotiated in an open and transparent manner. In this respect we may think of a more balanced IPR regime as an orphaned discussion, looking for a new home.

## Bibliography

Braithwaite, J. and Drahos, P. 2000. *Global Business Regulation*. Cambridge: Cambridge University Press.

Flyverbom, M. forthcoming. *Organizing the Internet – Multi-Stakeholder Processes and the Global Politics of the Digital Revolution*. Cheltenham, UK; Northampton, MA: Edward Elgar.

Lessig, L. 2004. *Free Culture: How Big Media uses Technology and the Law to Lock Down Culture and Control Creativity*. New York: Penguin Press.

Siochrú, S. Ó. and Girard, B. 2002. *Global Media Governance: A Beginner's Guide*. New York: Rowman and Littlefield.

# 6.3
# Health Traditions in Kerala and Local Intellectual Property Rights

*Rubeena Aliar*

Local health traditions, henceforth referred as to LHT, as used in Kerala State, south western India, consist of health traditions passed over generations through oral discourse. They fall under the broad realm of traditional medicine though they are distinctly different from classical medicine, which is codified and supported by written text. These health traditions include home remedies ("grandma's medicine") and country or folk medicine (Kumaran 2002) and are generally devoid of any spiritual beliefs.

LHT are a collective community resource and the benefits accruing from the use of LHT belong to the local village communities or the community of local health practitioners, who have been employing these traditions for generations to prevent and cure ailments. Associated with the lack of obligations to the community for its ownership of the LHT, or any form of indigenous knowledge, are the problems relating to intellectual property rights, patents and benefit sharing.

However, intellectual property rights fail to ensure the rights of the local communities over their cultural resources such as LHT, since it is impossible to identify an individual inventor owing to the collective nature of traditional knowledge. In addition, traditional knowledge often cannot be attributed to a particular geographical location. The ownership of varieties of plants is alien to many social and cultural beliefs and the required criteria of "'novelty" and "inventive step" are not always possible, particularly in cases where the traditional knowledge has been in existence over a long period of time. Also the costs of applying for a patent and pursuing patent infringement cases are prohibitive (GRAIN and Kalpavriksh 2002).

Kerala State has excellent modern and traditional health care facilities. Modern medicine is preferred and used widely by the majority in the state, though other forms of medicine also enjoy a covetable position. Kanjoor, a small village where this case study was carried out, is no exception to this. The study on the extent of access and use of LHT threw light on the ineffectiveness of applying intellectual property rights to these traditions.

151

Historical profiles and timelines created as part of the study suggested that the use of LHT has declined as a result of the loss of traditional knowledge systems and associated absence of traditional medical practitioners. Ignorance and lack of awareness of LHT and a decline in resources that support them also accounted for the fall in popularity. Drastic changes to life styles and high scale promotion of modern medicine were other factors that led to the decline of LHT.

The role of traditional "barefoot doctors" or local health practitioners in preventing and curing the health problems of the poorest people of India and the knowledge they possess is remarkable (Hafeel and Shankar 1999). Though local health practitioners are often non-existent in the urban and suburban areas, they play a significant role in ensuring the health of the rural folk. In Kanjoor, there are four traditional practitioners who treat patients from the "panchayat" or local village government area, and from outside. The major reason for the declining number of traditional practitioners is the lack of recognition of their knowledge and contribution towards enhancing the health of local communities. According to Michael, a traditional healer:

> People from far and near have been treated here for hepatitis. I have saved many ill people who have come here in highly critical condition and whom the practitioners of modern medicine could not save. Even then we do not get much recognition. The only recognition and the best one is the recognition I get from many of my patients. I have sent letters to many organizations about our achievements in curing hepatitis and also about the effectiveness of my medicine and proofs for having cured so many people. But I have seldom received any positive responses from these organizations.

In this regard, it is essential that the practitioners are given due recognition and only such a step can ensure that the knowledge base of these health traditions is not effaced completely. The reluctance of the present generation to practice traditional medicine as a livelihood option emerges from this sense of insecurity. Besides local practitioners, the community itself possesses knowledge on curatives and preventives in the form of home remedies that are employed in their day-to-day life. It is doubtful how ownership of such knowledge possessed by the local communities or households can be defined in terms of intellectual property, or how intellectual property rights can be used in such context, since the knowledge possessed by these groups are a collective resource and cannot be confined to a single geographical area.

The increasing popularity of modern medicine and the decline of the herbal resource base on which LHTs depend has led to the decline in their use. Kerala, being in the tropical monsoon regime, is blessed with rich biodiversity, which however is fast declining owing to population growth and associated stress on land use. Though biological prospecting is seen as an efficient strategy for conducting research into the usefulness of herbs and preserving them, the rights of the community and the local health practitioners as source of this knowledge base are seldom recognized. Biological prospecting tends to promote commercialization of health care and makes herbal medicine unaffordable to the local communities. Here again the problems of benefit sharing need to be addressed. The benefits accruing from intellectual property rights need to be distributed equitably and effectively. There is a need for stringent legislation to curb the undue exploitation of the knowledge possessed by the local communities by any other organization. The government has a key role to play in enacting legislation and stringently implementing this. Many people present during the preparation of the stakeholder analysis matrix expressed their deepest fear that enacting any such stringent legislation will be resisted by many vested interest groups. It should be noted that the intellectual property rights of the community could be protected only through governmental intervention. However, it remains uncertain whether the local intellectual property rights of the community and the local practitioners will receive due recognition. To protect these local intellectual property rights it is necessary to enact laws benefiting the community and practitioners and thereby ensuring their legitimate rights.

## Bibliography

GRAIN and Kalpavriksh Environment Action Group 2002. *Traditional Knowledge of Biodiversity in Asia-Pacific: Problems of Piracy and Protection – A Briefing, India*, Barcelona: GRAIN.
Hafeel, A. and D. Shankar. 1999. "Revitalizing Indigenous Health Practices," *Compas Magazine* 1, Leusden: Compas.
Kumaran, T. V. 2002. "Mobilizing Local Knowledge – Local Health Traditions." Presented at the Workshop on the Ecosystem Approach to Human Health in Chennai (August), organized by the McMaster-York-Madras University Collaborative Research Program at the Chennai Metropolitan Development Authority, Chennai.

# Part VII
# The Digital Commons

# 7
## Socrates Back on the Street: Wikipedia's Citing of the Stanford Encyclopedia of Philosophy

*John Willinsky*

In this chapter I present a study of how a new generation of "open access" scholarly resources is being taken advantage of by those contributing to popular sources of knowledge. The study explores Wikipedia's citation of the *Stanford Encyclopedia of Philosophy* (SEP), as an innovative and open project of the academic community. SEP was begun in 1995 by Edward N. Zalta at Stanford's Center for the Study of Language and Information (Perry and Zalta 1997). Zalta set out to create a "dynamic" encyclopedia that combined peer review with ongoing updating and revision. Entries offer internal and external linkages to related materials, as well as access to the author's homepage and email. SEP is intended to be "useful both to professional scholars and the general public" (Perry and Zalta 1997), and, as an important part of that, it is entirely free to read online. This has been made possible by a variety of grants, with a current program to have it funded by an endowment, with support from research libraries and philosophy departments.

What makes SEP a new sort of knowledge resource is not only that it is free for online readers, but that it is at once peer-reviewed, periodic, and yet systematically working through an encyclopedic coverage of philosophy.[1] If some entries are heavily footnoted in the manner of a scholarly article in the humanities, others are as free of such notes as an entry in *Encyclopedia Britannica*. SEP is not published by a press, but is copyrighted by the Metaphysics Research Lab in the Center for the Study of Language and Information at Stanford University, as well as by its authors. At the time of this study, SEP had 1026 entries on philosophical figures and concepts (while adding a half-dozen or so a month) and was receiving roughly half a million hits a week.

This study asked how Wikipedia contributors and readers are using SEP. It specifically examines the links in Wikipedia that lead to entries in SEP and the extent of the use of those links over a two-week period. I would argue that SEP is being used by Wikipedia contributors and readers in ways that bear directly on the educational quality of both works. Yet if Wikipedia is "the free encyclopedia that anyone can edit," as Wikipedia labels itself,

SEP is the free encyclopedia that anyone who has established a reputation within the academic community as an authority on a particular philosophical topic can contribute to, following a process of rigorous peer review.

This study of the relationship between Wikipedia and the SEP follows an earlier study I conducted on the use in Wikipedia of research that had been made freely available online or through open access. This earlier study established that a very small proportion (2 percent) of Wikipedia entries had links to research that could be read by readers who were not members of a research library. The study went on to show that the proportion of entries that could provide reference links to research that was freely available to readers could have been much higher (60 percent) based on the existing degree of research that has been made open access.[2]

## Method

This study of Wikipedia and SEP has been made possible by the close cooperation of the editors at the SEP. After discussing the intent, scope, and design of the study with Principal Editor Edward N. Zalta and Senior Editor Uri Nodelman, both at Stanford University, they provided helpful suggestions to improve the study and the weblogs for SEP over a two-week period, in which the Internet Protocol (IP) addresses had been encrypted to protect the identity of users. The information provided to us on the weblogs identified that set of users who had arrived at SEP by clicking on a link in Wikipedia. The weblogs, for the two-week period between June 22 and July 5, 2008, revealed both where in Wikipedia people clicked on a link leading to SEP, and where exactly in SEP that link led.[3]

This data provides a picture of how Wikipedia readers were being led to SEP in the period covered. This information provided a basis for analyzing the frequency of use, the most popular linkages, and the nature of the connections. It did not tell us how long readers spent in SEP nor, of course, what sense they were making of the SEP entries they encountered. But it did tell us that readers of an entry on Aristotle or politics in Wikipedia were clicking on links leading to SEP. At the very least, we might assume that a click through to SEP indicated an interest in the reference that was being made in the text of the entry or under "External Links," if not an interest in learning more about the topic. It did tell us that in cases where the SEP link in Wikipedia was clearly identified as Stanford Encyclopedia of Philosophy, readers were pursuing a more specialized source of knowledge, if not an official academic resource.

In addition, I was able to consider how Wikipedia contributors were making use of SEP by taking advantage of Wikipedia's advanced search capacities through Wiki Searching External Links. I was able to search for every reference to SEP, whether it was labeled Stanford Encyclopedia of Philosophy or not, by using SEP's URL "plato.standord.edu." These two sources of information—contributor's decisions to include links to SEP in Wikipedia and how those links were used by readers (and contributors)—provided a starting point for considering the relationship between these two different types of work on the web.

While Wikipedia is a work of astounding numbers, whether for entries, languages, readers, or contributors, the numbers that describe the linkages from Wikipedia to SEP pale in comparison. Explicitly philosophical topics make up a very small proportion of this giant encyclopedia. Yet how SEP is cited in Wikipedia by contributors and which of those citations are used by readers throws further light on the process of learning to which these new works contribute, and the something more to that learning that can arise because of these connections.

## Results—SEP references in Wikipedia and the Aristotle file

At the time of this study, 1741 of Wikipedia's entries contained one or more links to 942 entries in SEP. To put that in proportion, the editors of Wikipedia had founds ways of connecting Wikipedia to slightly over 80 percent of the 1026 entries that made up SEP at that time (with additional links to archived earlier editions of SEP). In total, Wikipedia possessed 2263 links leading to SEP.[4] The entries that were linked between the two encyclopedias were most often, by a factor of two, about ideas such as "truth" and "causality," as well as "pleasure" and "the meaning of life," as opposed to entries about philosophers such as Kant and Descartes. The Wikipedia entry with the largest number of links to SEP was the one on Aristotle, with 14 links to SEP. It was followed by the Wikipedia user page devoted to the editor Simfish who had embedded 13 links to SEP on his page (Table 7.1).

The Wikipedia entry for Aristotle is a substantial piece of work, at close to 9000 words long, with edits on this page continuing to take place daily during the course of this study. Wikipedia rates the entry B-class by the

*Table 7.1*   Wikipedia pages with the most links to SEP

| Wikipedia Pages | SEP Links |
| --- | --- |
| Aristotle | 14 |
| User: Simfish | 13 |
| Truth | 10 |
| Immanuel Kant | 9 |
| Causality | 8 |
| Computational Epistemology | 8 |
| Philosophy of Physics | 8 |
| Politics | 8 |
| Rene Descartes | 8 |
| Epistemology | 7 |

Wikipedia Version 1.0 Editorial Team, which means "No reader should be left wanting, although the content may not be complete enough to satisfy a serious student or researcher." The entry itself dates back to April 21, 2003 when, at 3000 words, it was largely devoted to the philosopher's biography, with the text largely lifted without attribution and only minor changes from the *Catholic Encyclopedia* (Turner 1907). A small amount of space was given to his method and a summary of three criticisms of his work. The bibliography did contain links to what was already a rich set of Aristotle's works freely available online, principally through Virginia Tech. There was little commentary, then, with biography as background and the links to the actual works of Aristotle the extent of it. There was little commentary on Aristotle's work and no references to commentary or interpretations of this work, or of Aristotle's important place in the history of philosophy.

The current Wikipedia article on Aristotle includes well-referenced summaries of his work in physics, metaphysics, biology, and medicine, as well as practical philosophy. At the top of the entry, it offers readers a baseball card-like sidebar with his picture, birth and death dates, along with his "school" and "notable ideas." At the end of the entry, after the list of Aristotle's works, references, further reading, and see also entries, there is a list of external links made up of "Collections of Aristotle's Works," which are available in English and the original Greek from five websites, and "Articles on Aristotle," which includes links to the Aristotle entries in the *Internet Encyclopedia of Philosophy* and the *Catholic Encyclopedia*.[5] These two links are followed by a set of links to 13 entries in SEP that bear on Aristotle's work: scholarly surveys of focused topics from the Stanford Encyclopedia of Philosophy, articles on Aristotle in the Renaissance, biology, causality, and commentators on Aristotle, ethics, logic, mathematics, metaphysics, natural philosophy, non-contradiction, political theory, psychology, and rhetoric.

Notice that this set of SEP links in Wikipedia did not include the obvious SEP entry for Aristotle himself, but rather sought to extend Wikipedia's reach by providing links to SEP entries about this peripatetic philosopher's work, as each of the links leads to a SEP entry about Aristotle's ideas, as in Fred Miller's SEP entry "Aristotle's Political Theory," or Paula Gottlieb's SEP entry "Aristotle on Non-Contradiction."

SEP also turns up in one of the 51 "References" that operate as hyperlinked footnotes in the Wikipedia entry on Aristotle, at the end of the statement: "In a similar vein, John Philoponus, and later Galileo, showed by simple experiments that Aristotle's theory that the more massive object falls faster than a less massive object is incorrect."[6] Reference 12, which simply reads "Stanford Encyclopedia of Philosophy," leads directly to the section on Philoponus' theory of impetus in the SEP entry on John Philoponus, by Christian Wildberg. Wildberg points out that while Philoponus' theory is based on a misguided sense of a kinetic force being imparted to falling or thrown objects, it did lead Philoponus to experimentally test and disprove

Aristotle's conclusion about the differing speeds of falling bodies, much as Galileo did centuries later. Wildberg's entry for Philoponus provides a substantial list of primary and secondary sources, while nothing is listed under "Other Internet Resources" except "please contact the author with suggestions."

The other 50 references for the Wikipedia entry cite not only the specific passages in Aristotle's works that are being quoted in the entry, but a range of the scholarly literature on Aristotle, from an 80-year-old study of Platonism to a recently published Aristotelian study in Italian published in Milan. A few of the reference notes extend the discussion, as only footnotes can do, into wonderfully arcane points on how, for example, as in reference 41, Aristotle's work is divided into "exoteric" and "esoteric," although the Greek philosopher did not himself use the term "esoteric" in his work.[7]

## Citing SEP in Wikipedia discussion pages

One pattern that emerged in Wikipedia's use of SEP was how the Stanford Encyclopedia was cited 216 times in the discussion pages that accompany each page in Wikipedia (representing 10 percent of the links leading from Wikipedia to SEP). In the discussion pages, what SEP offers is an authority for grounding positions, adding weight to stances taken on debates over how Wikipedia entries should be handled. Foremost among the discussions in which SEP was invoked was the one centered on the entry Atheism. Here SEP was cited 18 times, amid an immense body of discussion that overshadows the entry itself, at a little over 10,000 words. By contrast, the discussion is made up dozens of archived pages, with some of them running well over 20,000 words.

In the course of this extensive "talk" about the entry Atheism, it is J. J. C. Smart's SEP entry Atheism and Agnosticism that comes up repeatedly. When "Brian" reaches the point of exasperation ("this is getting ridiculous, repeating the same arguments over and over"), he reaches "one of the most well known atheist philosophers, and his entry in the Stanford Encyclopedia of Philosophy (comprehensive as expected)" who has his own entry in Wikipedia, as it turns out. Smart is then cited for his statement "Atheism means the negation of theism, the denial of the existence of God."[8] At another point in the discussion on atheism, Adraeus, who works with the group WikiProject Philosophy, describes how she or he "spoke with 84-year-old Dr. John Smart of Philosophy who wrote the Atheism-Agnosticism entry in the Stanford Encyclopedia of Philosophy" to clarify a point in dispute about whether "the weak definition of atheism is the product of the free thought movement." Smart "responded that at the time of writing he was unaware of the distinctions."[9]

Another discussant objects to the removal from Wikipedia of something from Smart's SEP entry Atheism and Agnosticism, and emphatically rejects that text is removed from Wikipedia when it is found to be poorly written and clumsy, stating: "The removed text was 'a direct quote' from *Stanford Encyclopedia of Philosophy*" (with a link to Smart's SEP entry Atheism and Agnosticism). Someone then points out that the citation of SEP was a breach of copyright (which is unlikely if a citation was provided, I would add, given the rules of "fair use"), while yet another discussant makes the interesting point that "SEP entries are written from the perspective of the authoring philosopher" while "our job as Wikipedia editors is to use whatever resources [are] available to us to provide objective contributions to our encyclopedia." This discussant does allow that "you can, however, quote SEP entries if you provide citation."[10]

More than once in these discussion pages, Smart's SEP entry Atheism and Agnosticism is directly cited for how it defines atheism ('atheism' means the negation of theism, the denial of the existence of God)[11] with one discussant pointing out how Smart introduces Wittgenstein's notion of "family resemblance" (for agnosticism and atheism) in explaining the word's use.[12] As a result of this discussion, Smart's entry in SEP stood, with the entry Classical Logic, as the second most-cited entries in Wikipedia, only one citation behind the SEP entry Libertarianism, which was cited 15 times in Wikipedia (Table 7.2). The prominence of classical logic has a way of tempering what might otherwise seem to be the libertarian interests in atheism and their opposites, given the degree of discussion.

A debate about whether "orthomolecular medicine represents a new paradigm" includes a citation of SEP entry Scientific Progress to support the claim that "A new paradigm can be true, false, or commonly, in different stages of development."[13] In the discussion around the Wikipedia entry for

*Table 7.2*  Stanford Encyclopedia of Philosophy entries most often cited in Wikipedia

| SEP Entry | Wikipedia citations |
| --- | --- |
| Libertarianism | 15 |
| Atheism and Agnosticism | 14 |
| Classical Logic | 14 |
| Copernicus | 13 |
| David Hume | 13 |
| Ontological Arguments | 13 |
| William Godwin | 12 |
| Karl Marx | 12 |
| Friedrich Nietzsche | 12 |
| Zeno's Paradoxes | 12 |
| Karl Popper | 12 |

*Note*: The URL for SEP itself, although not counted as a SEP entry for purposes of this table, is included 72 times.

2003 US Invasion of Iraq, SEP is cited as one of a half-dozen reference works that provide a definition of "sovereignty."[14]

But then SEP is also used in the course of these discussion pages in an effort to put an end to what might otherwise seem an exercise in scholarly pedantry (which I clearly have a weakness for). During May 2008, there was a 5000 word discussion over whether "the" should be capitalized in Karl Marx's essay "On the Jew Question," which was a critique of Bruno Bauer's 1843 book "The Jewish Question."[15] After Wikipedia contributor Schwalker is challenged for using the Marxist Internet Archive as a point of reference—"a website is not WP: RS [Wikipedia: Reliable Sources] in comparison with books"—Schwalker comes back with "an independent renowned web-page" in the form of Jonathan Wolff's SEP entry for Karl Marx, which uses both "On the Jewish Question" and "On The Jewish Question" in the course of the entry. The discussion continues, covering among other things, the history of the standardized editions of Marx, and ends in a muddle. The result of it all? The Wikipedia entry for Marx's essay is "On the Jewish Question," while the Wikisource version of Marx's essay in English is entitled "On The Jewish Question" ("Out of its own entrails, bourgeois society continually creates Jews").[16]

In the course of this long discussion, SEP is one of the rare instances of directly drawing on the parallel world of sanctioned scholarship, apart from the detailed consideration of the works of Marx himself. The ready availability of SEP contributes not so much to a pale imitation of scholarly squabbling, but to a thing that cannot be distinguished from such debate except in its remarkable public documentation, which, if such a record has no other imaginable value (with its time-stamped entries), serves to make this case about the "learnedness" nature of the learning that Wikipedia involves. Discussants invite challenges, seek out new sources and further evidence, argue from policy, challenge policy, carefully consider opponents' stands, calls for order ("Please remain civil towards Malik, who has made an extra effort to try and assist you through processes that you have trouble following correctly"), hold votes, and seek consensus, even if they appear rarely to change their minds about the issue at hand.

Yet what must stand as the ideal instance of SEP's contribution to the construction of Wikipedia entries comes, oddly enough, in the discussion page for Bobby Jindal, current governor of the state of Louisiana. Considerable discussion (7000 words) is given over to how best to represent and reference Jindal's stance on abortions. He is, in his own words, "100 percent against abortion, no exceptions," and the question for his Wikipedia entry is how to represent his stance in relation to its consequences for women's health. The discussion deals at some length with the teachings of the Catholic Church, of which Jindal is a member.[17] Wikipedia's policy is "that editors must take particular care about adding biographical material about a living person" and that there is to be "no original research" (reiterated in the "biographies of living persons policy"), which becomes an issue when Jindal's well-known stance is cited without backing by, to further quote the policy, "reliable, third-party, published sources."[18]

At one point in this discussion, Daniel Zimmerman backs up the idea that Jindal could support an abortion if it was the result of a "double effect" in which the abortion took place as a result of another vital, life-saving procedure, by stating that "here is another source that describes the double effect and shows that under Catholic law, direct abortions are prohibited even if it is to save the life of the woman."[19] He then includes a quote describing a hypothetical example in which a doctor who opposes abortion nonetheless operates on a pregnant woman with cancer, which he knows will result in the loss of the fetus. The quoted example is from Alison McIntrye's SEP entry Doctrine of Double Effect ("an action that causes a serious harm, such as the death of a human being, as a side effect of promoting some good end").[20] Daniel Zimmerman refers to the SEP entry as a "verifiable source," and while a hypothetical example, of the sort McIntrye provides second-hand, makes for an odd instance of verifiability, the example does speak poignantly to the distinction that he is making in this case.

## Readers' use of Wikipedia's links to SEP

To first summarize the use of the SEP links that are embedded in Wikipedia over the two-week period in question, it appears that Wikipedia readers clicked on SEP links 13,363 times, coming from 10,960 IP addresses.[21] Machiavelli was the SEP entry most frequently visited by a Wikipedia reader. Of the 2262 SEP links that had been placed in Wikipedia by various editors, it turns out that over half (1248) were clicked at least once during the two-week test period (and 355 of them were clicked ten or more times during this two-week period). The 216 SEP links used in the midst of the talk on the discussion pages to make one point or another received less attention but were clicked 16 times during the test period, suggesting that readers and editors are returning to this aspect of Wikipedia a few months or years after the point was originally made, which in two cases had to do with using a SEP entry as a model for Wikipedia entries on affirmative action and qualia ("properties of sensory experiences").

A little more than a third of the IP addresses (possibly representing more than one reader) recorded more than one leap from Wikipedia to SEP, with a small number of readers' addresses (30) recording ten or more clicks from Wikipedia to SEP. Of the 1026 entries in SEP at the time of the study, 780 were visited by Wikipedia readers at least once, with close to half of them visited once a day on average. However one looks at it, readers of those Wikipedia entries that had SEP links used a good proportion of those links, and did so, one may presume, to either extend what they could learn about the topic at hand or at least see what authority Wikipedia rested on and what more was available on this topic.

To look at this use in more detail, SEP itself was the site most frequently visited by Wikipedia readers, and the link leading to SEP's homepage (see Table 7.2) is, by far, the most common SEP link in Wikipedia (Table 7.3). That readers went to have a look at this encyclopedia of philosophy itself might be taken as an expression of interest about educational resources on the web. The weblogs indicate that most of these readers came to SEP from the Wikipedia entry for the Stanford Encyclopedia of Philosophy, with a good number also coming, as well, from the Wikipedia entry Philosophy, in which SEP is the first external link (although it is not listed among the 26 reference works listed as further reading for the entry Philosophy).

Otherwise, the closely grouped array of the top five visited SEP entries display an interesting cluster that ranges from the politics and historical figure of Machiavelli, to the far more abstract definition of art and the concept of ethics, back to the political philosophy of critical theory. Among the top visited SEP entries (apart from the SEP homepage) none is to be found in the SEP entries with the most links in Wikipedia (Table 7.2), thus readers' use of these SEP links was somewhat independent (with the exception of SEP itself) of how often editors Wikipedia cited particular SEP entries. The majority of links to the SEP homepage came from a Wikipedia entry for SEP. This only strengthens the notion that readers were pursuing particular interests, perhaps sparked by how the links are used in the entry.

The entry Machiavelli in Wikipedia contains two links to SEP's entry on Machiavelli, which is also cited in the Wikipedia entries Politics and Virtù. With Virtù, Wikipedia has a short 80-word entry describing the concept which Machiavelli used to cover those qualities needed to advance the state (not always virtuous by any means). The link to SEP under Virtù leads directly

*Table 7.3*  SEP pages most frequently visited from Wikipedia over a two-week period, with number of links to the SEP pages found in Wikipedia

| SEP Page | Visits | Links |
|---|---|---|
| Stanford Encyclopedia of Philosophy | 496 | 72 |
| Machiavelli | 243 | 4 |
| Definition of Art | 103 | 1 |
| Ethics | 100 | 1 |
| Philosophy of Economics | 99 | 4 |
| Critical Theory | 98 | 5 |
| Propositions | 88 | 1 |
| Respect | 87 | 5 |
| "Principia Mathematica" | 85 | 7 |
| Existentialism | 80 | 4 |

to the relevant section of the Machiavelli SEP entry, where Carey Nederman explains the concept in relation to Machiavelli's concepts of power and fortune. Wikipedia's entry Politics makes considerable use of SEP, with editors providing linking to its entries Authority, Confucius on Politics, Plato's Ethics and Politics in *The Republic*, Aristotle's Political Theory, Machiavelli, Locke, Mill, and Marx. The discussion on Machiavelli in the Wikipedia entry Politics includes a quotation from the opening of the SEP entry Machiavelli on a clearly intriguing point about the man's stance: "Machiavelli did not invent 'Machiavellianism' and may not even have been a 'Machiavellian' in the sense often ascribed to him."

The same delicate point comes up in the opening paragraph of the Wikipedia entry Machiavelli: "His work, particularly in 'The Prince,' made his name synonymous with ruthless politics, deceit and the pursuit of power by any means. The validity of that reputation is disputed." And while that ends the paragraph all too abruptly, the disputed statement is footnoted with a reference to the albeit mislabeled 'Stanford Dictionary of Philosophy,' which leads to Nederman's SEP entry. It is the same SEP entry that is listed under external links in Wikipedia's entry Machiavelli.[22] Yet I find the use of SEP in footnotes especially effective, as it offers readers what is still a rare experience of being able immediately to follow up one's search with a current and esteemed consideration of the point being made. With the entries Politics and Machiavelli, SEP is used to support a point of disputed reputation. The unjustified criticism of Machiavelli is discussed again later in the Wikipedia entry on the statesman, his reputation largely redeemed by the editor quoting Anthony Parel (1972) in a footnote: "the authentic Machiavelli is one who subordinates personal interests for the common good ... setting aside personal interests in making sacrifices for the common good" (or, more precisely, I would think, "for the state"). The Parel quote is an excellent choice as it highlights the needed distinction between Machiavelli's reputed and actual stance. In contrast, the SEP footnotes, although they would be improved by the briefest of annotations that spoke to the question at hand, still hold an advantage over the Parel footnote by providing a gateway to Nederman's much fuller consideration of Machiavelli. As to why the links to the Machiavelli entry in SEP tops the list of links that readers pursued, it may seem obvious enough in these desperate political times when state interests seem to overrule public concerns. Not the least of it was the publication of *Machiavelli's Shadow: The Rise and Fall of Karl Rove* by Paul Alexander on June 10, 2008.

The third most frequently visited SEP page by Wikipedia readers in this study was an external link in the entry Art that read "The Definition of Art entry in the Stanford Encyclopedia of Philosophy by Thomas Adajian." This model of reference clarity was not the only link to the definition of art in the entry. Under Further Reading, a small boxed-off section reads "Look up 'Art' in the Wiktionary, the free dictionary." The Wiktionary definition of

"Art," begins, "Human effort to imitate, supplement, alter, or counteract the work of nature." The Wikipedia entry Art itself includes a definition in the second paragraph of the entry, followed immediately by a disclaimer around definitions: "Generally art is a (product of) human activity, made with the intention of stimulating the human senses as well as the human mind; by transmitting emotions and/or ideas. Beyond this description, there is no general agreed-upon definition of art." Still, on reaching the end of the Wikipedia readers have looked for further insight into what defines art. The other entries under external links seem to lack the focus of the SEP one: "Art and Play from the Dictionary of the History of Ideas" seems narrow by comparison, and it is not immediately apparent what an "In-depth directory of art" is, to take two of the seven entries.

The SEP entry Definition of Art by Thomas Adajian does not provide any easy answers for art collectors or students. It begins, "the definition of art is controversial in contemporary philosophy," and concludes with a quotation by Kendall Walton stating, "it is not at all clear that the words—'What is art?'—express anything like a single question." Of course, we do not know how far readers went with this 6000-word SEP entry, after arriving there via Wikipedia. But if they only read the first sentence and looked away, they would have had a taste of what philosophy makes of the world. And to suggest that 50 people a week somewhere in the world just might do that in the course of their day seems slight enough but not insignificant from an educational perspective.

A final instance of relatively frequent SEP use comes from the entry René Descartes, whose entry in Wikipedia led to 229 visits to SEP. What distinguishes the SEP links in this entry is how well the bridge from the Wikipedia entry to the wealth of materials on Descartes in SEP was set up (Table 7.1). Wikipedia's external links for Descartes begin with a set of "General" entries, which include Descartes' works online, followed by a second set grouped under Stanford Encyclopedia of Philosophy. Each link leads to a substantial essay by a different philosopher, not least of which is the entry Descartes and the Pineal Gland (the seat of the soul and the organ of thought for Descartes) by Gert-Jan Lokhorst. However, the most popular of the Descartes pages in SEP for Wikipedia readers was the entry "Descartes' Epistemology" (97 visits), followed by "Descartes' Theory of Ideas" (62).

## Discussion

What I am referring to as the educational quality of Wikipedia has much to do with the stance that its editors are taking on knowledge. It is a stance that sets Wikipedia apart in what I find an encouragingly educational way. The learning that I am concerned with here takes place peripherally, out of the corner of the reader's eye, barely noticed, perhaps only after reading about this or that Wikipedia feature in another setting. Wikipedia's increasing emphasis on verifiability and documentation only further exposes the learning apparatus.[23] It exposes the degree to which knowledge is the product of learning, and is often subject to "editing," "discussion," "history,"— those three Wikipedia muses that hang as tabs over every entry. This learning has to do with Wikipedia's openness not just to editing, but to discussing the editing at a level of detail that far exceeds the entry being discussed, and openness to the history of the editing so that the process of making an entry is subject to review and further discussion. This amounts to a statement about knowledge. It represents an epistemic stance.

As Fallis has argued in some detail, Wikipedia is not about the newly established wisdom of crowds (2008: 1670). Wikipedia entries are not some sort of unconscious expression of what is collectively known, of the sort prediction markets use to accurately second-guess election results (Sunstein 2006). Rather, the Wikipedia entries considered here, at least, represent the hard and concerted work of people, many of them members of the group WikiProject Philosophy. Still, there is a further educational risk arising from this appreciation of Wikipedia as a labor of the love of knowledge. It could well lead people to assume that Wikipedia represents this open and always-formative, often-contested approach to knowledge, in a way that stands apart from knowledge proper—from the certain, fixed, and well-formed knowledge arrived at by academics and other experts, and publicly fronted by, for example, *Encyclopedia Britannica.* Thus I find that even the favorable judgments of Wikipedia, as when Fallis finds it "sufficiently reliable" and "more verifiable than most other information sources" (2008: 1667), miss the parallels in the tentative and formative nature of learning in both public and academic realms, as well as their increasing openness. The focus on verifiable reliability does little to encourage the educational value of strengthening connections between the two realms.

It is fortunate indeed that efforts are underway on a number of fronts, from federal legislation to university policies, to increase the degree to which scholarly literature is freely available for such purposes, and many articles are now available in open access archives or through open access journals (Harnad et al. 2008; Willinsky 2006). While I do not want to discourage anyone from going to the public library to pile up the books from the further reading lists for Aristotle, Machiavelli, and Descartes, I have previously argued, in ways that I hope this study of SEP reinforces, that Wikipedia editors would do well wherever possible to use references that readers can begin reading at a click (Willinsky 2007).[24]

In this case, Wikipedia editors have already drawn on 80 percent of the entries in the Stanford Encyclopedia of Philosophy, and readers are using the vast majority of those links to consider a particular aspect of a philosopher's work, as we saw with Aristotle, as well as to see how philosophers deal with such puzzling questions as "What is art?" If it is not always a good thing for reference works to cite other reference works, in this case it makes perfect sense. SEP's comprehensive reviews of philosophical topics and figures provides Wikipedia readers with the next step up, letting them in on how philosophers talk to other students of philosophy. The openness and intersection of these two enterprises, however that openness differs, only increases the opportunities to learn about the nature of what we know and how we know it. What could be more philosophical than that? It is what I most find most encouraging about the uncertain pursuit of knowledge in this new realm.

## Notes

1. Although it is tempting to contrast the scientific journal and the encyclopedia as very different publishing forms, from the first year of the *Philosophical Transactions* in 1665, there has a repackaging of a year's issues into a volume with an index that turned it, in effect, into an encyclopedia.
2. Björk, Roos, and Lauri, for example, have calculated that 20 percent of the scholarly literature's yearly output is freely available online through open access journals, archives and personal websites (2008).
3. The SEP weblogs originally contained 17,724 records of users coming from Wikipedia to SEP. Then, for purposes of this research, the logs were rid of computer-generated traffic from bots and crawlers (836 records), any doubling of records through a redirect that took users from one URL to another in SEP (3297 records), and records without an identifiable source and/or target (188 records). This left a total of 13,363 records of users moving from Wikipedia to SEP.
4. Although this is a study of how readers of Wikipedia come to SEP, it can be pointed out that SEP contains 17 entries with linked references to Wikipedia entries, largely under "Internet Resources," with a link, for example in the SEP entry "Time" leading to Wikipedia entries "Eternalism," "Philosophy of Space and Time," and "Presentism."
5. The Aristotle entry for the Internet Encyclopedia of Philosophy has as its concluding line: "The author of this article is anonymous. The IEP is actively seeking an author who will write a replacement article," while the Catholic Encyclopedia entry starts out, "The greatest of heathen Philosophers" and was originally published in the 1907 edition of the encyclopedia.
6. These items are listed in the entry as "Notes," but are referred to as "footnotes" in the Wikipedia style guideline Citing Sources, and will be called such in this article: "These [references, footnotes, parenthetical reference] are the most common methods of making articles verifiable. A Wikipedia editor is free to use any of these methods or to develop new methods; no method is preferred."
7. Wikipedia co-founder Jimmy Wales has recently spoken of how Wikipedia entries have been improved as they are "more detailed, more accurate, hopefully better written, fleshed out more, with ... two or three footnotes to tell you where to go and check it" (Young 2008). In May 2005, for example, the Wikipedia entry on Aristotle, which was by then over 5000 words, was without footnotes or any links to SEP. The 51 footnotes that currently accompany the Aristotle entry may tell you where to go and check, but with the reference to SEP (as well as to Aristotle's will and Cicero's *Academica*), Wikipedia takes the reader directly there in a click, whether to a discussion of Aristotle's disproven theory of falling bodies, or to the primary sources (courtesy of Google Books and Project Gutenberg).
8. http://en.Wikipedia.org/wiki/Talk:Atheism/Archive_33.

9. http://en.Wikipedia.org/wiki/Talk:Atheism/Archive_14.
10. http://en.Wikipedia.org/wiki/Talk:Atheism/Archive_22.
11. http://plato.stanford.edu/entries/atheism-agnosticism/.
12. http://en.Wikipedia.org/wiki/Talk:Atheism/Archive_27.
13. http://en.Wikipedia.org/wiki/Talk:Orthomolecular_medicine/Archive_3.
14. http://en.Wikipedia.org/wiki/Talk:2003_U.S._invasion_of_Iraq.
15. http://en.Wikipedia.org/wiki/Talk:On_the_Jewish_Question.
16. http://en.wikisource.org/wiki/Selected_Essays_by_Karl_Marx/On_The_Jewish_ Question.
17. http://en.Wikipedia.org/wiki/Talk:Bobby_Jindal.
18. http://en.Wikipedia.org/wiki/*Wikipedia*:Biographies_of_living_persons.
19. http://en.Wikipedia.org/wiki/Talk:Bobby_Jindal.
20. http://plato.stanford.edu/entries/double-effect/.
21. A terminal in a library would have a fixed IP address with many users in the course of a day, just as some users would have a single IP address in their homes, for example.
22. At least one reader expressed appreciation for the SEP link in the discussion page for the entry, after much discussion of the poor quality of writing in the entry: "I am glad the Stanford text/link is here, at the bottom—the best I've seen in a while ... Now, that's a well written text on M.!!!"
23. Wikipedia has policy and style guidelines on verifiability, verification methods, citing sources, referencing for beginners, WikiProject for citation cleanup, and a number more.
24. Recently, Google Scholar made finding an open access version all the more easy by indicating in its search results which links lead directly (without subscription or credit card) to a PDF or HTML file.

## Bibliography

Aristotle. 2008. In *Encyclopedia Britannica*. Retrieved September 3, 2008, from http://search.eb.com/eb/article-33161.

Aristotle (384–322 BCE): Overview 2006. *Internet Encyclopedia of Philosophy*. http://www.utm.edu/research/iep/a/aristotl.htm, accessed May 9, 2008.

Björk, Bo-Christer, Annikki Roos, and Mari Lauri 2008. "Global Annual Volume of Peer Reviewed Scholarly Articles and the Share Available via Different Open Access Options." Proceedings ELPUB2008 Conference on Electronic Publishing, Toronto, Canada. Retrieved from http://www.oacs.shh.fi/publications/elpub-2008.pdf, accessed December 21, 2008.

Cicero, Marcus Tullius 2005. *Academica*, ed. James S. Reid. London: Macmillan, 1874. Gutenberg Project. http://www2.cddc.vt.edu/gutenberg/1/4/9/7/14970/14970-h/14970-h.htm#BkII_119, accessed April 21, 2009.

Harnad, S., T. Brody, F. Vallieres, L. Carr, S. Hitchcock, Y. Gingras, C. Oppenheim, C. Hajjem and E Hilf. 2008. "The Access/Impact Problem and the Green and Gold Roads to Open Access: An Update." *Serials Review* 34 (1): 36–40.

Parel, Anthony. 1972. "Introduction: Machiavelli's Method and His Interpreters." In *The Political Calculus: Essays on Machiavelli's Philosophy*, ed. Anthony Parel. Toronto: University of Toronto Press, 3–28.

Perry, John and Edward N. Zalta. 1997. "Why Philosophy Needs a 'Dynamic' Encyclopedia." Unpublished paper. Center for the Study of Language and Information. Scott, Jaschik. 2007. "A Stand against Wikipedia." *Inside Higher Education*

(26 January). Retrieved from http://www.insidehighered.com/news/2007/01/26/wiki, accessed June 21, 2009.

Stanford University. http://plato.stanford.edu/pubs/why.html, accessed November 10, 2009.

Sunstein, Cass R. 2006. *Infotopia: How Many Minds Produce Knowledge.* New York: Oxford University Press.

Turner, W. 1907. Aristotle. In *The Catholic Encyclopedia.* New York: Robert Appleton.

Willinsky, J. 2007. "What Open Access Research Can Do for Wikipedia." *First Monday* 12(3). Retrieved from http://www.uic.edu/htbin/cgiwrap/bin/ojs/index.php/fm/article/view/1624/1539, accessed January 26, 2009.

Willinsky, J. 2006. *The Access Principle: The Case for Open Access to Research and Scholarship.* Cambridge, MA: MIT Press.

Young, Jeffery R. 2008. "Wikipedia's Co-Founder Wants to Make It More Useful to Academe." *Chronicle of Higher Education*, 54(40), A18 (13 June).

# 7.1
## The Challenges of Digitally Mapping Marginal Sub-Regions and Localities: A Case Study of South India

*Aloka Parasher-Sen*

The intention of this project by the French Institute Pondicherry (http://www.ifpindia.org/histatlas/) is to prepare a historical atlas of South India, from pre-historic times to 1600 CE, in digital format, through a combination of maps, photographs, illustrations, texts and geographical information system (GIS) functionalities. The challenge was to focus on relatively marginal localities in a sub-region of the present day linguistically unified state of Andhra Pradesh, which has historically been called Telangana (Reddy 1998).

Digital technology has helped to illustrate these economically underdeveloped localities that had significant historical data, frequently ignored in the meta-historical narratives of the larger regional identity formation, because such data did not fit into the dominant historical narrative. Some of the region's monumental aspects had been mapped earlier (Subbarao 1958) but a systematic accumulation of the not so striking had not been part of the pan-Indian and pan-regional accounts. However, the process of not only highlighting the monumental has permitted a layering of the small but significant pieces of information that have inscribed the past of these localities in decidedly different ways.

Mapping was carried out using templates for different periods under a thematic schema of political, social, cultural, and economic knowledge, so that static maps plotting only one kind of historical information for each period and theme were avoided (Schwartzberg 1978). Being by definition interactive, these digital maps were simultaneously inclusive, as they exhaustively exploited diverse sources, in some cases for the first time. This made the historical and cultural past accessible to a large audience while concomitantly satisfying the innate desire of peoples of the sub-region to celebrate their past in all its glory.

The danger of eulogizing a forgotten past within an essentialized and primordial frame to assert a dogmatic ideology of uniqueness of the sub-region and its people had to be avoided (Parasher-Sen 2009). The digital mapping process explained how over time regions and sub-regions developed a tenuous identity that changed in the context of the demands of the economic and political challenges of the times (Parasher-Sen 2006). In other words, all periods did not exhibit similar boundaries for a particular sub-region under discussion, and in some periods the local identity was necessarily more prominent than the total identity (Parasher-Sen 2007). By creating several parameters for the retrieval of the data through the use of multiple symbols for tools, coins, writing samples, buildings, burials, and religious edifices, and within each further demarcating several types by using written material found in local contexts, it became possible to highlight in greater depth the intricate nature of the way historical mapping can emerge.

In straddling the diversity of time, space, and data through this technology one had to contend with its pitfalls. Virtual maps are not replicas of maps as in printed atlases. The fluidity of the former initially perturbs first time users, especially those unfamiliar with the region. By their very definition, the user as a critical participant has to be comfortable with grasping these virtual maps that allow for multiplicity and simultaneity in comprehending complex data and its visualization. For those seeking permanence and certainty, this can be inhibiting. The process of digital mapping immediately makes it possible for the local to be rapidly appropriated into the global. The danger of essentializing and romanticizing the local is thus imminent, or highly likely. To avoid this, the larger project of mapping South India has enabled us to highlight not only differences but also commonalities between sub-regions.

While critiquing older methods of mapping for the Indian subcontinent as a whole (Green 1937), we had noted that the structure of political and economic power had imposed categories of understanding that were alien to those that had been thus systematized (Parasher-Sen 2006). This danger remains. The baggage that accompanies computer technology is even more overpowering and overarching in its reach to control, organize, and map the subject. Though we cannot escape the portals of knowledge creation and generation that govern our lives today, we can sensitize ourselves to the fact that all regions of the globe cannot and need not react to this technology in a similar way. Ian Hodder's (1986) assessment that dissemination of the past is absolutely dependent on power, control of theory, method and communication is even more pertinent and relevant in the digital age of today and we must remain forewarned of this.

## Bibliography

Green, E. W. 1937. *An Atlas of Indian History*, London: Macmillan.
Hodder, Ian. 1986. *Reading the Past, Current Approaches to Interpretation in Archaeology.* Cambridge: Cambridge University Press.
French Institute of Pondicherry. *Historical Atlas of South India* Website: http://www.ifpindia.org/histatlas/, accessed January 11, 2010.
Parasher-Sen, Aloka. 2006. "The Making of Digital Historical Atlas: Some Examples from South India." *The International Journal of Technology Knowledge and Society* 2 (4).

Parasher-Sen, Aloka. 2007. "Localities, Coins and the Transition to the Early State in the Deccan," *Studies in History*, 23 (2): 231–69.

Parasher-Sen, Aloka. 2009. "Perceptions of Time, Cultural Boundaries and Region in Early Indian Texts." *Indian Historical Review*, 36 (2): 183–207.

Reddy, E. Siva Nagi. 1998. *Evolution of Building Technology in Andhradesa Early and Medieval Periods.* 2 vols. New Delhi: Bharatiya Kala Prakashan.

Subbarao, B. 1958. *The Personality of India: Pre- and Proto-Historic Foundations of India and Pakistan.* Baroda: M.S. University.

Schwartzberg, J. E. (ed.) 1978. *A Historical Atlas of South Asia.* Chicago: Digital South Asia Library, University of Chicago. Website: http://dsal.uchicago.edu/reference/schwartzberg, accessed January 15, 2010.

# 7.2
## How Open Source Software and Wireless Networks are Transforming Two Cultures: An Investigation in Urban North America and Rural Africa

*David Yates and Anas Tawileh*

Many people in developed countries take access to information and communication technologies for granted. Such individuals live and work in a well-connected information society with access to digital resources that is almost universal. However, a global analysis of the contemporary information society paints a much more complex picture (Chinn and Fairlie 2007; Fuchs 2009; Yates, Gulati and Weiss 2010).

For businesses and the more fortunate, digital inclusion enables access to resources that provide information, communication, education, commerce, and entertainment. For the less fortunate and more isolated members of contemporary society, access to such resources can help to overcome the many barriers to participation in the global information society. In the early stages of development, however, new technologies may reinforce or even widen existing economic, political, and social inequalities between the haves and have-nots (NTIA 2000; United Nations 2001; van Dijk 2006). Much of the previous research supports the view that technological advances mostly have created new or exacerbated existing inequalities between the information rich and poor, both within nations and between nations (Guillén and Suárez 2005; Fuchsand Horak 2008; Mossberger, Tolbertand Gilbert 2006).

This research, summarized here (Yates, McGonagle and Tawileh 2008), explores how open source software and wireless networks enable digital inclusion in the US and Africa. It began by measuring the digital divide in these very different regions of the world. Our research demonstrated quantitatively and qualitatively how the digital divide places populations in both regions at a disadvantage. Next we examined the role of technologies in bridging the digital divide along three complementary dimensions. First, we showed that both affordable technology and sound policy are necessary for digital inclusion. Second, we looked at how these two technologies are extended, integrated, and customized in information and communication technology (ICT) solutions that are both creative and effective. Third, we described how the hardware and software in networked systems have been tailored to support applications that are as diverse as the people using them.

Because the digital divide is wider in most regions in Africa than it is in the US, bridging the digital divide in Africa often requires more comprehensive ICT implementations. We surveyed wireless networking and open source software that enabled ICT solutions and analyzed their critical role in bridging the digital divide. In our research we presented more detailed investigations into a few important technologies, specifically IEEE 802.11 wireless networks, wireless mesh networks, and GNU/Linux-based open source software. We saw the immediate and widespread impact that these technologies have in communities as different as Castle Square in Boston, Massachusetts and Karagwe, Tanzania. We also saw how these technologies are being used to educate children today, so that they will acquire the knowledge and develop the ingenuity needed to design global ICT and ICT for development (ICT4D) in the future.

## Bibliography

Chinn, M. D. and R. W. Fairlie. 2007. "The Determinants of the Global Digital Divide: A Cross-Country Analysis of Computer and Internet Penetration." *Oxford Economic Papers* 59(1): 16–44.

Fuchs, C. 2009. "The Role of Income Inequality in a Multivariate Cross-National Analysis of the Digital Divide." *Social Science Computer Review* 27(1): 41–58.

Fuchs, C. and E. Horak. 2008. "Africa and the Digital Divide." *Telematics and Informatics* 25(2): 99–116.

Guillén, M. F. and S. L. Suárez. 2005. "Explaining the Global Digital Divide: Economic, Political and Sociological Drivers of Cross-National Internet Use." *Social Forces* 84(2): 681–708.

Mossberger, K., C. J. Tolbert and M. Gilbert. 2006. "Race, Place, and Information Technology." *Urban Affairs Review* 41(5): 583–620.

NTIA 2000. *Falling Through the Net: Toward Digital Inclusion.* Washington, DC: National Telecommunications and Information Administration, US Department of Commerce.

United Nations. 2001. *Human Development Report 2001: Making New Technologies Work for Human Development.* New York: Oxford University Press.

van Dijk, J. A. G. M. 2006. "Digital Divide Research, Achievements and Shortcomings." *Poetics* 34 (4–5): 221–35.

Yates, D. J., G. J. Gulati and J. W. Weiss. 2010. "Towards Universal Broadband: Understanding the Impact of Policy Initiatives on Broadband Diffusion and Affordability." *Proceedings of the European Conference on Information Systems (ECIS)*, Pretoria, South Africa.

Yates, D. J., T. McGonagle and A. Tawileh. 2008. "How Open Source Software and Wireless Networks are Transforming Two Cultures: An Investigation in Urban North America and Rural Africa." *International Journal of Technology, Knowledge and Society* 4(6): 145–57.

# 7.3
# The Secrets of Biblioland: A Case Study

*Elena Moschini*

The Secrets of Biblioland (http://www.biblioland.org) is a Flash-based educational interactive game designed to support university students in developing academic information literacy skills. The game is also intended to instigate a sense of belonging to the academic community, and to develop awareness of the historical value of academic work and the role of new generations in the interpretation and preservation of knowledge repositories. The game is designed to be embedded in a blended learning approach to information literacy; students and tutors should use it to explore the historical development of scholarship, issues on the role of libraries and information, as well as referencing and plagiarism, using the gaming elements but also the online materials available via the website. The choice of developing a Flash game is linked to accessibility and usability issues, as the game should be playable by a wide audience. The illustration and game design styles are also meant to address an audience of casual gamers (Juul 2010) and do not have the ambition of emulating or competing with commercial platform games.

In The Secrets of Biblioland, the gamers play as students embarking on an adventure to find a "lost reference." The world of Biblioland has lost one of its most precious references and it is in danger of disappearing altogether. Axel, one of the game characters and a Member of the Fellowship of the Seekers, asks the gamers to help Biblioland in the quest for the Lost Reference. This challenge takes the gamers/learners through a number of fantasy libraries, each library constituting a level in the game. The game has nine levels: Introduction, Socrates and Ancient Greece, the lost library of Alexandria, the Bobbio medieval library, Gutenberg and the printing press, the contemporary British Library, the Plagiarism Cellar, the BiblioFuture library, and Conclusion. Each level introduces a set of topics that relate to various issues about academic literacy. The final phase of the game reveals a password to access further learning resources published on the accompanying website.

This project stems from the experience of trying to engage the generation of the "Digital Natives" as defined by Mark Prensky (2001b); this requires an understanding of the language of contemporary learners and the development of innovative approaches to information literacy for higher education. The educational potential of games, and their role in developing literacy and competence skills for a new generation of learners, has been widely explored in the literature (Gee 2003, 2007; Gibson, Aldrich and Prensky 2007; Prensky 2001a; van Eck 2006). Gaming and mobile platforms and virtual environments are now part of the mainstream digital landscape and offer new opportunities for the development of significant educational experiences (Kirriemuir 2007; Ritterfeld, Cody and Vorderer 2009; Peachey et al. 2010; Klopfer 2008). The Secrets of Biblioland is also a research project; the ongoing evaluation of the game is focused on exploring the role of educational games in adult learning strategies and the role of game-based learning in the development of academic digital literacy skills (Moschini 2008). Furthermore, the project aims to explore the challenges, advantages, and limitations of educational in-house game development in higher education settings.

The Secrets of Biblioland has been supported by an ESCalate Development Grant from the UK Higher Education Academy and a small grant from the CAPD Learning & Teaching Projects Fund from London Metropolitan University (UK). The Secrets of Biblioland has been developed by a team of academics with expertise in game design and information literacy: Elena Moschini (team leader), Che Guevara John and Vanda Corrigan from London Metropolitan University, and Enrico Benco from Middlesex University (UK).

## Bibliography

Gee, J. P. 2003. *What Video Games Have to Teach Us about Learning and Literacy*. New York, Basingstoke: Palgrave Macmillan.

Gee, J. P. 2007. *Good Video Games + Good Learning: Collected Essays on Video Games, Learning, and Literacy*. New York: P. Lang.

Gibson, D., C. Aldrich and M. Prensky (eds.) 2007. *Games and Simulations in Online Learning: Research and Development Frameworks*. Hershey, PA; London: Information Science Publications.

Juul, J. 2010. *A Casual Revolution: Reinventing Video Games and Their Players*. Cambridge, MA: MIT Press.

Kirriemuir, J. 2007. "Digital Games in Libraries and Information Science." Presented at Tampere University, Finland, November 29. http://www.slideshare.net/silversprite/digital-games-in-libraries-and-information-science, accessed October 5, 2010.

Klopfer, E. 2008. *Augmented Learning: Research and Design of Mobile Educational Games*. Cambridge, MA; London: MIT.

Moschini, E. 2008. "The 'Secrets of Biblioland' Interactive Game: An Innovative Approach to Information Literacy." *International Journal of the Book* 5 (3): 1–6.

Peachey, A., J. Gillen, D. Livingstone and S. Smith-Robbins (eds.) 2010. *Researching Learning in Virtual Worlds*. London: Springer.

Prensky, M. 2001a. *Digital Game-Based Learning*. New York, London: McGraw-Hill.

Prensky, M. 2001b. "Digital Natives, Digital Immigrants." *On the Horizon* 9 (5): 106.

Ritterfeld, U., M. J. Cody and P. Vorderer (eds.) 2009. *Serious Games: Mechanisms and Effects*. New York; London: Routledge.

van Eck, R. 2006. "Digital Game-based Learning: It's Not Just the Digital Natives Who Are Restless." *Educause Review* 42(2) (March–April): 16–30.

# Part VIII
# New Infrastructures of Science

# 8

# Towards a Science 2.0 based on Technologies of Recommendation, Innovation, and Reuse

*Karim Gherab-Martín*

Web 2.0 is a term that has come into widespread use in recent years to describe the emerging social dynamic that characterizes the creation and consumption of web content (O'Reilly 2005). The concept of Web 2.0 involves a new way of understanding the relationship between agents that play a role in the production, marketing, distribution, and utilization of digital content and highlights the steadily converging relationship between producer and consumer. Sociologist Alvin Toffler (1980) perceived this convergence and coined the term "prosumer."

Following in the footsteps of Web 2.0, science is embarking on a course that will transform it into Science 2.0. A detailed analysis of the characteristics of a Science 2.0 is beyond the scope of this chapter. I will focus only on the scientific publications and communication system and point out why the current proposals of open access to academic journals are not ambitious enough and how it is that many of the Internet's current technological and methodological trends can manifest themselves in a seemingly more protected field such as science. Thus, only a general idea of the proposal will be given here—a model that could be called Open Reuse.

## Technologies of recommendation

Many Web 2.0 projects are characterized by their encouragement of user participation. The idea is to create social networks that bring about the emergence of a collective intelligence that, in turn, improves the products and services offered online. This collective intelligence is fueled by the recommendations users make to each other in exchanging their experiences. These recommendations have been technologically implemented in various ways, sometimes by means of tags with keywords or brief descriptions, other times through more extensive annotations and comments and, on other occasions, using clever techniques to relate products, behaviors, likings, or trends.

A well-known example of success in the use of this type of technique is Amazon.[1] A tool that allows users to write reviews or make comments about their favorite books has been added to Amazon website's automatic recommendation system, which is based on user purchases and behavior. Amazon also trawls books in search of similar text strings and interprets which similar texts convey similar ideas, thus creating additional links between books.

Among the Internet search engines, Google has clearly stolen the lead as a result of PageRank, an algorithm that, among other things, counts the hyperlinks between websites. The more people link with a specific website, the more confidence the search engine has in this site and, consequently, the higher it will appear in the lists that Google presents on-screen to show the results. What is more, Google gives greater weight to websites that are linked by sites that already have a high confidence value because there are many other sites that link to them. It is important to realize that all these hyperlinks are "intentionally dependent" on people. In other words, they depend on the will, knowledge, beliefs, wishes, and intentions of the users. Therefore, if we have a specific project or business, we should know how to manage these subjective qualities correctly in order to improve our service to the user or client group we are targeting.

Google's founders took the idea from a model that chemist and information scientist Eugene Garfield (1955) applied to science. Scholars are very familiar with citation indexes and the impact factor of the different academic journals because tenured positions, grants, subsidies, and even their salaries depend on them. Citation indexes function in a similar way to Google's PageRank algorithm, as explained above. The difference is that, instead of counting hyperlinks, Garfield counted the number of times a specific article or journal was cited in other articles written at a later date. This is usually known as the "cited by" list.

Garfield's inspiration came from routine procedures of US courts. Indeed, this may stand as the historical origin of the technologies of recommendation. US attorneys cite prior cases because they set precedent: a prior case with a favorable outcome might serve as a convincing argument for the one who brings it up. This case is referenced by the volume and page on which it is recorded, and a particular statute is cited by the article, chapter, and section of the publication in which it is found.

In the final quarter of the nineteenth century, a publication entitled Shepard's Citations appeared. It belonged to the Frank Shepard Company, a consulting firm that undertook to provide attorneys with such citations— cases, statutes, and sentences. In particular, it showed the history of a case by documenting all subsequent cases in which it was cited and by reporting on the outcome of sentencing in those subsequent cases. The index of legal cases even detailed the history of each case using a simple letter code: "e" for those cases in which the court gave an explanation of the original case cited; "a" to denote that on that occasion the court affirmed the precedent; "d" to specify that distinctions regarding the original case were introduced; and so on. Let us consider an example (Adair 1955):

|   |     |       |
|---|-----|-------|
|   | 101 | Mass 210 |
|   | 112 | Mass 65 |
| e | 130 | Mass 89 |
|   | 165 | Mass 210 |
| d | 192 | Mass 69 |
|   | 205 | Mass 113 |
| e | 212 | Mass 173 |
|   | 221 | Mass 210 |
| a | 281 | US 63 |
|   | 35  | HLR. 76 |

The attorney begins his research with case 101, the original case, appearing on page 210 of the reports section corresponding to the State of Massachusetts. The cases that appear in the above list are all those in which the original case was cited. Thus, in case 130 Mass 89, the court gave an explanation of the original case (as indicated by the letter code "e"); in case 192 Mass 69, the court introduced a clarification or distinction ("d") with respect to the original case, thereby limiting the area in which the case was valid as precedent; and in case 281 US 63, the US Supreme Court asserted ("a") that the case constituted sound legal precedent, which was published in the *Harvard Law Review*(HLR.).

Garfield quickly perceived the validity of this scheme for indexing scientific knowledge and immediately applied it to reference works in the area of chemistry patents (Garfield 1957). The way patent offices function is similar to the scheme we have been describing in that, to confirm that an invention is truly original, examiners must check the bibliography. Their decision about the originality of an applicant invention must be supported by references to previous research and patents in such a way that the entire history of a field of invention is accurately recorded and may be retrieved efficiently and without gaps.

I have turned to historical acts to show that technologies of recommendation have not come into being as a result of digital technology. They were ideas that already existed and were applied with greater or lesser difficulty. The advantage of digital technology is that it offers us unparalleled flexibility in comparing all types of data, trends, likings, citations, hyperlinks, downloads, clicks, and so on. It offers us possibilities our forefathers could not even dream about. And with the Internet, we have the ideal setting to offer users around the world any innovation of this type that may occur to us. We are a long way from exhausting all ideas for creating innovative technologies of recommendation. The so-called "folksonomies," sometimes known as social tagging, are an example of a collaborative tagging technique that allows users to freely edit tags, thereby categorizing and organizing the huge amount of content being published every day on the Internet—a recent and remarkable example that goes beyond the traditional taxonomy.

A tag is a keyword attached to a piece of information that usually describes a digital object, which could be a website, a document, a blog, a photograph, a sound, a piece of software, among others, and allows it to be found again by browsing, filtering, or searching. Tagging means assigning some words to an item and is a technique that has been popularized by many websites associated with Web 2.0. Launched in 2003, Del.icio.us[2] coined the term social bookmarking and seems to have pioneered social tagging as we know it today.

In providing for the use of nicknames, Delicious could respect its users' privacy and, at the same time, openly show all tags that all users had assigned to each link saved in Delicious. This enabled any user to see the tags most often used by another user, the most popular tags associated with a particular website (a link), or all links associated with a specific tag. To put it another way, in gambling on a collaborative tagging effort, Delicious did more than enable users to share links or bookmarks (social bookmarking); it enabled them to share tags associated with those bookmarks (social tagging), an ingenious addition that launched a sensible improvement. It did not take long for the idea of assigning tags to links to be extended[3] to other types of objects such as blogs (Technorati[4]), photographs (Flickr[5]), videos (YouTube[6]), and citations (Diigo[7]) published online, as well as books (LibraryThing[8]) and scientific articles (CiteULike,[9] Connotea,[10] 2Collab,[11] and BibSonomy[12]).

In a folksonomy, the item's creator is free to use whatever words she or he likes as tags; the item's viewers can also tag it using whatever words they choose. Thus a folksonomy is a collection of tags on a website in which many users informally tag many items; it is essentially a social tagging system in which the user is free to write keywords or phrases that have a particular meaning for him or her. As a result, the expression "social tagging" has a broader and more flexible meaning than "folksonomy" because the first takes in both the free-tagging and controlled-tagging systems (in which there are more restrictions on the choice of keywords), while the second leaves the user free to choose his or her own keywords. We can visualize a folksonomy, then, as a social free-tagging system.

It must be stressed that the primary purpose of social bookmarking, social tagging, and folksonomies is to enable the user to store links and create a personal catalog of bookmarks and their associated tags to facilitate their subsequent retrieval. For example, all bookmarks for websites with similar content could have one tag in common and thereby be grouped together. Tagging an item consists primarily of committing it to memory for personal reasons using the available technology.

Once tags are assigned to a specific bookmark (or other digital object), however, both the tags and the bookmark (or digital object) may be placed at the disposal of all other users who may, in turn, locate the tagged bookmarks (or objects) via the tags created and also identify other users with similar interests. Because tagging is done in a public forum, the social dynamic forces users to choose relevant tags.

## Free flow of information in an industry of innovative reuse

If we disregard the term "innovative," then what do we mean by the expression "industry of reuse?" In reality, the expression refers implicitly to the free reuse of third-party content in creating derivative works with added value. This added value may be very subtle and can be a source of controversy at times. Let us take the case of the music industry: imagine for a moment that, instead of a copyright license, all songs in the world had a Creative Commons[13] or copyleft[14] type of license that allows a CD to be created using a strange and apparently absurd combination of songs never before put together for sale. Simply combining the original creations of others without rhyme or reason could generate controversy, obviously, but what if the CD is a success among music critics and experts? It could happen that what a layperson thought was a random combination of songs turns out to show tremendous intuition on the part of that person because the songs have some characteristic in common that is not readily perceived.

Let us now suppose that this CD is not a success among music critics. Would this invalidate the argument in favor of the free reuse of content? Would the author of the CD be committing a crime because there is no added value in a simple combination of songs that already existed? Let us imagine another scenario: suppose that, three years after the release of the CD, a company launches a data mining technological device on the market that is capable of finding similar patterns in the chords of all the songs—patterns that had previously been detected only by the CD author through a stroke of genius. In this case, the author of the CD would be spared the accusations, and his or her CD would be considered a valuable and unexpected contribution to music, perhaps the work of a genius. But what if the recently invented music data mining device does not discover these patterns? Should it be inferred that there is no added value in the CD? Or should the music data mining device be considered not good enough to discover the underlying patterns of the songs? Obviously, the answer eludes us. Of course, we would be inclined to believe that there is no added value in the CD, but we would not be completely certain.

The question of added value is an important issue in the foregoing example. We are referring to whether the author has, in fact, contributed any innovation in combining the songs in a CD. We cannot wait for someone to invent a machine that indicates whether there has been an innovation or not in the previously mentioned initiative. What is obvious is that the authors of the individual songs would not like to see someone who has contributed nothing new receiving economic gain from a product that cost them considerable effort. Naturally, intellectual property rights make it impossible to think about a situation like this, and perhaps the copyleft type of license is not the best bet for the music industry or the progress of music. In any case, this issue is the music industry's concern, and we will leave it to the experts in that field.

We will focus on the best choice for science. Our expertise lies in how science functions and what the expectations of scientists are when they publish an article in an academic journal. The foregoing example helps us to see a crucial difference between science and cultural industries. Let us explain why the case of science is the opposite of music, in this regard, and why science is a good field for exploiting resources in the way I have indicated—through an industry of innovative reuse.

Meanwhile, let us mention three examples of successful applications of content reuse, where content can refer to software, data, articles, and the like. But first I would like to point out that what I have named "industry of innovative reuse" is called "innovation in assembly" by Tim O'Reilly (2005), author of "What is Web 2.0?" O'Reilly wrote:

> When commodity components are abundant, you can create value simply by assembling them in novel or effective ways. Much as the PC revolution provided many opportunities for innovation in assembly of commodity hardware, with companies like Dell making a science out of such assembly, thereby defeating companies whose business model required innovation in product development, we believe that Web 2.0 will provide opportunities for companies to beat the competition by getting better at harnessing and integrating services provided by others.

I will speak briefly about the first example before focusing in more detail on the second and third ones. The first example worth mentioning is Amazon, the famous online bookstore. As in the case of Barnes & Noble,[15] the original Amazon database came from the ISBN registry provider R. R. Bowker. But Amazon continued to improve its data and increase the value of its content by adding other complementary information such as cover images, tables of content, indexes, and sample materials. In order to give even more value to the offered service, Amazon encouraged its users to introduce comments and reviews. In this way, and after several years of making improvements, Amazon has overtaken Bowker and become the reference for many scholars and librarians in their consultations of bibliographical data.

The second is a well-known example: the Free Libre Open Source Software (FLOSS) industry. Here, the reused content is software. In this case, depending on the type of license, developers who use a free software application are often obliged to keep the software they are constructing from it free, as well. This simple procedure, made possible by a GNU General Public License—a particular type of copyleft license—has resulted in innumerable innovations and is threatening the dominion of the large, multinational, proprietary software corporations. Notice that, in English, the word "free" has two meanings—"free" as in "freedom" and "free" as in "free beer"—and this has often led to misunderstandings relating to the free software philosophy. The FLOSS community is not opposed to the marketing of software or the economic benefits, as long as free access to the software is preserved. For example, one business model might be to develop free software components and then reap the benefits by adapting these components to the client's needs or installing them in the client's servers and carrying out routine maintenance. Another might be to accept donations. In any case, open access (public access) to the software is the critical factor, not that it be no-cost software. This is why the Spanish word "Libre" was included in the acronym FLOSS, to emphasize that "free" means "freedom" rather than "no-cost."

An interesting feature of the FLOSS community is that it is divided into two groups which have exactly the same objective but for different reasons. On one side are those led by Richard M. Stallman (2002) who takes an ethical stance on this issue, believing that the emphasis must be on the freedom to reuse software; on the other side are those who incline towards a pragmatic position, seeking only to promote the development and use of open source software because its innovative dynamism is of technological and economic benefit to society's industrial framework and to society as a whole. This is why the latter group decided to substitute the term "open source software" for "free software."

The concept of freedom of reuse goes well with the idea I want to clarify here, that the open reuse of content (or components or parts of a whole) and the freedom to combine it in various creative ways with other open content (or components or parts of another whole) may lead to new and useful products and services that are appealing to users. It seems reasonable to expect, of course, that a free policy, in the sense of "no cost," would enhance the practice of reuse. As I will show later, this is the case in science.

The third example tackles public sector information (PSI). Public entities are usually the largest producers of information in Europe, and European governments gain income from fees for commercial licenses that allow private investors to access and reuse this information. The goal of this licensing-based model is to recoup as soon as possible some of the investment of public funds. A study commissioned by the European Commission (PIRA International 2000) some years ago showed that this is not the best way to increase the return on investment, however, because these charges mean artificial barriers to the private sector's creation and development of value-added services and products for consumers.[16] Removing these barriers to the access and reuse of PSI yields higher taxation and employment benefits because of the higher volumes of commercial activity.

The private sector finds ways of exploiting PSI for commercial gain by delivering products and services that benefit their consumers:

- By supporting the original mandate of public sector institutions but doing so more cost-effectively and more efficiently than the public sector itself;
- Through aggregating and linking raw information from diverse sources into one location;
- By creating innovative services, processes and products such as indexes, catalogs and metadata;
- By adapting information for each specific academic field or commercial sector for a variety of purposes, for instance, by using analytical data software;
- By delivering information through new channels;
- By displaying information in creative and attractive ways such as viewer-friendly presentations, graphics, animations, simulations, interactive interfaces, and so on;
- By merging this information with other sectors' services and products.

For instance, the US makes its meteorological data available at no cost and, as a result, the US meteorological data reuse industry is ten times larger than Europe's in resulting profits and generation of employment. In European countries, meteorological data is often subject to the licensing-based model. Moreover, in some European Union (EU) member states, as was the case in Spain (PIRA International 2000: 76), license fees were so high that many private meteorological service providers preferred to get their information from the US rather than purchase it from their own domestic meteorological service institutions, even though the US meteorological service information was not well suited to their objective of providing specific information about the weather expected in different locations in Europe.

The US government scenario, which has been summarized as "a strong freedom of information law, no government copyright, fees limited to recouping the cost of dissemination, and no restrictions on reuse" (Weiss and Backlund 1997: 307), is currently being promoted by the EU, which has measured the loss of industrial value that PSI reuse limitations mean for Europe, especially if the situation is compared with the US. The above-mentioned study (PIRA International 2000: 16) showed that PSI in the EU generated a seven-fold return on the investment whereas a 39-fold return on the investment is seen in the US. What makes the difference, specifically, is the freedom US institutions offer the private sector to commercially reuse public information. For that reason, a recent European Community directive[17] is aimed at establishing a legal framework favorable to the creation of an industrial network – favoring private sector creation of value-added products and services. This directive took effect in EU Member States on July 1, 2005. After holding a public consultation, the European Commission published a Communiqué[18] on May 7, 2009 that appends a number of additional recommendations and actions to the original PSI Re-use Directive. A summary of the process is available in Corbin (2009).

## Scientific publishing and open access

Let us now apply all these ideas to science and, more specifically, to the open access movement (OA). This initiative promotes no-cost access to scientific articles via the Internet, at least to articles that are based on research and results financed with public funds.

The origins of OA date back to the period in which scholars realized that the immediacy afforded by the Internet for the dissemination of their discoveries among their colleagues was very important, because establishing priority of discovery in scientific endeavor was very important. Subsequently, the serials pricing crisis (Guédon 2001)—the exaggerated and unjustified increase in the prices of commercial journals—became the principal argument to promote OA initiatives. Thus, many librarians and a growing number of scholars are in favor of OA. Naturally, one can find opposition among many publishers of traditional (and electronic) subscription journals and also, again, many scholars.

Under normal conditions, the main interest of librarians is to provide scholars with the publications they need at the lowest possible cost, and the interest of commercial publishers is to offer the best articles, which maximizes their prestige and economic profits. On the other hand, young scholars, who are always more inclined than veteran scholars to change the established order, are more enthusiastic about OA and formulas such as open peer review, which is a more open and democratic peer review. And it is obvious that the editorial boards of the most prestigious journals have the most distinguished academics, who are usually more advanced in age and have more privileges. Therefore, these are generally not scholars who are eager to promote either OA or open peer review, although there have been notable exceptions (Knuth 2003).

Two roads have been identified to achieve full OA (Guédon 2006): the "golden" road, which refers to open access journals; and the "green" road, in which scholars self-archive their articles directly into either institutional repositories (IRs) or subject-based repositories (SRs). A number of "golden" road supporters advocate financing this with the so-called author pays model[19] (House of Commons 2003–4): if the author wishes to have his or her article open, then she or he should pay a certain amount of money to the journal publisher. But let us avoid the economic debates about whether the "green" road is sustainable or whether the "golden" road must follow the "author pays" model, and so on. Our objective is to propose an alternative model, which will be described below.

## Open reuse in science

Merton (1973) showed that scientific endeavor is ruled by a set of *sui generis* sociological parameters—behavior norms that shape the ethos of science. These Mertonian norms, sometimes dubbed CUDOS for communalism, universalism, disinterestedness, and organized skepticism, apply in scientific communication.

As a by-product of the communalism[20] and disinterestedness[21] norms, scientists are not concerned about monetary compensation when they intend to publish an article in a respected journal. On the contrary, they often have to pay, and this would be even truer under the "golden" road's "author-pays" model. The goal of scientists is to obtain the greatest possible impact by being cited the greatest number of times by others. Their professional prestige and, in turn, income, power, and influence depend on this, and, in order to achieve their goal, they are willing to surrender their articles at no cost, with the sole condition of being cited. Therefore, if their articles are published in various journals and monographs, they will not raise any objection. On the contrary, every time one of their articles is published, their chances of being read and subsequently cited increase.

Certainly, in some cases, articles are cited to refute a specific idea instead of to support it; it is just as certain, however, that a citation generally represents a reading recommendation and, therefore, intellectual support. In this way, different computing platforms have been developed, based on the Thomson Reuters Web of Knowledge[22] model, that show how many and which articles have cited a specific text we are interested in. Examples of this type are Citebase,[23] Citeseer,[24] Spires,[25] or Google Scholar.[26]

Following Stevan Harnad (2001), the model that I propose also argues that scientists self-archive their papers in their institution's OA repositories (IRs) or in successful subject-based repositories (SRs) such as arXiv[27] that have reached critical mass. The self-archived articles and critical reviews or counter-criticisms that are sometimes uploaded in SRs[28] can form a network of citations that add luster to different discussions,[29] whether these are groups of articles, experimental data, observations, or commentaries.

In the future, the repositories that have better technologies of recommendation will better sift "the wheat from the chaff"; the good articles will be separated from the bad ones using a combination of different methods of counting downloads, citations, co-citations, tags, and so on. In this way the network of articles will become denser and denser as more and more articles are added; at the same time, however, similarities and relationships between the different texts that were not readily apparent at first will emerge. The ancient tree of knowledge with its various branches will become a veritable jungle where the disciplines (branches) will begin to disappear within the thick undergrowth and discussions focused on resolving problems from various viewpoints will emerge.

My proposal is that authors be permitted to publish the articles they have self-archived in IRs or SRs as many times as they wish and in as many journals as they consider appropriate, so that what these journals publish need not be seen as discipline-specific articles that generally have no relationship to each other, but rather as unified discussions pointing to the resolution of a specific problem. In this way, even though authors will continue to submit their articles to academic journals through the traditional procedure, the most astute journals will "fish" the digital repositories for what they consider the best articles for their readers. Consequently, their added value will lie more in presenting discussions than in publishing a mass of (often) unrelated articles. Innovative journals whose editorial boards or knowledgeable elite make the right choice and group articles and comments related to resolving common problems, thereby forming optimal discussions on a particular theme, will be rewarded. Competition among journals would be enhanced.

Different journals will be able to choose identical articles, if they so desire, since their added value will reside in showing the relationships between them. In other words, there will be no exclusivity contracts for these articles. There will be Creative Commons licenses, a kind of copyleft instead of copyright. Electronic journals could reuse the articles as many times as they wish, if this is of benefit to the discussion they publish. Peer review of each article or commentary could continue as it has until now or could open up to new and more democratic formulas, but the editorial board would have the additional responsibility of constructing an epistemological building with those articles. These are the advantages that digital technology and the Internet offer us, and this is the value we should demand of twenty-first-century science.

My proposal, then, challenges the Ingelfinger Rule, a policy promulgated in 1969 by the editor of the *New England Journal of Medicine*, Franz J. Ingelfinger (1969), to protect journals from publishing material that had already been published and, thus, lost its originality. As Lawrence K. Altman (1996b: 1459) has pointed out, however, "many people overlook the fact that Ingelfinger's economic motivation for imposing the rule was, as he said, a 'selfish' concern for protecting the copyright." Altman (1996a, 1996b) has shown that, far from being epistemically motivated, F. J. Ingelfinger's primary objective was commercial in nature, namely, to keep the mass media and other journals from publishing articles that the authors wanted to publish first in the *New England Journal of Medicine*. As a result, under his mandate, subscriptions to the journal doubled between 1969 and 1977 and almost tripled in 1996.

Because my Open Reuse proposal is based on OA to scientific articles published in digital repositories that are accessible at no cost via the Internet, it would survive only if the journals are innovative both in information technology and in "technologies of knowledge," so as to add value when reusing the content. Therefore, they should purge their technologies of recommendation, so as to choose good discussions, and they should know how to reuse or assemble the articles in digital repositories in an innovative manner. Readers will then look for discussions rather than unrelated articles. The articles, commentaries, data, videos,[30] blogs, or wikis that make up these discussions are the various "fish" that would be "fresh caught" in the digital repository fishing ground (at no cost) which, once treated and preserved in the journals' market (OA or paid), would become a properly cooked "fish dish." In this way, even though the repositories are OA, journals would be compensated financially through the subscriptions of scholars who wish to read what the gurus in their field consider relevant discussion for resolving a problem under consideration. Thus, under this Open Reuse proposal, articles could be used on more than one occasion and by more than one journal—no longer being loose objects but becoming part of the discussions of a twenty-first-century science that will be written and read differently from how we write it and read it today.

## Notes

This work has been undertaken thanks to funds coming from the Spanish Ministry of Science and Innovation as part of the research project FF12008-03599 – Philosophy of Human and Social Technosciences.

1. http://www.amazon.com/.
2. Now called Delicious.com.
3. Not all initiatives have followed the same model. There are websites that allow links or digital objects to be tagged and read by anyone, and there are other websites where tags may be read by the general public but only authors are permitted to tag their own objects.
4. http://technorati.com/.
5. http://www.flickr.com/.
6. http://www.youtube.com/.
7. http://www.diigo.com/.
8. http://www.librarything.com/.
9. http://www.citeulike.org/.
10. http://www.connotea.org/.
11. http://www.2collab.com/.
12. http://www.bibsonomy.org/.
13. Creative Commons defines the spectrum of possibilities between full copyright and the public domain, from *all rights reserved* to *no rights reserved*. Therefore, it is a "some rights reserved" copyright licence. More information is available here: http://creativecommons.org/.
14. For a definition of *copyleft*, see http://www.gnu.org/copyleft/. From here on, I will use the term "copyleft" meaning all these kinds of "loose" licenses as opposed to copyright.

15. http://www.barnesandnoble.com/.
16. For an executive summary, see PIRA International, University of East Anglia and KnowledgeView (2000).
17. Directive 2003/98/EC of the European Parliament and of the Council, of November 17, 2003, on the reuse of public sector information, *Official Journal of the European Union*, December 31, 2003, L345/90–L345/96, available (in various languages) at http://www.epsiplus.net/reports/european_directive_on_psi/directive_2003_98_ec.
18. Commission staff working document accompanying the Communiqué from the Commission to the European Parliament, the Council, the European Economic and Social Committee, and the Committee of the Regions on the re-use of Public Sector Information, Review of Directive 2003/98/EC [COM(2009) 212 final], available at http://eur-lex.europa.eu/LexUriServ/LexUriServ.do?uri=SEC:2009:0597:FIN:EN:PDF.
19. Actually, the model should be called the "author-proxy-pays" model.
20. Communalism means that scientific results are the common property of the entire scientific community.
21. Disinterestedness means that the results scientists present should not be mingled with their financial interests, personal beliefs, or activism for a cause.
22. http://isiwebofknowledge.com/.
23. http://www.citebase.org/.
24. http://citeseer.ist.psu.edu/.
25. http://www.slac.stanford.edu/spires/hep/.
26. http://scholar.google.com.
27. http://arxiv.org/.
28. See, for instance, http://arxiv.org/abs/quant-ph/0606092.
29. See http://arxiv.org/abs/quant-ph/0607101 and http://arxiv.org/abs/quant-ph/0610174.
30. See, for instance, the Journal of Visualized Experiments: http://www.jove.com/.

## Bibliography

Adair, W. C. 1955. "Citation Indexes for Scientific Literature." *American Documentation* 6 (1955): 31–2. Available at http://www.garfield.library.upenn.edu/papers/adaircitationindexesforscientificliterature1955.html, accessed on May 15, 2010.

Altman, Lawrence K. 1996a. "The Ingelfinger rule, Embargoes, and Journal Peer Review – Part 1." *The Lancet* 347 (9012): 1382–86.

Altman, Lawrence K. 1996b. "The Ingelfinger rule, Embargoes, and Journal Peer Review – Part 2." *The Lancet* 347 (9013): 1459–63.

Corbin, Christopher. 2009. "EC Communication on the PSI Re-Use Directive: PSI Re-Use Stakeholder Reaction." *European PSI Platform* (Topic Report 3). Available at http://www.epsiplus.net/topic_reports/topic_report_no_3_ec_communication_on_the_psi_re_use_directive_psi_re_use_stakeholder_reaction, accessed on May 15, 2010.

Garfield, Eugene. 1957. "Breaking the Subject Index Barrier: A Citation Index for Chemical Patents." *Journal of the Patent Office Society* XXXIX (8), (August): 583–95. Available at http://www.garfield.library.upenn.edu/es- says/v6p472y1983.pdf, accessed on May 15, 2010.

Garfield, Eugene. 1955. "Citation Indexes for Science: a New Dimension in Documentation through Association of Ideas." *Science* 122, (3159) (July 15): 103–11.

Available at http://www.garfield.library.upenn.edu/essays/v6p468y1983.pdf, accessed on May 15, 2010.

Guédon, Jean-Claude. 2001. "In Oldenburg's Long Shadow: Librarians, Research Scientists, Publishers, and the Control of Scientific Publishing." Available at http://www.arl.org/resources/pubs/mmproceedings/138guedon.shtml, accessed on May 15, 2010.

Guédon, Jean-Claude. 2006. "Open Access: a Symptom and a Promise." In *Open Access: Key Strategic, Technical and Economic Aspects*, ed. Neil Jacobs. Oxford, UK: Chandos Publishing, 27–38.

Harnad, Stevan. 2001. "The Self-archiving Initiative." *Nature* 410: 1024–25.

Ingelfinger, Franz J. 1969. "Definition of 'Sole Contribution.'" *The New England Journal of Medicine*, 281: 676–7.

House of Commons—Science and Technology Committee. 2003–04. *Scientific Publications: Free for all?* Tenth Report of Session I: Report. London: The Stationery Office Limited. Available at http://www.publications.parliament.uk/pa/cm200304/cmselect/cmsctech/399/399.pdf, accessed on May 15, 2010.

Knuth, Donald E. 2003. "Donald Knuth's Public Letter to Fellow Members of the Editorial Board of the *Journal of Algorithms*." October 25. Available at http://www-cs-faculty.stanford.edu/~knuth/joalet.pdf, accessed on May 15, 2010.

Merton, Robert K. 1973. *The Sociology of Science: Theoretical and Empirical Investigations.* Chicago: University of Chicago Press.

O'Reilly, Tim. 2005. "What Is Web 2.0? Design Patterns and Business Models for the Next Generation of Software." Available at http://www.oreillynet.com/lpt/a/6228, accessed on May 15, 2010.

PIRA International. 2000. *Commercial Exploitation of Europe's Public Sector Information: Final Report for the European Commission.* Leatherhead: PIRA International. October 30. Available at ftp://ftp.cordis.lu/pub/econtent/docs/commercial_final_report.pdf, accessed on May 15, 2010.

PIRA International, University of East Anglia and KnowledgeView. 2000. *Commercial Exploitation of Europe's Public Sector Information: Executive Summary.* Luxembourg: Office for Official Publications of the European Communities. September 20. Available at ftp://ftp.cordis.lu/pub/econtent/docs/2000_1558_en.pdf, accessed on May 15, 2010.

Stallman, Richard M. 2002. *Free Software, Free Society: Selected Essays of Richard M. Stallman.* Boston, MA: Free Software Foundation. Available at www.gnu.org/philosophy/fsfs/rms-essays.pdf, accessed on May 15, 2010.

Toffler, Alvin. 1980. *The Third Wave.* New York: Bantam Books.

Weiss, Peter N. and Peter Backlund, 1997. "International Information Policy in Conflict: Open and Unrestricted Access versus Government Commercialization." In *Borders in Cyberspace: Information Policy and the Global Information Infrastructure*, eds. Brian Kahin and Charles Nesson. Cambridge, MA: MIT Press, 300–21.

# 8.1

# Democratizing the Science of Risk Management – An End-User-Driven Approach to Managing Risks to Drinking Water Systems in First Nations Communities

*Khosrow Farahbakhsh and Benjamin Kelly*

First Nations communities across Canada have historically faced numerous challenges. Over the past several years, the issue of reliable and safe drinking water has been particularly stressful for a growing number of aboriginal populations. According to a report from Health Canada (2009), the frequency of drinking water advisories in First Nations communities has risen from approximately 130 in 2003 to over 250 in 2007.

"Experts" have traditionally conceptualized risk reduction associated with water contamination by way of scientifically driven notions of risk and what they conceive as a manageable and acceptable community tolerance of risk. Approaching risk from the expert's (scientists, policy makers, etc.) point of view often ignores the fact that risk as a whole is inferred, complex, multidimensional, and ambiguous, and plagued with uncertainty. Such views often focus on "rational" and "objective" aspects of risk, ignoring more subjective and community-specific knowledge systems. The risk assessment tools generated by experts, influence policy makers and these standards then determine how limited financial resources are best distributed among "at risk communities." Usually local knowledge is not integrated into these models, thus increasing the potential for improper assessment of risk. This lack of inclusion may significantly harm First Nations communities.

Risk assessment tools generally suffer from a number of limitations; most evident are the conflicting priorities and the social-cultural disconnect that emerges between those who develop and administer the tools and those mostly affected by them. Specifically, these limitations represent conflicting definitions of what is considered "acceptable risk." What may be acceptable for administrators and experts may not be acceptable to community members and vice versa. Most risk assessment tools focus on "hazards" and their potential routes of community exposure. Hazards are usually defined by experts as chemical and biological contaminants that can lead to some

195

type of waterborne illness. For community members, taste and odor may be a far more important indicator of risk than chemical and biological contaminants.

In a study conducted by one of the authors, various risk assessment tools were evaluated from the perspective of First Nations end-users—operators, public work personnel and others. Although criteria identified by First Nations end-users corresponded reasonably well with those identified in the scientific literature, they did diverge on a number of important culturally specific non-technical issues. Other uses of water such as traditional and spiritual uses and fishing that were identified as important by First Nations participants are generally absent in the scientific literature. As a result, none of the risk assessment tools include assessment of water availability and safety for traditional uses. Most risk assessment tools treat drinking water in isolation, whereas according to the First Nations participants, drinking water should be assessed in conjunction with watershed, wastewater, housing, and land use.

Considering that end-users' perspectives and priorities are diverse, inclusion of these priorities into a risk assessment tool will result in an overly complicated and cumbersome tool. Even using a fully integrated tool, assuming one exists, may not ensure safety of water supplies, as risk assessment does not necessarily result in effective risk management. Risk management that recognizes expert–lay collaboration requires a model of knowledge translation that builds capacity among end-users to manage their own risk (see Kelly and Farahbakhsh 2008). In general, risk assessment tools are not concerned with capacity development. A bottom-up approach to risk management that includes end-users' priorities and perspectives and involves building the capacity of the end-users may be a more beneficial way of ensuring safer water supplies in First Nations communities. Such an approach should empower communities to understand risk as it relates to both lay and expert epistemologies, therefore promoting positive decisions based on a sound understanding of drinking water, both technically and culturally. An end-user-driven bottom-up approach may help identify risk management processes that are flexible and representative of the diversity of First Nations communities. The "one size fits all" approach may not be suitable in managing water-related risks and ensuring safety of water supplies.

A more democratic approach to improving drinking water safety may eventually involve a shift from more rigid risk assessment tools to an approach that enables communities to develop their own plans for water safety. This seems to be the direction now promoted by the World Health Organization. A democratic, bottom-up approach to the development of a water safety plan would empower communities to choose acceptable levels of risk instead of having risks imposed from above. If adequate resources were supplied to build capacity in communities and enable development of proactive water safety plans, these plans could be used to direct funds towards building safer water systems based on each communities' vision for the future.

## Bibliography

Health Canada. 2009. 'Drinking Water Advisories in First Nations Communities in Canada – A National Overview (1995–2007). http://www.hc-sc.gc.ca/fniah-spnia/alt_formats/pdf/pubs/promotion/environ/2009_water-qualit-eau-canada/2009_water-qualit-eau-canada-eng.pdf. Accessed February 15, 2010.

Kelly, Benjamin and Khosrow Farahbakhsh. 2008. 'Innovative Knowledge Translation in Urban Water Management: An Attempt at Democratizing Science'. *International Journal of Technology, Knowledge and Society* 4: 73–83.

# 8.2
# The Public Debate on Science and Technology: Transgenic Corn in Mexico

*Julio E. Rubio*

The recent debate over transgenic corn in Mexico has brought to light the different social agents that participate in the public debate on science and technology.

The first social agent is the scientific community. The public debate over transgenic corn in Mexico arrived to international discussion forums with an article on the presence of transgenic corn in Oaxaca, which was published in 2002 by Berkeley scientist Ignacio Chapela in the prestigious journal *Nature*. The article has sparked two fundamental debates. The first concerns the validity of Chapela's conclusions. The second concerns the potential effects of transgenic plants on Mexico's cereals and native species. In 2002, *Nature* revoked the findings of the article on the basis of methodological inconsistencies. In spite of this, the article's findings had already had an impact on the scientific community. The Autonomous National University of Mexico (UNAM) and the Center for Research and Advance Studies of the National Polytechnic Institute (Cinvestav), for instance, published articles confirming the presence of transgenic corn in Oaxaca.

The second social agent is the government. Politics provides a space for negotiating society's public decisions. Government's role in science and technology is to mediate between social agents and regulation by means of laws. The Mexican government has assumed a cautious role in the debate over transgenic corn. The government's first intervention was the creation in 1999 of the Interdepartmental Commission on the Biosecurity of Genetically Modified Organisms (CIBIOGEM) as a means of regulating the use of biotechnology, especially genetic technologies. The legislative branch has also taken measures to regulate the use of biotechnology in Mexico. In 2005 the Mexican Congress approved the Biosecurity of Genetically Modified Organisms Act.

The third social agent is civil society. Civil society has played an important role in the debate over transgenic corn. In fact national and international NGOs have taken the lead. The principal agent representing the civil society has been Greenpeace, which has demonstrated a great capacity for coordinating actions of diverse types in different cities around the country, including making statements, and spreading information and civilian resistance. For example, Greenpeace has published in newspapers that Mexico imports 6 million tons of corn per year, 25–30 percent of which is transgenic. Greenpeace has also elaborated and published a list of foodstuffs containing transgenic corn. Civilian resistance has also been a very effective been strategy for Greenpeace. For instance, some members once tied themselves to the anchors of ships bound for Mexico that contained transgenic products, causing the boats to return to the US. Finally, in order to fortify its social capital, Greenpeace has used its alliance with other civil organizations like the National Association of Farm Products Merchandisers (ANEC), National Union of Indigenous Regional Farming Organizations (UNORCA) and the Action Group on Erosion, Technology and Concentration in order to sue for CIBIOGEM not to lift the moratorium on transgenic experimentation.

The final social agent is the business sector. Business has played an important part in the debate over transgenic corn, although it has maintained a discrete presence. The most important companies in the field, which together make up the association Agro-Bio, are Monsanto, DuPont, Syngenta, Aventis, and Savia. These companies have made their position clear. For example, in Colombia, Agro-Bio published an extensive investigation of the use transgenic corn, 'Genetically Modified Maize,' which defends the use of transgenic foods on the basis of serious scientific studies but also reveals the organization's special interests.

So these four domains of social actors play important and different roles in the public debate on science and technology, each with their own agenda and interests. For a full account of the relationship between society, science, and technology all four domains need to be included, and critically analyzed in any analysis.

## Bibliography

Barkin, David. 2006. "Building a Future for Rural Mexico." *Latin American Perspectives* 33 (2): 2006–12.

Chaturvedi, Sachin. 2002. "Agricultural Biotechnology and New Trends in the IPR Regime: Challenges before Developing Countries." *Economic and Political Weekly* 37 (13) (March 30–April 5): 1212–22.

Chaturvedi, Sachin. 2001. "Continued Ambiguity on GMOs." *Economic and Political Weekly* 36 (42) (October 20–26): 3981.

Guadagnuolo, R., J. Clegg and N. C. Ellstrand. 2006. "Relative Fitness of Transgenic vs. Non-Transgenic Maize × Teosinte Hybrids: A Field Evaluation". *Ecological Applications* 16 (5) (October): 1967–74.

Herrera-Estrella, Luis. 1999. "Transgenic Plants for Tropical Regions: Some Considerations about their Development and their Transfer to the Small Farmer." *Proceedings of the National Academy of Sciences of the United States of America* 96 (11) (May 25): 5978–81.

Kaiser, Jocelyn. 2005. "Calming Fears, No Foreign Genes Found in Mexico's Maize." *Science* 309 (5737) (August 12): 1000.

Lele, Uma. 2003. "Biotechnology: Opportunities and Challenges for Developing Countries." *American Journal of Agricultural Economics* 85 (5), Proceedings Issue (December): 1119–25.

Lofstedt, Ragnar E., Baruch Fischhoff and Ilya R. Fischhoff. 2002. "Precautionary Principles: General Definitions and Specific Applications to Genetically Modified Organisms." *Journal of Policy Analysis and Management* 21 (3) (Summer): 381–407.

Mann, Charles C. 2002. "Has GM Corn 'Invaded' Mexico?" *Science*, New Series 295 (5560) (1 March): 1617–19.

Mann, Charles C. 2002. "New Law Could Turn Scientists into Outlaws." *Science*, New Series 296 (5573) (May 31): 1591.

Martínez-Soriano, Juan Pablo Ricardo and Diana Sara Leal-Klevezas. 2000. "Transgenic Maize in Mexico: No Need for Concern." *Science*, New Series 287 (5457) (February 25): 1399.

Ortiz-García, S., E. Ezcurra, B. Schoel, F. Acevedo, J. Soberón, A. A. Snow and Barbara A. Schaal. 2003–4. "Absence of Detectable Transgenes in Local Landraces of Maize in Oaxaca, Mexico." *Proceedings of the National Academy of Sciences of the United States of America* 102 (35).

Pilson, Diana and Holly R. Prendeville. 2004. "Ecological Effects of Transgenic Crops and the Escape of Transgenes into Wild Populations." *Annual Review of Ecology, Evolution, and Systematics* 35: 149–74.

Qaim, Martin. 1999. "Potential Benefits of Agricultural Biotechnology: An Example from the Mexican Potato Sector." *Review of Agricultural Economics* 21(2) (Autumn–Winter): 390–408.

Raven Source, Peter H. 2005. "Transgenes in Mexican Maize: Desirability or Inevitability?" *Proceedings of the National Academy of Sciences of the United States of America* 102 (37) (September 13): 13003–4.

Snow A. A., D. A. Andow, P. Gepts, E. M. Hallerman, A. Power, J. M. Tiedje and L. L. Wolfenbarger. 2005. "Genetically Engineered Organisms and the Environment: Current Status and Recommendations." *Ecological Applications*, 15 (2) (April): 377–404.

Stone, Glenn Davis. 2002. "Both Sides Now: Fallacies in the Genetic Modification Wars, Implications for Developing Countries, and Anthropological Perspectives." *Current Anthropology* 43 (4) (August–October).

Vogel, Gretchen. 2004. "Europe Takes Tentative Steps toward Approval of Commercial GM Crops." *Science*, New Series 303 (5657) (23 January): 448–9.

# 8.3
## Evolving Publishing Practices in Mathematics: Wiles, Perelman, and arXiv

*Manuel González Villa*

The 25th International Congress of Mathematicians held in Madrid in 2006 confirmed that the Russian mathematician Grisha Perelman, who was awarded a Fields Medal, had solved the Poincaré Conjecture (PC). In the process of solving this major mathematical problem, Perelman has also helped to create significant changes in publishing practices for mathematicians.

The proof of the PC was without doubt the greatest mathematical breakthrough since the proof of the Fermat's Last Theorem (FLT) by the British mathematician Andrew Wiles in the mid-1990s. As some commentators have noted (Jackson 2007; Nasar and Gruber 2006), there are several parallelisms between the stories of these two achievements.

Both problems were stated long before their solutions and the problems were already settled for many particular cases. The basic strategy of proof had already been proposed. However, both proofs required the highest originality and astonishingly deep technical skills. Both Wiles and Perelman had to test their tenacity and perseverance, working in isolation and almost secretly, for seven years. Finally, Wiles and Perelman became firm candidates for the Fields Medal, the most famous mathematical award that recognizes mathematicians younger than 40, when they announced their work.

On the other hand, the decade that separates FLT and PC had seen dramatic changes in the way mathematicians disseminate their results.

Wiles presented his proof in a series of lectures at the Newton Institute (Cambridge, UK) on June 1993. Later he submitted a 200-page manuscript to the journal *Inventiones Mathematicae*. However, the referees found a gap and Wiles had to rework the proof. The revision lasted longer than a year and expectations grew everyday. Wiles then communicated his decision not to circulate the manuscript until he had fixed up the issues raised by the referees. Wiles solved all difficulties in September 1994 and published his results in two papers in *Annals of Mathematics* in May 1995.

In June 1993 it was widely believed that Wiles would pick up a Fields Medal the next year at the International Congress of Mathematics (ICM) because, even without a full proof of FLT, his amazing progress might deserve recognition. However, the committee did not award Wiles. The reason might be found in the delay of the final proof and the lack of a formal publication. Four years later Wiles was too old for a Fields Medal, namely, over 40.

Perelman has been presented as the classic absent-minded genius, who retired and devoted seven years to resolving an obscure mathematical conundrum. He has been described as a hermit who feeds himself only on bread, milk, and cheese and let his hair and fingernails grow without limit. Newspapers have highlighted his disinterest in money, prizes, and jobs.

At the time he was awarded the Fields Medal, he had resigned from his job at the Steklov Institute, and was living with his mother in a sparsely furnished apartment in a modest suburb of Moscow. He rejected the Fields Medal and did not seem to care about the $1 million Millennium Prize the Clay Institute has offered him in March 2010.

Perelman's behavior is so striking that psychologists have been speculating that he might suffer from a disorder called Asperger's Syndrome (Gessen, 2009), which is also compatible with his brilliant intelligence.

Nevertheless, Perelman had been extremely modern in the way he communicated his work. Starting in November 2002, Perelman uploaded three preprints on arXiv, an Internet repository for preprints on physics and mathematics founded by Paul Ginsparg in 1991, and contained more than 600,000 preprints up to May 2010. It is funded by the National Science Foundation and Cornell University, and has an annual budget of $400,000 for 2010.

Surprisingly, Perelman did not submit any of these preprints to journals. Some authors have interpreted this as a sign of his self-imposed isolation.[1] But Perelman knew the identity of those mathematicians who were interested in the PC and were able to recognize the importance of his contributions. When he posted his first preprint, he warned them by sending an e-mail. Meanwhile, uploading documents onto arXiv was becoming a standard way of communication among mathematicians,[2] but mathematicians still used to send their preprints to journals besides uploading them on arXiv.

Perelman's decision was considered outrageous. Taking into account the importance of the results, Perelman took a considerable risk: "If the proof was flawed, he would be publicly humiliated, and there would be no way to prevent another mathematician from fixing any errors and claiming victory" (Nasar and Gruber 2006).

In April 2003 Perelman went back to US to lecture on his work at several universities. He posted his third preprint in June and let people study and judge his papers.

Usually, the warranties for the correctness and the value of a paper rely on its publication in a research journal. The editors of the journal ask peer experts to referee and judge the manuscript to help them in making a final decision. This is why many mathematicians wondered who was going to check Perelman's proof and whether the mathematical community was going to accept it, provided that the proof turned out to be right.

After Perelman posted his first preprint, at least three groups (Kleiner and Lott 2008; Morgan and Tian 2007 and Cao and Zhu 2006) spontaneously started the task of understanding Perelman's work and checking all its details.[3] It was a long and hard project, partly because of the technical difficulty and partly because of the brief style of Perelman's preprints. These projects were so efficient that the Fields Medal Committee saw no problem in awarding the medal to Perelman in 2006.

So we have seen major differences in the way Wiles and Perelman communicated their results. We do not aim to discuss the reasons behind them. We prefer to point out how Perelman has revolutionized the practice of mathematical publication. Perelman showed us how the new ways of publication, based on the web and outside of the traditional journals, might be both efficient and reliable.

## Notes

1. Gessen (1998) has suggested another point of view: "His decision to post his proof on the arXiv had been an intentional revolt against the very idea of scientific journals distributed by paid subscription."
2. According to arXiv (http://arxiv.org/Stats/hcamonthly.html), more than 400 new preprints on mathematics were posted monthly during 2002.
3. Kleiner and Lott wrote notes as they studied the manuscripts and updated them on the following webpage: http://math.berkeley.edu/~lott/ricciflow/perelman.html.

## Bibliography

Cao, H. D. and X. P. Zhu. 2006. "A Complete Proof of the Poincaré and Geometrization Conjectures—Application of the Hamilton-Perelman Theory of the Ricci Flow." *Asian Journal of Mathematics* 10 (2): 165–492.

Gessen, M. 2009. *Perfect Rigor: A Genius and the Mathematical Breakthrough of the Century*. New York: Houghton Mifflin Harcourt.

Hamilton, R. 2006. "The Poincaré Conjecture," *Plenary Lecture 25th ICM Madrid*. Abstract at http://icm2006.org/v_f/AbsDef/Invited/hamilton.pdf, accessed January 21, 2007.

Jackson, A. 2007. "Two Landmarks, Two Heroes," *Notices of the AMS* 54 (9): 1117, at http://www.ams.org/notices/200709/tx070901117p.pdf, accessed July 6, 2008.

Kleiner, B. and J. Lott 2008. "Notes on Perelman's Papers," *Geometry and Topology* 12; 2587–2855, doi:10.2140/gt.2008.12.2587, arXiv:math/0605667.

Lott, J. 2006. "The Work of Grigory Perelman," in M. Sanz-Solé, J. Soria, J. L. Varona and J. Verdera (eds.) *Proceedings of the International Congress of Mathematicians Madrid 2006*. Volume I. Plenary Lectures, European Mathematical Society Publishing House, 66–76.

Morgan, J. W. and G. Tian 2007. "Ricci Flow and the Poincaré Conjecture," *Clay Mathematics Institute*. See also "Ricci Flow and the Poincaré Conjecture," arXiv: math/0607607.

Morgan, J. W. 2006. "The Poincare Conjecture," in M. Sanz-Solé, J. Soria, J. L. Varona and J. Verdera (eds.) *Proceedings of the International Congress of Mathematicians Madrid 2006*. Volume I. Plenary Lectures, European Mathematical Society Publishing House: 713–36.

Nasar, S. and D. Gruber. 2006. "Manifold Destiny," *The New Yorker* (August 20) at http://www.newyorker.com/printables/fact/060828fa_fact2, accessed December 12, 2006.

Perelman, G. 2002. "The Entropy Formula for the Ricci Flow and its Geometric Applications." arXiv:math/0211159 (November).

Perelman, G. 2003. "Ricci Flow with Surgery on Three-manifolds." arXiv:math/0303109 (March).

Perelman, G. 2003. "Finite Extinction Time for the Solutions to the Ricci Flow on Certain Three-manifolds." arXiv:math/0307245 (July).

Sanz-Solé, M., J. Soria, J. L. Varona and J. Verdera (eds.) 2006. *Proceedings of the International Congress of Mathematicians Madrid 2006*. Volume I. Plenary Lectures, European Mathematical Society Publishing House.

Sigh, S. 1998. *Fermat's Enigma: The Epic Quest to Solve the World's Greatest Mathematical Problem*. New York: Anchor Books.

# Part IX
# Digital Aesthetics

# 9
# Fabrication

*Sean Cubitt*

There is no single digital aesthetics. The term is plural. There is an aesthetics of code, and industrial design (Gelernter 1998), of specific software packages like Flash (Munster 2003), of programming and interface design (Fishwick 2006), of sound (Dyson 2009), of viruses (Parikka 2007). And yet there is, intuitively, something which seems to draw all these together. Now the majority of domestic media are digital, from radios to TVs, cameras to mobiles, and the vast majority of printed materials; now our vehicles, houses and furniture are designed in CADCAM systems; now we no longer make images look digital as a special effect; now we are ready to think about what is common, even if it is not essential to digital media.

One term, admittedly but usefully ambiguous, that may help understand some of these qualities is "fabrication." Although computers are still manufactured in factories by factory hands (the word "manufacture" meaning "made by hand"), many of their components, including the most important functioning parts, the chips, are no longer manufactured but fabricated. After a short introduction, this chapter moves on to consider the implications of this process of making digital devices for the way we use them. Fabrication also has the sense of a story made up to cover over something else: we accuse politicians of fabricating stories. Here is a second meaning of fabrication in digital aesthetics: our species' (*homo faber*) tendency to narrate, to stitch data together in time-based structures which explain them to us. A third sense comes from a false etymology: the weaving of fabrics. From web-weaving to the archeology of the Jacquard loom, the concept of the digital as a net embroidered invisibly over the world presents a third vantage on the aesthetics of digital media. There is always the problem, however, of concentrating so hard on the present that we cannot see its continuities and discontinuities with the immediate and the more distant past. To begin with then, a thesis on history.

## The mode of representation

Although the world has moved on since the question of representation was the burning topic of 1980s media studies, the question still remains.

It is a truism to say that representing inevitably implies a reduction of the thing represented. A tree is a complex, evolving organism of many scales: a photograph is a relatively simple, flat, unchanging object of only one size and shape. Erecting a philosophy of tragic loss on this banal observation, as thinkers from Heidegger (1977) to Baudrillard have done, is perhaps an excessive reaction: the world remains, however many photographs we take. Aesthetics, the topic of this chapter, historically referred to the realm of the senses. In our visual arts, where aesthetics has flourished in recent centuries, the senses are reduced to one or two: seeing, and to a lesser extent, hearing, with language hived off into a separate domain. The pedigree of the wider "sensorium" should alert aestheticians to the historical problem inherent in the unexceptional difference between the world and our pictures of it. There is no single difference, but a vast sea of them: a pencil drawing is not a drawing with brush and ink; a 35mm slide is not a digital photograph; and the instantly recognizable gestures of a Klee, a Picasso or a Cézanne distinguish them as much as the motifs they depict or the intentions we infer and the meanings we extract from their works.

At the same time, there are tides in the history of humankind which lift the individuals, their media, and their very intentions, as boats rise with the tide. However different from one another, Cézanne, Picasso and Klee clearly belong to a different period than Dürer and Raphael: there are commonalities of technique, motif and ways of earning a living shared by people of a generation, or an age. There is, in short, not only representation, nor a simple plural: representations. There have been major modes of representation, capable of extraordinary richness, but nonetheless exhibiting a certain coherence over periods of hundreds of years. Periodization is a perilous path, and any essay in periodizing throws up exceptions to the rule. But it is worth adventuring some such broad map in the present context, because what is at stake, digital aesthetics, is a specifically digital mode of representation. Therefore I hazard the following: the medievals, with their heraldry (Pastoureau 1987) and doctrine of signatures (Foucault 1970), their religious calendar and closeness to the seasons, inhabited and worked within a system governed by semantic relations between depictions and the world. Note that we are not speaking exclusively of those things we now store in art galleries, but of a whole visual culture comprising architecture, banners, uniforms, carved furniture and graffiti. The Enlightenment occupied a visual culture of realism. In science and art, it pursued representations of the world as it appears, and later, with the Romantics, of the appearance of the world in a human eye. Since the nineteenth century, with its new photographic and printing techniques and its culture of technical drawing, we have been struggling towards some new mode of representation, which appears to come clear in the digital era: an arithmetic accounting of the world.

A typical chip in widespread use for digital imaging is the CCD or charge-coupled device. The CCD chip of a digital camera comprises a p-doped (positively charged) thin crystalline lattice deposited on a transmitting layer. Light arrives from the lens onto the lattice, each cell of which acts as a capacitor accumulating an electric charge according to the amount of light, or more specifically the luminance, arriving at that cell. The charges at each pixel are in effect the latent image, like the undeveloped filmstrip in a traditional camera. The array is linked to a control circuit which, after exposure, instructs each capacitor to pass its charge on to its neighbor. The last capacitor in the array then passes its charge to an amplifier, which converts the charge into a voltage. The process is repeated till all the charges have been converted to voltage, digitized, sampled and stored by the underlying CCD semiconductor. The majority of modern CCDs use a buried-channel design, where areas of the silicon substrate are implanted with phosphorous ions giving them an n-doped (negatively charged) designation. These areas act as channels through which the electric charge generated by the light-sensitive upper layer will travel. The actual capacitor layer lies on top of this buried-channel layer. On top of them both is a layer comprising polysilicon gates, perpendicular to the channels. The channels are separated by oxides, which stop charge flowing from one channel to the next, while the gates control the flow of charge from the capacitors towards their destination.

Lying on top of the layers, immediately above the focal plane, lies a Bayer mask, which for every four pixels filters one red, one blue and two green. The fundamental set-up then is this: light is organized by the lens and filters, and gathered on a grid during the exposure time. The information in the grid is passed through the system of gates and channels in ordered array to its conversion, via voltage, to stored data. The charge-coupled device operates as a kind of clock. The exposure charges the lattice, but the charge is drained from it down ordered channels in lockstep units. The chip moves its data from spatial to temporal and back to spatial ordering. Without the clock function allied to the interlocking grids, the charges would mingle and pour out in no order at all, chaotically, as noise. The CCD imposes a very specific order, or a pair of orders, on the light which it gathers. This is characterized by whole-number steps of equal unit duration and area. The result is an array of discrete, ordered units. This is the arithmetic structure of digital imaging, its unit-steps the basis of all binary digital logic, and all digital interfaces with the world. Such hardware architectures structure digital image gathering and reproduction in scanners, microphones, still and video cameras; and in LCD, LED and plasma screens, data projectors and almost all digital printers. Even more than earlier standards, such as aspect ratios in the cinema, these formal standards are today hardwired into not only displays but also the motherboards of computers and wireless devices.

## Fabrication

There is a curious design feature of CCD chips. To make the orientation and structure of the crystal lattice identical to that of the underlying chip, the crystals are grown on the chip itself, whose carefully doped structure acts as a seed. This is the reason why their production is called "fabrication" rather than "manufacture": the scales are far, far smaller than human hands or tools. As growing snowflakes are catalyzed by a mote of dust, the crystalline perfection of the CCD requires an imperfection to start its growth. The orientation of the chip's crystal is a result of that invisible nanoscale flaw at a scale that becomes observable in its behavior. This step from contingent speck to formal organization is characteristic of digital media. It is a process of abstraction, not of divorce.

There is a modulation of the observational realism which dominated the Newtonian era from Galileo and the Renaissance through to Mendel and Darwin, all profoundly based on seeing. In digital instruments, among which we should count the CCD and similar fabricated chips, the objects to be accounted for often lie far beyond the visible spectrum: quantum events, astronomically ancient light, and radiation at extreme wavelengths. The typical displays are no longer the illusionistic field drawings of the great naturalists, or the meticulous pen-and-wash accounts of sunspots left by Copernicus and Galileo, but diagrams and graphs. The fundamental information being displayed is numerical, and where it is presented in pictorial form, it is the result of algorithms designed to squeeze non-visible light into the vocabulary of Red, Green and Blue that digital displays can handle.

It must be reiterated: two processes are involved here as they are in CCD chips. First, despite the tiny scale of individual pixels on a chip and the short duration of exposures – one twenty-fifth of a second for a video frame, for example – we are in the presence of quantum events occurring, literally, at the speed of light. The wavelengths and frequencies of incoming photons in their thousands are averaged over the exposure time and across the area of the pixel. This process of sampling is intrinsically one of averaging. But it is also, and simultaneously, a process of deriving from that average a unit number denoting the address of the pixel and the summed color to be associated with it. The process of averaging is probabilistic: fluctuations far from the norm of the incoming light are discounted as random (in the same way such chips must discount quantum effects occurring at their own surface) or assimilated to a statistical median. The random has no place in this accounting of the world. It has the same shape – perhaps Deleuze might have said the same diagram – as the mode of rule which Foucault described as biopolitical: the management not of individuals through discipline but of populations through actuarial principles of probability. Digital tools do not record, as photography and cinematography record: they measure and they manage light.

With this in mind it is easier to understand the realism inherent in digital media. The dominant media of the previous epoch may have been the illusionistic image and narrative. The dominant media of the twenty-first century are spreadsheets, databases and geographical information systems (GIS). None of these are anything but realist in aspiration, in the sense that they intend to give accounts of the world which can be used to do things in it. But what each has in common is a very particular arithmetic abstraction which turns time into space. The origins of the spreadsheet and database lie in ledgers used to lost transactions, or to give an account of holdings (as in the Domesday Book) over time or at a moment in time. The very form of the ledger meant that items were entered chronologically starting at page one. Reconstructing a year's accounts implied going through them in the order in which they occurred. But a spreadsheet no longer has this problem, and databases shed the old narrative order with the invention of the vertical filing cabinet in 1898, allowing records to be ordered not by time but by the alphabet or any other system. Electronic databases multiplied the modes of organization in an already organized system. Maps meanwhile have been with us in their modern form since Mercator. But the old maps belonged to the geometry of projection, which also dominated perspectival art: the new maps, especially since the introduction of postal (ZIP) code systems since the 1930s, have been associated with all other geographically organized data, like censuses, and the resulting GIS systems have become vital tools in population management, especially when wedded to the surveillance networks facilitated by satellite observation.

Digital images are of a kind with these workplace media. As we have seen, the CCD chip makes a record of average illumination over a pixel's expanse during an exposure, and stores the resulting charge in the form of digital integers. Most of the codecs (compression-decompression algorithms) in use in contemporary media use comparisons between frames to decide which areas (blocks, groups of blocks) in the frame are effectively the same from exposure to exposure, using that information to reduce the amount of information that has to be transmitted, and so the time and energy required to transmit it. Some codecs, like H.261 used by YouTube, and many mobile screens add a key frame function, nominating the beginnings and ends of "scenes" automatically by measuring major changes between two frames. Extrapolating from these beginning and end points, H.261 and similar technologies can reduce even further the fidelity of the image, on the principle that sports fans have come to see players and balls, not grass. This is why sometimes you will see cricket balls disappear momentarily, or birds appear suddenly in mid-pitch as if from nowhere; and why dark or otherwise consistent areas of color often appear blocky: the aggregations of pixels known as blocks have been averaged to the same color, so that the edges of the grid show where they meet another averaged tone.

The enumeration and averaging of light is of one kind with both workplace media and the fundamentals of digital hardware and software. Combined, they produce "good-enough" images, typical of the good-enough aesthetic of mass consumerism, and images designed on the informatic principle that information is difference. There is however a critical distinction between this principle and its earliest enunciation by Gregory Bateson (1973: 351): "a difference that makes a difference in some later state of affairs." The differences involved are not temporal but spatial, as in the case of spreadsheets, databases and GIS. Their differences are effectively equivalent one to another: one block or group of blocks, one pixel or one frame, can be exchanged for any other by simple mathematical functions. This foundation in exchange illuminates a further critical aspect of fabrication: its proximity to the commodity form.

## Fabricating

In extreme environments, such as high-orbital space, where the Hubble Space Telescope (HST) gathers light from ancient stars, extreme instruments are in use. The Hubble's photon counter does exactly what it says: it counts the smeared, old photons coming from the faintest and furthest objects in the sky as they arrive one by one. It is a strange destiny: light that has travelled thousands of light years arrives at the HST only to be transformed into charge, voltage and streams of transmitted data. Instruments like these are so finely tuned that they can, as in this case, and in the case of many devices designed to gather data about sub-atomic events, change the course of the events they measure (Heisenberg 1989). One definition of high-definition might then be: media which obliterate the events they record. The more familiar meaning of the term is the professional (and in a bastardized form the domestic) equipment used to produce imagery over 1 K resolution—where the K, for thousand, indicates the number of horizontal lines in a scan.

The most successful commercial "hi-def" camera is the Red One, which boasts a maximum 4.5K resolution, or 4480 × 1920 pixels (although in practice much of the data stream is required for non-visual information). This device is built around a 35 mm CMOS chip. Each pixel in a CMOS (complementary metal oxide semiconductor) chip has a light-sensitive JFET (junction gate field effect transistor) whose activation by incoming photons "pinches" the current flowing through the transistor from source to drain. Less noisy than CCD, because the JFET is completely drained of charge after each shot, and because of the different architecture of the pixel array, CMOS uses an Active Pixel Sensor (APS) architecture, in which the charge moves from the JFET light-sensitive element direct to a buffer transistor which acts as an amplifier, before passing it on to a third row-select transistor, which allows whole rows of pixels to be transferred to the read-out electronics where it is stored in memory as a digital array. This triggers the fourth transistor, a reset gate, which drains residual charge from the pixel prior to the next exposure. This combination of the movement of charge within the pixel itself along with the inclusion of an amplifier in every individual pixel was produced by Eric Fossum (2005) and his collaborators on the back of the mass-manufactured CMOS chip, used extensively in webcams and mobile phones. Far more widespread and thus more cheaply manufactured than CCDs, CMOS chips are characteristically not power-hungry, and therefore run cool and use less battery power, and have a low signal-to-noise ratio. CMOS chips use both positive and negative (p-type and n-type) semiconductors in pairs, which reduces the waste heat produced by other types of logic chips. In effect each pixel operates as if it were a whole CCD chip, but with less noise, and less of the blooming and smear effects associated with the quantum effects and transfer of charge across CCD chips. Perhaps equally significant is that CMOS delivers digital output, which requires only buffering before being delivered to its output, and the output is external to the chip itself, rather than integrated as it is in CCDs. But it remains the case that the size of pixels—currently well above the diffraction-limit of visible wavelengths—means that they must sample the light across their surface area and over the duration of exposure; that like CCDs, they are manufactured in discrete unit arrays; and that under low light conditions they will produce artifacts of virtual light through their built-in amplification of incoming photon signals.

The Red One captures data at a maximum rate of 42 megabytes per second: that is about one tenth of the sensitivity of the chip itself. The chipset produces, as mentioned above, digital data rather than voltage: the work of compression is accomplished in the chip itself, reducing the flow of data to rates manageable by inbuilt or external storage, since data has to be transported somehow, to make room for the next frame. This is the RAW format, nominally uncompressed. Circumventing this unwanted compression, to access the "latent image" gathered before the chip processes the incoming data (Flaxton 2009), is a demanding task which few undertake, but which points towards possible higher resolutions, some available in high-end professional equipment, more in military applications, up to at least 28K. The potential for deeply immersive depiction is intoxicating: true high-definition projection in the 4K-plus region gives startling verisimilitude, so much so that it raises once again the question of representation.

We need at this juncture to reconsider digital image gathering as image processing, and more generally as data processing. What are gathered are photons, whose wavelengths and frequencies are effectively measured, as in the Hubble photon counter, by being converted into charge. The subsequent in-camera processing (and further automated processing occurring as the data is first transferred to a computer, then opened in a particular editing software package, even before being manipulated by an end-user) abstracts from light a set of numerical data. This act of abstraction is the central fact of digital processing, which takes the ancient Pythagorean concept of mathematics as the language of nature that underpinned the realist epoch of classical science and makes of it a method for converting the flux of the physical universe into a stable matrix of integers.

Writing shortly before the mass marketing of digital cameras, Vilém Flusser defined photography (here I give only part of this definition) as "an image created and distributed automatically by programmed apparatuses in the course of a game necessarily based on chance" (Flusser 2000: 76). Automation is intrinsic to Flusser's photographic apparatus. Once designed, the camera operates according to the program written into its structure. This automation not only abstracts values from the world, but reconstructs the world as information (Flusser 2000: 39). Following Shannon and Weaver's (1949) mathematical definition of information as a ratio between probabilities, Flusser sees the camera seizing not the world but abstract "state of things" data. Information depends on the balance between repetition and novelty. The human user and the world the camera observes only add improbability, chance, to the mix, increasing the amount of data which it can convert into photographs. In an increasingly organized world, in which neo-liberal economics have already turned almost every human activity into markets, capital finds itself less and less capable of perpetual innovation, and more and more drawing on the inventiveness of what were previously consumers.

So it is that fabricating our version of the world is no longer a tribal epic, no longer a Dickensian narrative stitching the disparate parts of society together, but a network of mutually supportive activities. It is a moot point whether these are an extension of existing political economy into new realms (as Fuchs 2008 argues, among others), or whether it is the first signal of a new political economy as Bauwens (2005) and Barbrook (1998) have suggested. What is clear is that the combination of automation with the employment (usually unpaid) of creative labor to act as a randomizing factor in the operation of the network (Flusser's "apparatus") produces an assemblage which abstracts from the physical world in pursuit of a kind of ordering, which conforms not simply to the world depicted but to the managerial and commodified forms of the contemporary information economy.

In one sense there is nothing new here. The universe is a flux of matter and energy. Human beings have always sought, with the passion of an instinct as basic as hunger and sex, to build order against the entropic drift of pure flows. What changes is the specific mode of ordering which we undertake in different historical and cultural epochs. My argument is that we now see, in the digital apparatus, the realization of a tendency during the last 150 years or so towards sampling, abstraction, enumeration and probabilistic averaging, a mode of order which underpins our emerging digital mode of representation. At its best, this represents, as it does in scientific instrumentation and data visualization, a near-magical process in which we grow images on the skin of the world, as we grow crystals on the seed of a chip. In this sense digital media, especially high-definition models, are *more* realistic than their analog realist forebears.

## Fabric

At the same time, digital data is entirely malleable. As we have seen, this malleability is presumed in image processing as much as in data-processing. Thus workplace media are used not only for describing the state of affairs, but for extrapolating future scenarios. This is the kind of simulation which is intrinsic to the digital media, not the simulacra posited by Baudrillard (1994). Simulations manipulating variables in a database in order to extrapolate possible future scenarios are the model for the processing of visual, audio and other sensory outputs from computers and other digital devices. There are two aspects that are significant for the theme of digital aesthetics. The first is the nature of the "futures" produced in simulations: they are extensions of the present. More radical approaches to the concept of the future stress that it is definitionally different from the present, definitionally non-existent (yet), and by definition unknowable (Bloch 1986). Used as tools for strategic planning and risk management, the corporate equivalents of the Stalinist five-year plan, simulations deny the difference of the future, establish knowledge about it, and prepare tactics for ensuring that the unforeseen either never happens, or can be accommodated into existing systems. Thus we weave a shroud for our own future.

The second aspect of simulation useful for our topic is manipulation. While it is true that our miscellaneous software packages are amazingly potent, it is also the case that they tend to obey rules of market capitalism. In 2005, Adobe and macromedia amalgamated to produce a suite of tools which now dominate 2D design for print and web applications. In the same year, Autodesk, which already dominated the market for CADCAM and architectural software, completed a suite of purchases that made it the market leader across 3D applications. These two firms maximize the efficiency of their software by providing both a single style of interface for all their programs, and by integrating sophisticated workflow management, allowing teams of creatives and engineers to work on a single project without tripping over each other. Such managed workflows and common interfaces enable networked projects, but they do so within carefully designed, standardized and therefore constrained working environments. Something similar may be said of browsers: whatever you seek on the Internet, it will always be surrounded by a browser window. Embroider as you will: you must always do your needlepoint inside a frame.

Web-weaving and its rich associations with textiles (Plant 1997, 2003) give us another approach towards manipulation and team-work: mutuality. The argument raised by Michel Bauwens is that a cashless network economy arises when, if I give my labor freely, I get in return the labor of many others. This is the principle of Wikipedia, of Linux and the Free/Libre Open Source Software (FLOSS) movement: my hour of labor reaps the benefits of hundreds of thousands of hours of the labor of my peers. This network aesthetic has ended the age in which production and consumption were distinct activities. Today, from booking travel to designing kitchens, consumers are actively engaged in the design and production of their goods, and nowhere more so than in computing, where installing software and customizing are familiar chores. While diaries, scrapbooks and family albums have a long history, their publication and sharing in richly interactive forms is new. Understandably in these circumstances, the most important companies have been those that either provide the necessary tools for participation, or control the distribution of the results. The proliferation of business models, from Apple's iTunes and iBooks stores based on copyright, to Google's advertising funded drive to "organize the world's information and make it accessible and useful" (Google 2010), would seem to indicate that we are far from exhausting the capabilities of the network to provide a new terrain for creativity, profit and power.

To the extent that there is one digital aesthetic, then, that aesthetic should be understood in the subjunctive mood, the mood of "would that it were," or the conditional of "if only it were the case that." It is subjunctive perhaps especially in the second sense of fabricating proposed here: that of narrating to ourselves our world and our places in it. Not only the enumerative drive but also the networking process, when added to the powers of manipulation, leads us to treat every digital datum as something never entirely true, as something on the lip of existence that never quite exists. This is perhaps why we experience the vastly accelerated world of real-time information flows as ephemeral, indeed the generation of data has already outstripped our capacity to store it (Gantz 2008), and why so many contemporary thinkers describe the digital media as weightless, friction-free and immaterial.

The final fact to note on digital aesthetics is that the digital is far from immaterial. Our machines, in their millions, are made from metals whose extraction, in many instances, ravages the environments they are sourced in and the lives of the people who mine them, and in several instances supplies are running low (Cohen 2007). The construction of our machines occurs for the most part in offshore sweatshops in what can only be described as the Third World (see for example Fusco 2001). The energy required to run the network, domestic and office computers, mobile devices and the rapidly growing server industry required for cloud computing was almost 5 percent of US power usage in the middle of the last decade (Koomey 2007). Recycling, where it takes place at all, destroys lives and environments with a virulence not even the extraction of raw materials can equal (Basel Action Network 2002, 2005), and as we approach peak oil, the cost of recycling will increase, perhaps beyond what can be afforded. More and better machines are not an available solution: there are not enough materials on the planet to equip the newly wealthy of India and China with the concentration of devices to which we have become accustomed in the West. Any attempt to understand digital aesthetics must touch this solid ground, the materiality of mediation. In doing so it should also discover that the terms "immaterial" and "immediate" cannot be applied to digital media. Perhaps more significantly in the long run, network logic is an opportunity not only for a new economy but also for a new polity, which includes machines not as slaves but as partners. Some signs of this already exist (Knorr-Cetina and Brügger's 2002 ethnography of finance traders shows them treating their networks as equal partners, for example). Beyond this now rapidly approaching horizon, concern for environmental impacts of digital media should also teach us that the old species-specific anthropology, politics and aesthetics of Kant are no longer viable. We will need to enter an unforeseeable future partnership with the physical world of our planet, or there will be no digital, no aesthetic, no future and no fabric.

## Bibliography

Barbrook, Richard. 1998. "The Internet Gift Economy," *First Monday.* Available at http://firstmonday.org/htbin/cgiwrap/bin/ojs/index.php/fm/article/view/631/552. Retrieved October 5, 2010.

Basel Action Network. 2002. "Exporting Harm: The High-Tech Trashing of Asia." Available at http://www.ban.org/E-waste/technotrashfinalcomp.pdf, retrieved March 17, 2009.

Basel Action Network. 2005. "The Digital Dump: Exporting High-Tech Re-use and Abuse to Africa". Available at http://www.ban.org/BANreports/10-24-05/index.htm, retrieved March 17, 2009.

Baudrillard, Jean. 1994. *Simulacra and Simulation*, trans. Sheila Faria Glaser. Ann Arbor: University of Michigan Press.

Bateson, Gregory. 1973. *Steps to an Ecology of Mind: Collected Essays in Anthropolgy, Psychiatry, Evolution and Epistemology.* London: Paladin.

Bauwens, Michel. 2005. "The Political Economy of Peer Production," in *C-Theory.* Available at http://www.ctheory.net/articles.aspx?id=499, accessed December 6, 2005.

Bloch, Ernst. 1986. *The Principle of Hope*, 3 vols, trans. Neville Plaice, Stephen Plaice and Paul Knight. Cambridge MA: MIT Press.

Cohen, David. 2007. "Earth's Natural Wealth: an Audit", *New Scientist* 23 (May). Available at http://www.newscientist.com/article/mg19426051.200-earths-natural-wealth-an-audit.html?full=true, accessed August 1, 2007.

Dyson, Frances. 2009. *Sounding New Media: Immersion and Embodiment in the Arts and Culture.* Berkeley: University of California Press.

Fishwick, Paul A. (ed.) 2006. *Aesthetic Computing* Cambridge MA: MIT Press.

Flaxton, Terry. 2009. Interview with the author, September 2. Available at http://www.digital-light.net.au/interviews. Retrieved October 5, 2010.

Flusser, Vilém. 2000. *Towards a Philosophy of Photography*, trans. Anthony Matthews, intro. Hubertus Von Amelunxen. London: Reaktion Books.

Fossum, Eric R. 2005. "Gigapixel Digital Film Sensor (DFS) Proposal." Nanospace Manipulation of Photons and Electrons for Nanovision Systems, the 7th Takayanagi Kenjiro Memorial Symposium and the 2nd International Symposium on Nanovision Science, University of Shizuoka, Hamamatsu, Japan, October 25–26.

Foucault, Michel. 1970. *The Order of Things: An Archaeology of the Human Sciences.* London: Tavistock.

Fuchs, Christian. 2008. *Internet and Society: Social Theory and the Information Age.* London: Routledge.

Fusco, Coco. 2001. *The Bodies that Were Not Ours and Other Writings.* London: Iniva/ Routledge.

Gantz, John. (project director) 2008. *The Diverse and Exploding Digital Universe*, IDC White Paper. Framingham MA: IDC (March). Available at http://www.emc.com/collateral/ analyst-reports/diverse-exploding-digital-universe.pdf, accessed April 21, 2008.

Gelernter, David. 1998. *The Aesthetics of Computing: Elegance and the Heart of Technology*. London: Phoenix.

Heidegger, Martin. 1977. "The Age of the World Picture," in *The Question Concerning Technology and Other Essays*, trans. William Lovitt. New York: Harper & Row.

Heisenberg, Werner. 1989. *Physics and Philosophy: The Revolution in Modern Science*, intro. Paul Davies. London: Penguin.

Knorr Cetina, Karin and Urs Bruegger. 2002. "Traders' Engagement with Markets: A Postsocial Relationship," *Theory Culture and Society* 19 (5/6): 161–85.

Munster, Anna. 2003. "Compression and the Intensification of Visual Information in Flash Aesthetics," in *Melbourne DAC 2003 streamingworlds*, ed. Adrian Miles. Long paper proceedings of the 5th International Digital Arts and Culture Conference Melbourne DAC (May) Melbourne, RMIT University School of Applied Communication, 150–9.

Parikka, Jussi. 2007. *Digital Contagions: A Media Archaeology of Computer Viruses*, New York: Peter Lang.

Pastoureau, Michel. 1987. *Heraldry: Its Origins and Meaning*, trans Francisca Garvie. London: Thames and Hudson.

Plant, Sadie. 1997. *Zeros and Ones: Digital Women + The New Technoculture*. London: 4th Estate.

Plant, Sadie. 2003. "Mobile Knitting," in *Information is Alive: Art and Theory on Archiving and Retrieving Data*, eds. Joke Brouwer, Arjen Mulder and Susan Charlton. Amsterdam: V2 / NAI Publishing, 26–37.

Shannon, Claude E. and Warren Weaver. 1949. *The Mathematical Theory of Communication*. Urbana: University of Indiana Press.

# 9.1

## Closing the Gap between Art and Life: Digital Art as Discursive Framework

*John Byrne*

Over the last two decades, culture has changed irreconcilably. No longer the ideological window dressing of cold war politics, it now finds itself at the core of a growing globalized economy. Politically aligned with local, national and international regeneration policies, culture often finds itself being used as a bridging mechanism between different and often conflicting social, political and economic interests. According to the theorist George Yudice (2003), culture is now expedient—there is no way of funding or producing small, medium or large scale art projects that do not bring into alignment a collection of inchoate and seemingly incompatible interests. Map on to this the digitally driven emergence of "convergence culture"—where, as Henry Jenkins (2006) argues, "old and new media collide," "grassroots and corporate media intersect" and media producers and media consumers "interact in unpredictable ways,"—and you have the most sophisticated, complex and volatile cultural arena in human history. Within this capricious and frequently arbitrary milieu, art has long since lost its own self-asserted right to set itself apart as some form of omnipotent commentator on the ills of mass culture. The shock tactics of the avant-garde are now the stock in trade of advertising agencies and music producers the world over, and art, as we knew it, now has to fight for the oxygen of publicity with sport, daytime television, online social networking and game consoles. It is a losing battle.

However, many emergent artists, art groups and art institutions are rising to this challenge by interfacing art, technology and popular culture in new and innovative ways. One such example is Grizedale Arts (http://www. grizedale.org) in the UK. Based in England's Lake District, a designated zone of natural beauty, the work of Grizedale Arts frequently addresses issues of culture, the tourism and leisure industry, economic sustainability and employment. Their work is also consciously embedded within a complex and shifting network of local, national and international agendas concerning the relationships between art and social regeneration.. Operating from the site of a farm overlooking Lake Coniston, Grizedale intentionally exploits an amalgamation of freely available low and high tech methodologies to engage and interact with communities around the globe. Here, the emphasis is no longer on the production of tangible art objects but rather on the production of ideas, solutions and new knowledge. One example of this is the project "Seven Samurai" (2007). For this project, Grizedale commissioned seven UK based artists to live and work in the remote Japanese village of Toge. The idea was not only for the artists to produce artworks that could be shared online, but also for them to work with the villagers to find new solutions for self-sustainability. The "Happy Stacking" (2009) project in China is a more recent version of this project.

For Adam Sutherland and Alistair Hudson, Director and Co-Director of Grizedale Arts respectively, digital technologies play a crucial role. Unlike most contemporary arts organizations—who frequently see "Second Life" as either an opportunity to archive pre-existing projects cheaply in the public domain or, less frequently, as a new arena for the production of rather dated and predictable "digital art"—they argue that digital technologies now provide the primary site for the experience of new forms of artwork, whatever they may be. This fundamental shift away from the digital as a mere site for the representation of existing artworks (or, what amounts to the same thing, for the representation of long held ideologies of significant art production under the guise of the digital) is crucial. By re-positioning the use of digital technologies as a fundamental part of the continual reintegration of the arts into the circuit of contemporary cultural production, Grizedale's projects, and others like them, are helping continually to redefine the role of art. By working within the converging field of art and life, Grizedale uses digital technologies to avoid the production of art as a simple leisure commodity, while at the same time eschewing the all too familiar temptation of claiming that artists can somehow change the world through radical uses of new technology. Instead of this, the making of radical art is seen as an active and meaningful participation in the production and distribution of our collective digital and physical futures.

## Bibliography

Grizedale Arts. http://www.grizedale.org, accessed October 7, 2010.

Happy Stacking. 2009. http://www.grizedale.org/projects/happy.stacking, accessed October 7, 2010.

Jenkins, Henry. 2006. *Convergence Culture: Where Old and New Media Collide*. New York: New York University Press.

Seven Samurai. 2007. http://www.grizedale.org/projects/seven.samurai, accessed on October 7, 2010.

Yúdice, George. 2003. *The Expediency of Culture: Uses of Culture in the Global Era*. Durham NC: Duke University Press.

# 9.2
## Digital Art: *Blowing Zen* in the City

*Melissa D. Milton-Smith*

In the new millennium digital artists have questioned the nature of human existence in complex spaces like the city. Themes of connection and alienation pervade digital art works, in the spirit of the times (Tofts 2005). In this case study I will discuss how digital art operates as powerful platform of articulation. With reference to Nebojša Šeric Shoba's film *Blowing Zen* (2003), I will explore a digital reconstruction of New York City's Times Square.

I use the term "digital art" to provide a language for locating creative works mediated by digital technology in one of three ways: as the product, process or subject thereof. Digital artworks often incorporate eclectic forms of media, from film and video, to gaming and wearable technology (Murphie and Potts 2003). The use of the word "art" acknowledges the influence of other artistic styles, while countering technologically determinist notions of "new media" (Stallabrass 2003).

Shoba's *Blowing Zen* is a critical example of new millenium digital art. The digital film was exhibited as part of "Metropolis," curated by Anonda Bell at Australia's National Gallery of Victoria in 2004. In *Blowing Zen* Shoba presents a dehumanized and post-apocalyptic Times Square. He empties the metropolis of human life, and replaces its sounds with the haunting strains of an oriental flute.

In recalling a visit to Times Square, Shoba described how "thousands of people are stepping over your feet, pushing you around. Noise is unbearable, [there is a] smell of burned meat from nearby food stands, hundreds of advertising panels decorated by millions of flashing lights, outbursts of shopping hysteria" (Shoba 2004). In *Blowing Zen* Shoba provides an alternative view, where movement comes from neon signage and the flow of traffic lights.

*Blowing Zen* converses with the uneasy climate of the new millennium, punctuated by the horrific events of September 11, 2001. The evacuation of New York City, and the ensuing silence, affected individuals worldwide. In *Blowing Zen*, the US flag reminds us of the people who once inhabited this space. In this city the flickering faces on digital billboards have replaced human life. The artwork arguably engages with Jean Baudrillard's notion that in an era of simulation "old distinctions and orientations are abolished: objects no longer relate at all to their processes of human production, there is a loss of emotional content and of 'objective' or critical distance" (1992: 22).

*Blowing Zen* was created during a time of global upheaval and change. Shoba's digital reconstruction of Times Square reflects on this particular moment of production. In *Blowing Zen* he shows us how digital art can operate as a powerful platform of articulation, and reminds us that although a city may be a hub for commerce and technology (Sassen 2000), it is nothing without its people.

## Bibliography

Baudrillard, J. 1992. "From Simulacra and Simulations." In *Modernism/Postmodernism*, ed. P. Brooker. London: Longman.

Bell, A. (ed.) 2004. *Metropolis* Exhibition brochure. Retrieved from http://www.ngv.vic.gov.au/metropolis/resources/rb_metropolis.pdf, accessed March 6, 2010.

Murphie, A. and J. Potts. 2003. *Culture and Technology*. New York: Palgrave Macmillan.

Sassen, S. 2000. *Cities in a World Economy, Sociology for a New Century*, second edn. Thousand Oaks: Pine Forge Press.

Shoba, N. Š. 2004. In *Metropolis* Exhibition Brochure, ed. A. Bell. National Gallery of Victoria. Retrieved from http://www.ngv.vic.gov.au/metropolis/resources/rb_metropolis.pdf, accessed March 6, 2010.

Stallabrass, J. 2003. *Internet Art: The Online Clash of Culture and Commerce*. London: Tate Publishing.

Tofts, D. 2005. *Interzone: Media Arts in Australia*. Melbourne: Craftsman House.

# 9.3
## Digital Aesthetics in Everyday Technologies: A Case Study of the NY Art Beat iPhone Application

*Tamsyn Gilbert*

Digital technologies are shaping cultural institutions in new and surprising ways. Much of the current research on the ways we look at the intersection of cultural institutions and technology focuses on the access to knowledge (through online databases and catalogues). This case study attempts to go further and suggests that digital technologies may, in fact, alter how we understand aesthetics more generally. It will do so via an examination of the NY Art Beat iPhone application and its use in the cell phone.

From the outset, it should be noted that digital aesthetics can take many forms, from the aesthetics *of* the physical (digital) object, for example, the flat screen television and the iPod, to the aesthetics of the content *on* the digital object such as the webpage, television program or cartoon graphics. Unlike previous discussions of aesthetics, digital aesthetics have to deal also with the function and role of reproductive technology in an unprecedented way. Objects that were traditionally functional in some way now take on a different aesthetic quality because of the way in which they are refashioned by the digital. This is not a completely new phenomenon as "function" and "art" have always merged (for example, the map). This case study is interested in new interactions of this "function–art" dynamic, specifically through the refashioning of the spatiality of the viewer and the art museum.

On the NY Art Beat website, the iPhone application is advertised with the phrase "With NYAB, there's always a great art show around the corner." The NY Art Beat application begins with a simple question, "would you like to use your current location?" From here, the application locates the user via the global positioning system (GPS) and organizes local art exhibitions according to distance. The application creates a list of the nearest exhibitions and on clicking each item in the list, the user can find out a number of different pieces of coded information, for example, the price of the institution, the opening hours, the scheduled dates for the exhibition, the telephone number of the institution, what type of exhibition it is (painting, photography, video and so on) and the location of the exhibition as well as an image of one of the art works that can be seen in the exhibition.

On clicking on the location of the gallery or museum, the user opens up a map that not only pinpoints the location of the art institution but also provides directions from the location of the user. Here the map is refashioned around specific, art-related data and not only the geographical locations of sites. Consequently the merging of the function of the digital technologies and the digital aesthetic experience becomes apparent. As Sean Cubitt states in his book *Digital Aesthetics*:

> The map is like an advertisement for the land: its constantly provisional nature is precisely what makes it available for public communication, as the incompleteness and incoherence of the system must direct outwards, towards its uses, for stability it cannot attain on its own.
>
> (1998: 55)

This case study seeks to highlight the ways in which the digital "aesthetic" experience sets up a unique dynamic of beauty, taste, the senses and the artwork, critically determined and embedded in the functions afforded by the mobile digital technology to everyday life, such as speed, memory and directions. The NY Art Beat iPhone application allows for portability of information in a mode that renders links to traditional art gallery experiences into a digitally mapped topography. That is, the work of art and the information that goes along with it is constructed in relation to a map of exhibition sites, the location of the user at a certain time, and in a specific place and a specific frame of aesthetic interests. The nature of the data "mashup" is such that every construction of aesthetic options is a unique configuration of time, space and aesthetic preferences.

## Bibliography

Cubitt, S. 1998. *Digital Aesthetics*. Thousand Oaks, CA: Sage Publications.
NY Art Beat, http://www.nyartbeat.com/apps,accessed May 15, 2010.

# Part X
# Digital Labor

# 10
# Redrawing the Labor Line: Technology and Work in Digital Capitalism

*Eran Fisher*

One of the most promising innovations predicated on information and communication technology, henceforth called network technology, is arguably the emergence of network production, enabled by a host of web applications and organizational forms such as open source, peer-to-peer (P2P) production, wiki, social production, social networking, and crowdsourcing. In contemporary technology discourse, network production is described as revolutionary because it offers a more democratic, participatory, and collaborative mode of social and cultural production, and empowers individuals by allowing them more meaningful and creative engagement with the productive process. Network production harnesses human facets which have been hitherto excluded from production: authenticity, personal expression, and creativity. It facilitates the emergence of "prosumption" as a hybrid of production and consumption, rendering both practices more engaging, participatory, and fulfilling. Network production facilitates the crystallization of emergent, self-governed, and self-regulated collaborative projects, and its intrinsic flexibility allows for a more sophisticated utilization of resources such as play, joy, and free time, which can be harnessed into wealth creation.

Network production is assumed in this public discourse to be predicated on network technology. According to that rationale, technological revolution brings about social revolution: network technology is seen as a new means of production that, having been democratized and popularized, revolutionizes the relations and mode of production, and society at large. It is therefore common to hail network technology as bringing about not only a benevolent new society, but one that is post-capitalist, shifting control and power over from capitalists and corporations to workers.

Such discourse represents the dominant view that new technology creates new social relations and a new society. But the discourse on technology is not simply a reflection of the centrality of technology in the operation of modern societies; rather, it plays a constitutive part in that operation, and enables that centrality. I therefore want to offer here another view of the discourse on network production, which sees this technologically centered discourse not simply as a transparent reflection of contemporary technological realities but as taking an active part in the constitution of such reality. It is a view which sees technology not only as a material reality that constructs society, but a discursive reality that legitimates it, specifically, legitimizing new realities of production, new employment arrangements, and, more generally, new power relations under a new, post-Fordist capitalism.

In what follows I outline some key narratives of the discourse on network production, and highlight their role in legitimating new realities of work, employment, and production, by substituting a technologistic presentation for an analysis of power relations. Lastly, I will suggest how these narratives constitute a new spirit of capitalism which fits its post-Fordist phase.

## Network production: fusing the system's rationality and emancipation

The analysis of the discourse on network production draws on articles from *Wired*, a popular monthly magazine published since 1993 with a current circulation of about 700,000, which features one of the most crystallized articulations of this discourse. *Wired* grew out of, and indeed concurrently with, the dot.com bubble, with the explosion of the Internet, and with the emergence of the high-tech industry as a central engine in the economy as well as a powerful source of identity and culture. The magazine is therefore utilized here as a case study of the digital discourse. Such a case study provides a holistic and exhaustive field suitable for examining the qualitative meaning of a phenomenon, rather than its scope or prevalence, and reaching an understanding of its broader social context (Kuper and Kuper 1996; Stake 1994).

Within the case study I use a method of discourse analysis in order to decipher key themes that underlie the multiplicity and particularity of stories. However, unlike the structuralist notion of a fixed underlying meaning to a text, the post-structuralist discourse analysis offered here sees the digital discourse not as a veil on reality—as ideology—but as taking part in the constitution of that reality. The purpose of the analysis then is to make explicit the connection between discursive narratives and new constellations of power (Fairclough 1995: 57–9; Kincheloe and McLaren 2003: 443–5; see also: Chouliaraki and Fairclough 1999; Bourdieu 1998a, 1998b).

The analysis of dominant discourses in popular magazines has already been utilized in the social sciences in such works as *Reading National Geographic* (Lutz and Collins 1993), which studies national, racial, gender, and geographical representations through a close analysis of the photography featured in the famous magazine, and *Making Sense of Men's Magazines*

(Stevenson, Jackson, and Brooks 2001), which offers an analysis of masculinity, capitalism, and consumption in contemporary Western society.

In the digital discourse, network production is hailed as "a new cultural force based on mass collaboration. Blogs, Wikipedia, open source, peer-to-peer—behold the power of the people" (Kelly 2005). The most exciting potential of the web, according to this, is in allowing individuals to come together and unleash a powerful, creative, and democratic force of production, "a new kind of participation that has ... developed into an emerging culture based on sharing" (Kelly 2005). This democratization means that it is no longer professionals and experts but simply web "users" who overtake the process of production and do most of the work.

This trend towards democratization is furthered, according to the digital discourse, by the fact that network production "permits easy modification and reuse, and thus promotes consumers into producers," transforming many endeavors "from spectator art to participatory democracy." The web, then, allows the transformation of passive, receiving audiences into content providers, it "assumes participation, not mere consumption," and ushers in a "great shift from audience to participants" (Kelly 2005). The web allows the sort of engagement with the world that follows the logic of interaction. It involves

> deep enthusiasm for making things, for interacting more deeply than just choosing options ... This impulse for participation has upended the economy and is steadily turning the sphere of social networking – smart mobs, hive minds, and collaborative action – into the main event.
>
> (Kelly 2005)

With network production, individuals are no longer simply peons of a consumerist mass society; instead, they regain agency and independence through production and consumption. The web topples the mass, homogenized society, and brings greater opportunities for individuals to be creative and self-expressive. It transforms "a world ruled by mass media and mass audiences to one ruled by messy media and messy participation" where "everyone alive will (on average) write a song, author a book, make a video, craft a weblog, and code a program" (Kelly 2005).

In such an environment no one is merely a consumer: "What matters is the network of social creation, the community of collaborative interaction that futurist Alvin Toffler called prosumption ... prosumers produce and consume at once" (Kelly 2005). Power is delivered to "the people" by the democratization afforded by network technology, which in turn allows participation and collaboration. As the means of production are democratized, so are the mode of production and social relations in general.

One of the technological emblems of network production is open source. The term "open source" refers originally to software products, but came in the digital discourse to epitomize the revolutionary potentialities entailed by network production. Open source is taken to constitute a radical break in the process of production in that for the first time since the industrial revolution the rationalization of the work process through technology emancipates, rather than alienates, workers from the productive process. Put differently, in contrast with previous advances in the mode of production, open source responds not only to the rationalization demands of production, but also to the emancipatory demands of workers. The discourse on open source therefore heralds the emergence of a more democratic and emancipatory form of rationality, fusing democracy and capitalism together:

> Open source embodies an ethos as fruitful and resilient as the closed capitalism Bill Gates represents: the spirit of democratic solutions to daunting problems. It's the creed of Emerson, who preached independent initiative and advocated a "creative economy" … It's the science of Frederick Taylor, who proved that distributing work could exponentially boost productivity and replace "suspicious watchfulness" with "mutual confidence". It's the logic of Adam Smith, whose notion of "enlightened self-interest" among workers neatly presages the primary motivation for many open source collaborators.
>
> (Goetz 2003)

Open source, then, represents a new kind of capitalism, characterized as open (as opposed to "the closed capitalism" of Bill Gates), democratic, conducive to individual independence and creativity, decentralized, collaborative, and distributive. The brilliance of open source, according to this narrative, lies not only in being "for the people," nor only in it being a new and powerful source of wealth creation for corporations; instead, its revolutionary potential lies precisely in it being both, in bridging and transcending contradictory interests: the systemic demands of capitalism, and the life world demands of freedom and creativity.

According to the digital discourse, then, open source is not "anticommercial or anticorporate" but rather a more democratic and emancipatory form of capitalism. It offers a superior capitalist tool and also holds the promise for workers' emancipation, a truly universal and benevolent instrument:

> While the assembly line accelerated the pace of production, it also embedded workers more deeply into the corporate manufacturing machine. Indeed, that was the big innovation of the 20th-century factory: the machines, rather than the workers, drove production. With open source, the people are back in charge. Through distributed collaboration, a multitude of workers can tackle a problem, all at once. The speed is even greater – but so is the freedom. It's a cottage industry on Internet time.
>
> (Goetz 2003)

The network mode of production, then, offers a way of responding to two demands which are at the heart of modernity: that of economic rationality, and that of human emancipation. The two demands were historically thought to be at odds with each other; moreover, the zenith of industrial society, with its technology and organizational corollaries came to epitomize the suppression of the demands for emancipation through technology. But now network technology is construed as the means by which the two demands can be met. Moreover, met not by a political, externally imposed compromise between contrasting interests (which was ultimately what the welfare state during Fordism was all about), but rather by their complete enmeshment into the workings of network technology and its corollary network production. As one commentator notes, "We are at a convergence moment, when a philosophy, a strategy, and a technology have aligned to unleash great innovation." "With open source," he further says, "you've got the first real industrial model that stems from the technology itself, rather than simply incorporating it" (Goetz 2003). Network production absorbs both rationalization and emancipation and resolves their hitherto contradictory and crisis-laden tendencies.

The discourse on network production then tells of the emancipation of individuals, the rationalization of production, and perhaps most importantly, how both emancipation and rationalization come to complement each other through network technology. The historically contradictory demands of capital, industry, efficiency, competition, and productivity on the one hand, and the demands of labor, satisfaction, creativity, cooperation, and personal fulfillment on the other hand, come into alignment with the introduction of network production, and as a result, persistent tensions that characterized industrial production are resolved. But what are the terms of this resolution? If capital and labor align along a shared line, what is the shape of this line, and what is the topography of the terrain it traverses?

## Crowds and enthusiasts: redrawing the labor line

One of the terms used to account for this new geography of work is "crowdsourcing." The term is a word play on the more familiar "outsourcing." Heralding "The Rise of Crowdsourcing," a *Wired* article explains the term: "Remember outsourcing? Sending jobs to India and China is so 2003. The new pool of cheap labor: everyday people using their spare cycles to create content, solve problems, even do corporate R&D" (Howe 2006). The rationale of network production, according to that view, is anchored in the ability to tap into a "pool of cheap labor" and mobilize it to one's production needs. This statement makes a distinction between outsourcing and crowdsourcing, but at the same time uncovers their similarities: in outsourcing, cheap labor is based on differentials in space; in crowdsourcing, cheap labor is based on differentials in time, time which is deemed "unproductive"—"spare cycles" that is, time which is not easily translatable to money and capital.

But what is the mechanism, or "trick," by which productive forces, harnessed for the project of capital accumulation, can nevertheless be conceived as unproductive (as "spare cycles") and hence be partially or wholly unpaid? Presumably the trick is technological: this is made possible by network technology. But the trick is also discursive: the trick is that this labor is construed in the digital discourse to be performed by "everyday people," rather than by professional, tenured, "in-house" workers. No wonder, then, that the front-cover teaser for the article proclaims: "Crowdsourcing: A billion amateurs want your job" (*Wired* 2006). The threat is clear: one's position in the workforce is threatened not so much by other workers, but by a new type of worker, a new reserve army of, this time around, amateurs. The discourse on network technology remaps the field of socially necessary labor, redrawing the lines between paid and unpaid labor. Network production entails the transference of parts of the productive process from paid, organized, professional labor force to unpaid, atomized, amateurish enthusiasts.

Network production entails a discursive redefinition of workers, with two key analytical innovations. The first is the dissociation of work—the investment of labor power in production—from employment—the institutional arrangement of personal livelihood.The redefinition of "workers" as hobbyists and dabblers legitimizes new modes of employment. The second analytical innovation is the construction of workers as prosumers. This renders the distinction between labor and capital obsolete and allows the equation of big corporations with small-time hobbyists; analytically, both are seen as homologous units of prosumption. With the redefinition of workers as prosumers, the category of workers as a distinct category is eliminated, and instead emerges the worker as entrepreneur (Huws 2003, Greenbaum 1995: 92; Sennet 2006: 52).

The introduction of these analytical novelties to contemporary discourse is what allows the astounding reference to the talent and work power of the crowd as "latent" in the discourse on network production, as in the following quote: "smart companies ... discover ways to tap the latent talent of the crowd. The labor isn't always free, but it costs a lot less than paying traditional employees. It's not outsourcing; it's crowdsourcing" (Howe 2006). According to this framework, network technology is uniquely capable of mobilizing this talent and work power into productive results; of course, with the condition that it is poorly paid for.

Network technology topples the bastion of what are by definition exclusionary categories: workers, professionals, and cadres. It undercuts their guild-like privileges and instead empowers "the people" in two ways: hardware and software become cheap enough for non-professionals to acquire, and web access renders virtually anyone a viable participant in the marketplace. As network technology democratizes economic opportunities it also blurs the boundaries between professionals and "enthusiasts." These new circumstances create a market heaven: consumers are able to buy more cheaply, and amateurs are able to enter into the "loop" of production and creativity and offer the fruits of their passion.

Notwithstanding the populist undertones of the digital discourse, network production characterizes not only the way that hobbyists contribute to Wikipedia, sell their products online, and author their blogs; increasingly, it is being integrated into the operation of large corporations, and of capitalism at large. As this happens, the relations of production between capitalists and workers are radically transformed. Network production allows for a significant reduction in production costs by creating a business model based on the labor power of uncompensated and undercompensated laborers. This, in turn, is dependent on a new social compact, according to which "the people" receive more opportunities to engage their skills, creativity, hobbies, and passions, and corporations receive hitherto unexploited sources for capital accumulation. In return, "the people" give up stable compensation schemes embedded in institutionalized employment, and corporations surrender their claims for authority, professionalism, and control over content.

A key example of this are the research and development (R&D) units of major corporations, that, with the aid of network production, are "finding a way to tap" into the "scientific talent and expertise" of the crowd (Howe 2006). Underlying this statement is the assumption that the old way of tapping into individuals' talent and labor power—by employing them—has fallen out of favor, and that network production offers new ways to do that. One mechanism of tapping into such "latent talent" is InnoCentive, a website where companies post R&D problems that people in the crowd are welcome to solve. Network production in this case amounts to a deep restructuring of R&D departments: "Exit the white lab coats; enter ... over 90,000 'solvers' who make up the network of scientists on InnoCentive" (Howe 2006). The shift, then, is from a standing army of full-time, tenured, and fully compensated scientists to a reserve army in the form of the networked crowd. The reference to the "white lab coats" is not incidental; admittedly, a staple of scientists' habitus but also a connotation (in Roland Barthes' terms) of a much deeper set of meanings. A white lab coat is the "uniform" of the worker; it stands for tenure, for being an integral part of a greater system; a peon perhaps, but a member nonetheless, without which the structure will not survive. The white coat, just like the white or blue-collar is an emblem of the bureaucratic organization of industrial society *par excellence*. The figurative "Exit the white lab coats" is therefore literal as well: exit tenure, mass employment, collective bargaining, working class, and so forth.

This theme is reiterated in a story about Lego's decisions to crowdsource one of its products to a "panel of citizen developers" (Koerner 2006). The term "citizen developers" makes a distinction between the reserve army of workers, professionals, and engineers on the one hand, and the bureaucratized, tenured ranks of Lego employees on the other hand. And them being citizens, rather than the rank and file of the Lego organization, supposedly unfetters in them qualities that are deemed missing from Lego employees. Network production is likened to bringing a breath of the fresh air of citizens, ventilating the stultifying, closed ranks of a military-like institution. Like the aforementioned "white lab coats," "citizen developers" too connotes the army as the archetypical bureaucratic organization, and sees in network production a "civilizing" mechanism (Weber 1978; Sennet 2006: ch. 1).

Network production, then, undermines the rigidity of a military-like bureaucracy and brings about more flexibility. According to the digital discourse, such flexibility liberates workers as well. For example, one "solver" in the InnoCentive crowd is described thus: after completing his graduate studies in physics, he went through "a succession of 'unsatisfactory' engineering jobs ... none of which fully exploited [his] scientific training or his need to tinker ... Not every quick and curious intellect can land a plum research post at a university or privately funded lab" (Koerner 2006). This mismatch between his talent and a traditional work position was accompanied by his reluctance to work in "a 9-to-5 environment." Crowdsourcing, according to this discourse, offers a solution to the limitations put forth by the traditional job market, which consists of "positions," located at definite and stable points in space and time, which are part of a "career path." These are seen as close-ended and restrictive, in contrast to the open-ended and flexible character of employment afforded by network production, which is specifically adept at exploiting the full potentiality of individuals.

InnoCentive was launched, according to the story "as a way to connect with brainpower outside the company – people who could help develop drugs and speed them to market," and more broadly as a tool for "firms eager to access the network's trove of ad hoc experts" (Koerner 2006). The flexible, ad-hoc employment schemes spawned by network production are explained in terms of the inherent superiority of the network as a mechanism for the mobilization of dispersed skills. These new schemes are based not only on "economic reasoning," that is, this type of employment costs less and is less burdening from the point of view of employers, thereby "curb[ing] the rising cost of corporate research" [Koerner 2006]), and not only on "emancipatory reasoning," in that it allows more flexibility from the point of view of employees who would rather not be locked in committed, but possibly unsatisfying and stifling, working relations, but also on a new "network reasoning," according to which the network is a repository of superior intelligence. Through the network, companies are able to tap into the expertise and knowledge of many more individuals than they can ever employ directly, and in turn, can achieve better, smarter results, materialized in the form of commodities. As one commentator says: "[t]he strength of a network like InnoCentive's is exactly the diversity of intellectual background" (Koerner 2006).

Network production, then, is presented as the condition for two trends, which are constructed as intertwined and co-dependent. On the one hand is the universal benevolence of a new superior smartness, which can lead—who knows?—to the development of a new life-saving drug by Eli-Lilly. On the other hand is the new labor regime of loose, ad-hoc, flexible employer–worker relations. The digital discourse constructs an inextricable link, according to which in order to reap the benefits of this new network rationality we must accept the more flexible, precarious, and privatized regime of employment, since both are predicated on the characteristics of the network. Network technology has the ability to make production more rational, but to do that successfully, it is implied, workers must be redefined as autonomous nodes in the network that can be flexibly mobilized to a specific project and then let go.

## Dismantling "Work"

The discourse on network production marks a shift in business culture, which sees "the current R&D model" of tenured research positions within corporations as "broken" (Howe 2006), and substitutes it with a networked employment model: flexible, ad-hoc, and project-based. As the locus of production becomes decreasingly the company, and increasingly the network itself (Castells 1996: 178ff.), network production is seen in this discourse as a solution to the constraints and limitation of what a "company" is. It is instructive to recall the etymology of the word, which is a derivative of companion, from the Latin *com* "with" and *panis* "bread," literally meaning "one who eats bread with another." It was first used in the twelfth century to mean "body of soldiers" and in the sixteenth century came to mean "business association." It is precisely these relics of communality, mutual responsibility, and exclusivity, etched in the word company, that the new ethos of network production wishes to have wither away, and revolutionize.

The new business culture, then, calls for companies to open up to the network of talent, intellect, and knowledge, and incorporate the new spirit of networks into their operation. Proctor & Gamble, for example, is reported to have gone from an "insular" corporate culture, which won it the nickname "the Kremlin on the Ohio," to a more open, network culture. This, according to an executive with the firm, "changed how we define the organization … We have 9,000 people on our R&D staff and up to 1.5 million researchers working through our external networks. The line between the two is hard to draw" (Howe 2006). In terms of production, the line is indeed hard to draw, but from the point of view of workers, network production entails not the elimination of the line between workers and the productive process but its redrawing. In this discourse, workers are redefined as free-lancers, compensated solely for the end-product, rather than for their labor, and bearing all the costs of their work: equipment, training, unproductive time due to illness, health and other types of insurances, and so forth. In other words: piece workers. All these are redefined as externalities, that is, socially necessary costs that are not accounted for in the compensation system between workers and employers or in the price system between sellers and buyers. For example, an individual's success story at InnoCentive encapsulates by construction another story which remains latent in the discourse: that of all the other "solvers" who worked on that particular challenge, but could not find a solution, or were too late in reaching its completion. Had this been part of the story, it would have pointed to the transfer of risks from companies to individuals, a transfer that can be conceived in terms of the privatization of the relations between companies and workers in network production, where companies meet workers at the marketplace to perform a specific task and then part ways. This more lax, less dependent relationship between capitalists and workers is ingrained in the very new neologisms put forth by InnoCentive for the various actors and practices involved in the production process: no longer "employers," "employees," and "employment," a linguistic affinity which points out their dialectics and mutuality; instead companies are referred to as "seekers," workers as "solvers," and work as "challenges."

While allowing more lax and flexible relations between employers and employees, network production, according to this discourse, paradoxically strengthens these relations:

> People mistake this [crowdsourcing] for outsourcing, which it most definitely is not ... Outsourcing is when I hire someone to perform a service and they do it and that's the end of the relationship. That's not much different from the way employment has worked throughout the ages. We're talking about bringing people in from the outside and involving them in this broadly creative, collaborative process. That's a whole new paradigm.
>
> (Howe 2006)

The new paradigm of network production, according to this, offers a tighter, more meaningful collaboration between company and workers. But this comes with the condition of reconstituting both company and workers as two autonomous nodes of network production. The privatization of work and the atomization of workers are presented as preconditions for a new paradigm of production relations, which are ultimately more intimate and satisfying. The collaboration of these independent nodes in the new capitalism's network production substitutes for the symbiotic relationship between company and workers in the old capitalism, a paradigm that had proved, according to this, to be no longer viable, if not completely dysfunctional.

"Work," then, is almost completely absent in the discourse on network production. The social category and subject formerly known as "worker" is referred to with a plethora of other labels, connoting "flexibility" ("ad hoc experts," "solvers" [Howe 2006]); "democracy" ("citizen developers" [Koerner 2006]; "everyday people," "users" [Howe 2006; Kelly 2005]; "the people," "ordinary people," "the audience" [Kelly 2005]; "regular folks," "peer[s]", "rowdy rabble" [Anderson 2006]; "fans" [Koerner, 2006]), and "personal creativity and passion" ("amateurs," "enthusiasts," "dabblers," "passionate, geeky volunteers" [Howe 2006]; "hobbyists" [Howe 2006; Koerner 2006]; and "fanatics," and "obsessed fans" [Koerner 2006]).

The discourse on network production is succinctly summarized by Chris Anderson, *Wired*'s editor-in-chief. Anderson upholds the power of network production as one of the most prominent "trends driving the global economy": "Blogs, user reviews, and photo-sharing – the peer production era has arrived" (Anderson 2006). Anderson situates network production within a framework of a historical materialism reminiscent of Alvin Toffler's and Daniel Bell's, one which locates technology at the heart of social change:

First, steam power replaced muscle power and launched the Industrial Revolution. Then Henry Ford's assembly line, along with advances in steel and plastic, ushered in the Second Industrial Revolution. Next came silicon and the Information Age. Each era was fueled by a faster, cheaper, and more widely available method of production that kicked efficiency to the next level and transformed the world.

(Anderson 2006)

The prominence of network production in the new economy, then, stems from its ability to harness hitherto unutilized sources of power to the production process. Ford's "assembly line" is transformed into network assemblages, and with it changes the whole process of production. "Now we have armies of amateurs, happy to work for free." And the economy is thus restructured to take advantage of this new power resource:

The tools of production, from blogging to video-sharing, are fully democratized, and the engine of growth is the spare cycles, talent, and capacity of regular folks, who are, in aggregate, creating a distributed labor force of unprecedented scale.

(Anderson 2006)

The new source of production, the new engine of growth, is the labor power that has been hitherto left unexploited, that is, outside the process of capital accumulation. It is comprised of "spare cycles,"—time—which is not harnessed to the productive process; time which, contrary to the famous maxim, is not money, and "talent"— skills, craftsmanship, and knowledge.

The new source of production is predicated on the technological ability to mobilize these "leftovers" of time and talent into the productive, and therefore profitable, process. The more this technological ability improves and the more it is possible to aggregate those sources of production, the more the "labor force" becomes, in Anderson's words, "distributed." Here the full meaning of the notion of "democracy" in the digital discourse is revealed. The more "fully" "the tools of production ... are ... democratized," the more the "labor force" becomes "distributed." Democratization is equated with the decentralization of labor, involving more individual autonomy and empowerment, and the evaporation of centralized, social-wide, and organized power.

## Conclusion: From safety net to the Internet

The discourse on network production is understood here not merely as a description of a new mode of capitalist production facilitated by technological networks, but also as a new analytical unit which naturalizes and legitimates new constellations of power which this mode entails, specifically, in regards to work. Underlying this argument is an assumption about the role of technology discourse in the construction of reality. To illuminate the status of technology discourse as a sociological object of study, it needs to be located within a broader body of knowledge that engages the relations between technology and society. There are three sociological approaches to studying these relations. The most prevalent theory of technology, both in the general public and in the social sciences is the "autonomous technology" approach, which frames the questions concerning technology and society in terms of the effects of the former over the latter, insisting that "technology shapes society." In media and cultural studies, technological determinism has been a productive working assumption, most notably in the works of Harold Innis, Marshal McLuhan, and Walter Ong (Webster 2005). In sociology, this approach guides such social theories as Daniel Bell's post-industrial society, and Manuel Castells' network society. Underlying this "technologistic" (Robins and Webster 1999) approach are the assumptions that technology is neutral (or a-social) (Bijker 1995; Feenberg 1995), and its effects on society are deterministic, inevitable, and benevolent (Winner 1977; Smith and Marx 1994; Feenberg 1991; Robins and Webster 1999: Ch. 2; Postman 1993; Mosco 2004).

A critique of these assumptions has sought to introduce social coordinates into the analysis of technology, arguing instead that "society shapes technology." The various threads within science and technology studies share "an insistence that the 'black-box' of technology must be opened, to allow the socio-economic patterns embedded in both the content of technologies and the processes of innovation to be exposed and analyzed" (Williams and Edge 1996). Likewise, Marxist analysis has been particularly fruitful in uncovering the extent to which technologies of production have been sites of class struggle, rather than the mere implementation of universal instrumental rationality (Braverman 1974; Noble 1984, 1995; Aronowitz and DiFazio 1994; Dickson 1988 and see Huws 2003 and Wajcman 2004 for a feminist analysis).

Whereas both approaches share an engagement with technology as an instrument, a third approach, and the one taken here, focuses on the social, political, ideological, and cultural dimensions of technology discourse, that is, analyzing "technology as discourse." According to this approach, the discourse on technology is not simply a reflection of the centrality of technology in the operation of modern societies; rather, it plays a constitutive part in that operation, and enables that centrality. The discourse on technology is understood to constitute a "projection" (Heffernan 2000) of social realities, a "technological vision" (Sturken and Thomas 2004) through which political, economic, and social transformations percolate and are processed and ideological struggles fought (Mayr 1986). Moreover, the discourse on technology assumes an active role in the construction of reality, which it presumes merely to describe, and is central in shaping the political, cultural, and social zeitgeist (Nye 1994; Herf 1984; Rabinbach 1992; Mosco 2004).

The strongest version of this approach has been put forth by members of the Frankfurt School as part of their broader critique of instrumental reason. According to this approach, in modernity, and specifically with the emergence of the modern nation-state and of capitalism, the discourse on technology has come to play a central role in the legitimation of a techno-political order where political questions are discussed as technical and technological problems (Fromm 1968; Marcuse 1991; Horkheimer and Adorno 1976; Feenberg 1991; Pippin 1995; Borgmann, 1988). This thesis is encapsulated in Habermas' notion of "technology-as-ideology" (Habermas 1970). Habermas lays out not only a general argument regarding the depoliticizing effects of technologistic consciousness, but also a historically specific critique regarding the legitimation of capitalism under the specific constellation of Fordism, specifically the construction of the Keynesian welfare state.

The four decades since Habermas laid out his theory have witnessed a deep structural transformation in capitalist societies from the Fordist phase of capitalism to its post-Fordist phase (Aglietta 2000). This transformation is from a welfare, Keynesian, social democratic state, to a privatized, deregulated, and neoliberal state (Harvey 2005; Yergin and Stanislaw 1998); from a national economy to a global one; from mass production and mass consumption to just-in-time production and mass customization (Harvey 1989); from a large scale, hierarchized, and centralized corporation that controls all aspects of production to lean, decentralized production, with a core company that outsources most facets of production to other, small and mid-size companies on a global scale (Castells 1996; Sklair 2002); and from industrial-mechanical technology to network technology.

This new social constellation was accompanied by a new legitimation discourse of technology. The discourse on network technology, then, is the new legitimation discourse of post-Fordism, marking a radical break from the discourse on technology which dominated the Fordist phase of capitalism. This break can be conceived in terms of the changing responses to the critiques of capitalism in those two phases on capitalism, and the changing role of technology discourse. While industrial technology, the assembly line, the bureaucratic corporation, and the Keynesian welfare state were conceived during Fordism as technologies and techniques that respond to the social critique of capitalism, and so were geared towards mitigating exploitation, network technology, flexible production and employment, and the neoliberal state are conceived during post-Fordism as technologies that respond to the humanist critique of capitalism, that is, are geared towards mitigating alienation (Boltanski and Chiapello 2005). In other words, while during Fordism technology legitimized the social-wide compact between capital, labor, and the state (Habermas 1970), during post-Fordism network technology legitimizes the decomposition of this compact and the constitution of its alternative: privatized relations within the context of a global market.

In the short scope of this chapter I have focused on merely one facet of the transformation to post-Fordism: the flexibilization of employment, that is, the shift from tenured, long-term career paths within a single organization to more precarious and flexible, ad-hoc, individualized employment schemes, or, put broadly, the de-socialization and privatization of socially necessary labor (Greenbaum 1995; Castells 1996; Sennet, 2000, 2006; Bauman 2000, Jessop 1994). In that context, network production is constructed in the digital discourse as amending, through the axial means of network technology, the pitfalls and shortcomings of the Fordist mode of production by responding to the humanist critique of capitalism and harnessing those facets that have been suppressed during Fordism—authenticity, creativity, personal expression, and so forth—into the new mode of production. At the same time that network production responds to these demands it also downplays and even rejects the demands put forth by the social critique of capitalism, the response to which was epitomized by tenure, job security, social security, and mutual responsibility that characterized the Fordist mode of production.

Through a "technologistic" framework (Robins and Webster 1999), which depoliticizes and naturalizes social transformations by analyzing them as technological inevitabilities, the discourse on network production embodies the shift in emphasis from the Fordist work ethics of structure (a cog in the machine), linearity (career advancement), tenure, and security (long-term, mutual commitment), to a post-Fordist work ethics of individualization, privatization, personal expression, creativity, and authenticity.

The discourse on network production, then, legitimizes a shift to a post-Fordist organization of labor and production. The sites where the social organization of work during Fordism and much of industrialism were contained and anchored—the company, the union, the professional, the cadre—are rendered obsolete with the emergence of the prosumer. At the same time that work becomes more meaningful and humane, allowing greater outlet for personal potential, and harnessing amateurish skills and leisure time into social reproduction, it also becomes more privatized and individualized, shifting more risks from capital to labor, and dismantling the social buffer zone offered during Fordism, a buffer zone which favored social equity and personal security over the development of individual potential (Harvey 2005). At the same time that the new spirit of networks offers more engaging roles to individuals in the process of production, it also accepts the individualization, atomization, and privatization of work life. As it promises more flexibility and creativity, such discourse also legitimates greater precariousness, instability, and vulnerability. The two are in fact presented as inextricably linked: a more fruitful and satisfying mode of production for individuals is "preconditioned" on their reconstitution as private, autonomous, and flexible units of prosumption.

The humanist demands for decentralization, dehierarchization, debureaucratization, more individual empowerment and satisfaction—criteria judged by the quality of the labor process—are shown in the digital discourse to go hand in hand with the traditional demands of capitalism for productivity, judged by the quality and quantity of the end result—products). Moreover, both demands are seen as mutually constitutive: more productivity demands that individuals' creativity and personal expression be harnessed to the productive process.

Network technology, in this discourse, plays a pivotal role as a source of social change, moving society closer towards overcoming class antagonism, which dominated industrial capitalism, because it is able to provide technological resolutions that speak to both system and life work. On the one hand, it answers the systemic demands of the economy for rationality, productivity, and growth by facilitating a mode of production which is more rational and efficient, and musters more creative and productive forces. On the other hand, network technology answers the demands for human emancipation and liberation through life activity, through work. In this respect, the discourse on network production represents not disengagement with the traditional critique of capitalism, but a continuation thereof. According to this discourse, then, network technology betters capitalism in two senses, responding to the inherent demands of capitalism for infinite accumulation, while at the same time transforming its operation to respond to the humanist demands regarding alienation, lack of satisfaction and creativity, and so forth.

The purpose of this chapter has been to uncover some of the hegemonic dimensions of the new spirit of capitalism as they pertain to network production. This perspective does not mean to downplay the novelties and liberating potentials encapsulated in this new mode of social production. It does, however, point to the degree to which this new ethos, commonly described as a challenge to capitalism, in fact represents the new spirit of capitalism itself. The narratives regarding the bottom-up, democratic, participatory, inclusive character of network production constitute "The New Spirit of Capitalism" (Boltanski and Chiapello 2005), or "the spirit of informationalism" (Castell 1996: 195ff.). This new spirit marks a watershed in the political culture of Western capitalist societies concerning work, employment, and other facets of contemporary capitalism, such as the new constellation of power between markets and state (Fisher 2007), moving us from a political culture which gains its legitimacy from the promise of a safety net to that of the Internet, promising individuals more engaging participation with the reproduction processes of society.

## Bibliography

Aglietta, M. 2000. *A Theory of Capitalist Regulation*. New York: Verso.

Anderson, C. 2006. "People Power." *Wired*, July.

Aronowitz, S. and W. DiFazio. 1994. *The Jobless Future: Sci-Tech and the Dogma of Work*. Minneapolis: University of Minnesota Press.

Bauman, Z. 2000. *Liquid Modernity*. Cambridge: Polity Press.

Bijker, W. 1995. *Of Bicycles, Bakelites, and Bulbs: Toward a Theory of Sociotechnical Change*. Cambridge: MIT Press.

Boltanski, L. and È. Chiapello. 2005. *The New Spirit of Capitalism*. London: Verso.

Borgmann, A. 1988. "Technology and Democracy." In *Technology and Politics*, eds. M. Kraft and N. Vig. Durham: Duke University Press.

Bourdieu, P. 1998a. "The Essence of Neoliberalism." *Le Monde Diplomatique*, December.

Bourdieu, P. 1998b. "A Reasoned Utopia and Economic Fatalism." *New Left Review*, 227.

Braverman, H. 1974. *Labor and Monopoly Capital: The Degradation of Work in the Twentieth Century*. New York: Monthly Review Press.

Castells, M. 1996. *The Rise of the Network Society*. Vol. 1 of *The Information Age: Economy, Society and Culture*. Oxford: Blackwell.

Chouliaraki, L. and N. Fairclough. 1999. *Discourse in Late Modernity: Rethinking Critical Discourse Analysis*. Edinburgh: Edinburgh University Press.

Dickson, D. 1988. *The New Politics of Science*. Chicago: The University of Chicago Press.

Fairclough, N. 1995. *Critical Discourse Analysis: The Critical Study of Language*. New York: Longman.

Feenberg, A. 1991. *Critical Theory of Technology*. New York: Oxford University Press.

Feenberg, A. 1995. "Subversive Rationalization: Technology, Power, and Democracy." In *Technology and the Politics of Knowledge*, eds. A. Feenberg and A. Hannay. Indianapolis: Indiana University Press, 3–22.

Fisher, E. 2007. "'Upgrading' Market Legitimation: Revisiting Habermas' Technology as Ideology in Neoliberal Times." *Fast Capitalism*, 2: 2. Available at: http://www.fastcapitalism.com, accessed October 1, 2010.

Fromm, E. 1968. "Where are We and Where are We Headed?" In *The Revolution of Hope*. New York: Harper & Row, 32–46.

Goetz, T. 2003. "Open Source Everywhere." *Wired*, November.

Greenbaum, J. 1995. *Windows on the Workplace: Computers, Jobs, and the Organization of Office Work in the Late Twentieth Century*. New York: Monthly Review Books.

Habermas, J. 1970. "Technology and Science as 'Ideology.'" In *Toward a Rational Society: Student Protest, Science, and Politics*. Boston: Beacon Press.

Harvey, D. 2005. *A Brief History of Neoliberalism*. Oxford: Oxford University Press.

Harvey, D. 1989. *The Condition of Postmodernity: An Enquiry into the Origins of Cultural Change*. Oxford: Blackwell.

Heffernan, N. 2000. *Capital, Class, and Technology in Contemporary American Culture: Projecting Post-Fordism*. London: Pluto Press.

Herf, J. 1984. *Reactionary Modernism: Technology, Culture, and Politics in Weimar and the Third Reich*. Cambridge: Cambridge University Press.

Horkheimer, M. and T. Adorno. 1976. *Dialectics of Enlightenment*. New York: Continuum.

Howe, J. 2006. "The Rise of Crowdsourcing." *Wired*, June.

Huws, U. 2003. *The Making of a Cybertariat: Virtual Work in the Real World*. New York: Monthly Review Press.

Jessop, R. 1994. "Post-Fordism and the State." In *Post-Fordism*, ed. A. Amin. Oxford: Blackwell, 251–79.

Kelly, K. 1995. *Out of Control: The New Biology of Machines, Social Systems and the Economic World*. Reading: Addison-Wesley.

Kelly, K. 1998. *New Rules for the New Economy: 10 Radical Strategies for a Connected World*. New York: Viking.

Kelly, K. 2005. "We are the Web." *Wired*, August.

Kincheloe J., and P. McLaren. 2003. "Rethinking Critical Theory and Qualitative Research." In *The Landscape of Qualitative Research: Theories and Issues*, eds. N. Denzin and Y. Lincoln. Thousand Oak: Sage Publications, 433–88.

Koerner, B. I. 2006. "Geeks in Toyland." *Wired*, February.

Kuper, A. and J. Kuper (eds.) 1996. *The Social Science Encyclopedia*. London: Routledge.

Lutz, C. and J. Collins. 1993. *Reading National Geographic*. Chicago: The University of Chicago Press.

Marcuse, H. 1991. *One Dimensional Man: Studies in the Ideology of Advanced Industrial Society*, 2nd edn. Boston: Beacon Press.

Mayr, O. 1986. *Authority, Liberty, and Automatic Machinery in Early Modern Europe*. Baltimore: Johns Hopkins University Press.

Mosco, V. 2004. *The Digital Sublime: Myth, Power, and Cyberspace*. Cambridge: The MIT Press.

Noble, D. 1984. *Forces of Production: A Social History of Industrial Automation*. New York: Knopf.

Noble, D. 1995. *Progress without People: New Technology, Unemployment, and the Message of Resistance*. Toronto: Between the Lines.

Nye, D. 1994. *American Technological Sublime*. Cambridge: MIT Press.

Pippin, R. 1995. "On the Notion of Technology as Ideology." In *Technology and the Politics of Knowledge*, eds. A. Feenberg and A. Hannay. Indianapolis: Indiana University Press, 43–61.

Postman, N. 1993. *Technopoly: The Surrender of Culture to Technology*. NewYork: Vintage Books.

Rabinbach, A. 1992. *The Human Motor: Energy, Fatigue, and the Origins of Modernity*. Berkeley: University of California Press.

Robins, K. and F. Webster. 1999. *Times of Technoculture: From the Information Society to the Virtual Life*. London and New York: Routledge.

Sennet, R. 2000. *The Corrosion of Character: The Personal Consequences of Work in the New Capitalism*. New York: Norton.

Sennet, R. 2006. *The Culture of the New Capitalism*. New Haven: Yale University Press.

Sklair, L. 2002. *Globalization: Capitalism and Its Alternatives*. Oxford: Oxford University Press.

Smith, M. and L. Marx (eds.) 1994. *Does Technology Drive History? The Dilemma of Technological Determinism*. Cambridge: MIT Press.

Stevenson, N., P. Jackson and K. Brooks. 2001. *Making Sense of Men's Magazines*. Cambridge: Polity Press.

Sturken, M. and D. Thomas. 2004. "Introduction: Technological Visions and the Rhetoric of the New." In *Technological Visions: The Hopes and Fears that Shape New Technologies*, eds. M. Sturken, D. Thomas, and S. J. Bell-Rokeach. Philadelphia: Temple University Press, 1–18.

Stake, R. 1994. "Case Studies." In *Handbook of Qualitative Research*, eds. N. K. Denzin and Y. S. Lincoln. Thousand Oaks: Sage Publications, 236–47.

Wajcman, J. 2004. *TechnoFeminism*. Cambridge: Polity.

Weber, M. [1921] 1978. *Economy and Society: An Outline of Interpretive Sociology*, eds. G. Roth and C. Wittich. Berkeley: University of California Press.

Webster, F. 2005. "Making Sense of the Information Age: Sociology and Cultural Studies." *Information, Communication & Society* 8 (4): 439–58.

Williams, R. and D. Edge. 1996. "The Social Shaping of Technology." *Research Policy* 25: 856–99.

Winner, L. 1977. *Autonomous Technology: Technics-out-of-Control as a Theme in Political Thought*. Cambridge: MIT Press.

*Wired*. 2006. Cover. February.

Yergin, D. and J. Stanislaw. 1998. *The Commanding Heights: The Battle between Government and the Marketplace that is Remaking the Modern World*. New York: Simon and Schuster.

# 10.1
# Work and Skills in the Telecommunications Industry

*Owen Darbishire*

The telecommunications industry has been transformed over the past 25 years. The switch from an analogue to digital and now IP-based architecture has created the core infrastructure of the information age. This is reflected in the growth in the share of national income it represents, from 2.3 to 3.0 percent in Britain. As a bellwether industry, it provides critical insights into the impact on society concerning changes in employment, its distribution among occupations, including technicians, and changing patterns of skills.

Having followed a U-shaped path, aggregate employment in the industry has ended unchanged over the past 25 years. However, this disguises dramatic falls at the incumbent fixed line operator, British Telecom (BT), following the digitalization of the network. Subsequently, from the mid-1990s, there has been a rapid, off-setting growth within the mobile phone sector, illustrating a substantial shift in corporate power, as BT's share of employment and revenue has fallen well below 40 percent.

Paradoxically for such a high-technology industry, there has been a substantial decline in the technical composition of the workforce, with a shift to customer service and, particularly, management employment. Between 1984 and 2005, the proportion of technicians at BT declined from 54 percent to 43 percent of employees, while the proportion of managers doubled to 30 percent. More dramatically, there has been a 60 percent decline in the absolute number of technicians, as previously routine (if nevertheless skilled) technical and maintenance tasks have been subsumed into the technology, coupled with remote network management, demanding expert thinking and non-routine "intellective" problem solving.

This decline has not been compensated by a corresponding growth in technicians within the mobile companies, which can be characterized as management and marketing organizations. Strikingly, although mobile firms are highly capital intensive (with a labor–capital ratio of 10 percent, compared with 25 percent for fixed-line companies) the proportion of managerial employees is typically 40–50 percent, with under 10 percent of technicians. These firms rely on highly skilled engineers to manage the networks remotely. Technical capability is built into the network equipment and only complex, non-routine tasks require manual intervention. The role of skilled, non-managerial technicians is minimized.

At the same time that this high technology industry has reduced its overall reliance on skilled technicians, there is no evidence of either deskilling or a polarization of skills. Indeed, the skill profile and classification of those that remain has increased, with top graded technicians increasing from 41 percent to 54 percent. Although there is some increased monitoring, substantial task autonomy remains, with long standard task times.

By contrast, work in the more female dominated customer service centers, where part-time and agency work has expanded significantly, has become increasingly segmented by market value. Taylorist approaches have been adopted in high volume residential sectors and quality assessed by "conformance to delivery of pre-defined specifications."

The development of IP-based next generation networks (NGNs) poses new challenges. For technicians, it is certain to result in a reduction in aggregate employment as 16 over-lay networks are replaced by a single architecture creating simpler network management. Problem solving techniques and complex communication may become more important and there is unlikely to be significant deskilling.

NGNs do, however, facilitate an expansion of the potential product portfolio considerably beyond that implied by concepts such as "triple play." The challenges within customer service centers are substantial. Given the importance of cross-selling, the strategic challenge is to ensure that customer service representatives have sufficient understanding of the fast-changing product base and can both interrogate the knowledge management systems and substantively understand the products as they interact with customers. Even though BT spends nearly 8 percent of its wages in direct and indirect training costs, the development of a dynamic skill-enhancement training system is required. This is in contrast to the traditional "periodic training" model, which had an estimated knowledge retention rate for complex products of only 20 percent.

## Bibliography

Batt, R. and H. Nohara. 2009. "How Institutions and Business Strategies Affect Wages: A Cross National Study of Call Centers." *Industrial and Labor Relations Review*. 62(4): 533–52.

Darbishire, O. 2008. "Next Generation Networks: Technological Change, Employment and Skills." *International Journal of Technology, Knowledge and Society* 2(7): 49–55.

Doellgast, V. 2009. "Still a Coordinated Model? Market Liberalization and the Transformation of Employment Relations in the German Telecommunications Industry." *Industrial and Labor Relations Review* 63 (1).

Levy, F. and R. J. Murnane. 2004. *The New Division of Labor: How Computers are creating the Next Job Market*. New York: Russell Sage Foundation.

Ross, P. and G. J. Bamber, 2009. "Strategic Choices in Pluralist and Unitarist Employment Relations Regimes: A Study of Australian Telecommunications." *Industrial and Labor Relations Review*. New York 63 (1): 24–41.

# 10.2
## US Policy Approaches to Digital Labor

*Madeline Carr*

Although digital labor is often discussed in the context of the private sector, governments have also been closely engaged with the development of this element of the workforce. In America, there have been strong political links between digital labor and ideas of American power—both material and social—and these ideas have been instrumental in shaping digital labor. A study of US political history of the late 1980s and 1990s, when the Internet was still in its infancy, reveals a proactive policy and legislative approach designed to enhance the US digital workforce in order to ensure future prosperity.

There were a number of important political transitions which contributed to the American government's early investment in digital labor. For many Americans, one of the key expectations to arise from the conclusion of the Cold War was that US military spending would be diverted to other areas[1] (Russett, Hartley and Murray 1994; Bartels 1994). Without the constant and imposing threat of nuclear war with the USSR, it was expected that funds would be freed up for other uses and speculation began on how best to spend the "peace dividend." While campaigning for president on a platform of economic renewal and strengthening global networks, Bill Clinton declared a preference for spending some of those funds on redefining the workforce. He argued that America needed "a plan to transfer the talents of our people from defense abroad to progress and prosperity at home" and that in the new information economy, American prosperity would depend on the capacity of workers to change, a change which the government was bound to facilitate.[2]

This link between new technology, the future of the workforce, and American economic power was intensified when Al Gore joined the administration as Vice President. He had, for some years, been part of a growing political movement known as the "Atari Democrats," who advocated large-scale government investment in new information and communications technology as a pathway to future American prosperity. As a Senator, Gore had introduced the High Performance Computing and Communication

Act of 1989, widely acknowledged as one of the key legislative platforms for America's later success in emerging technology. Introducing this Bill to Congress, Gore likened electronics in the information age to coal and iron in the industrial revolution, arguing that the "nation which most completely assimilates high-performance computing into its economy will very likely emerge as the dominant intellectual, economic and technological force in the next century."[3]

This combination of available funds, Clinton's emphasis on the economy, and the predominance of the Atari Democrats worked for a time to promote and support the growth of digital labor in the US. Vision and a proactive policy approach seemed to be richly rewarded by the huge economic and workforce benefits of the dot.com boom in the late 1990s. However, by the time the bubble burst in 2000, the American digital workforce was beset by a range of complications and obstacles. The IT recession resulted in a contraction of workplace opportunities; the steady trend of IT outsourcing to cheaper overseas labor markets was having an impact on low skilled digital labor; and by 2010, America's broadband performance had slipped to below that of 18 other OECD states, leading to speculation that the next wave of innovation may take place outside the US.[4]

The US Department of Labor projects continued high growth in demand for digital workers and as those skills not only permeate more and more of the workplace, but also to some degree make workers without them redundant, managing digital labor in the US will continue to present policy challenges for the government.[5] Understanding how governments regard the digital workforce as part of overall state power can provide insight into those decisions which ultimately help to shape and define digital labor.

## Notes

1. Professor Larry M. Bartels conducted a study of public policy preferences for defense spending in the post Cold War years. He found that "the dissolution of the Soviet Union produced marked changes in the defense spending preferences of politically informed Americans" and political elites but relatively little shift in opinion among the politically disengaged.
2. Remarks of Governor Bill Clinton, Wharton School of Business, University of Pennsylvania, Philadelphia, April 16, 1992. Available online at http://www.ibiblio. org/nii/econ-posit.html, accessed January 21, 2009.
3. Senator Albert Gore Jr., speech to Congress introducing the National High-Performance Computer Technology Act, reprinted in *Congressional Record* 135 (64), May 18, 1989. Full text available at http://w2.eff.org/Infrastructure/ Old/s1067_89_gore_hpc.bill, accessed February 11, 2009.
4. Statistics on this vary somewhat. This refers to a report recently funded by the Federal Communications Commission, *Next Generation Connectivity: A Review of Broadband Internet Transitions and Policy from around the world* (Harvard: The Berkman Center for Internet and Society, February 2010).

5. For statistics on this, refer to *Occupational Outlook Quarterly*, published by the US Department of Labor. Available online at http://www.bls.gov/opub/ooq/ooqhome.htm, accessed March 6, 2010.

## Bibliography

Bartels, L. M. 1994. "The American Public's Defense Spending Preferences in the Post-Cold War Era." *Public Opinion Quarterly* 58 (3) Autumn: 479–508.
Russett, B., T. Hartley and S. Murray. 1994. "The End of the Cold War, Attitude Change, and the Politics of Defense Spending." *PS: Political Science and Politics* 27 (1), March 1994: 17–21.

# 10.3
## Employability and Sustainability in the Graduate Job Market: A Case Study of ICT Graduates in Malaysia

*Suriyani Muhamad*

The emergence of digital technologies and the transformation to a knowledge-based economy has created a significant demand for workers who are highly skilled in the use of information and communication technology (ICT). One of the significant impacts of the rapid growth and diffusion of ICT is a shortage of such highly qualified workers. However, the shortage is not just associated with there not being enough ICT workers. Currently, in Malaysia, there is also a major concern expressed by employers that the shortage of ICT workers is due to a skills shortage and employability issues, particularly among ICT graduates.

As ICT has become an important enabler in Malaysia's development, efforts have therefore been made to increase the number of ICT personnel, especially graduates, in order to fulfill the increasing needs of the ICT workforce (Ed. Dev. Plan Malaysia 2001). On the other hand, most employers have reported difficulties in finding candidates with a degree in ICT, good transferable skills, and experience in ICT-related jobs. The lack of transferable skills is perceived as a key factor contributing to the high unemployment among ICT graduates in Malaysia, as seen from the employers' perspective (Czarina 2009; *Malay Mail* 2005a, 2005b).

Despite the increasing importance of transferable skills for employability, there has been little study of how important transferable skills are in ICT graduates' job placement or job profile, particularly in Malaysia. Some of the literature which discusses the importance of transferable skills in relation to graduates' employability includes Knight and Yorke (2002), Mason et al. (2003), Warn and Tranter (2001), Yorke and Harvey (2005) and Yorke (2006). This case study attempts to address this gap by examining the importance of transferable skills in ICT graduates' jobs as well as from the employers' perspectives.

The empirical focus of the case study involves quantitative study undertaken through a questionnaire survey.[1] Four different categories of skills (business knowledge, specialized technical skills, programming, and transferable or soft skills) were listed in the questionnaire. Transferable skills are the core of the analysis, as demanded by employers. The data was analyzed to observe and establish significant differences of opinion on the important of skills and competencies between ICT personnel in their job tasks and ICT students in their industrial training placement. In particular, the analysis aimed to identify the extent of the importance of transferable skills to the two groups. The analysis showed that all the listed transferable skills are statistically significantly different in importance between ICT personnel and students, and are more significantly important to ICT personnel. The obvious explanation is that ICT personnel who have worked in the industry for at least two years have gained real working experience. From this experience, they have realized that despite the importance of technical and programming skills, transferable skills are increasingly becoming fundamental in complementing their specialized ICT skills. Meanwhile, ICT students' industrial training placements are too short to give them as much experience as ICT personnel, and they have limited time for learning and attaining the important and relevance skills, including transferable skills. This is also related to the nature of learning and acquiring transferable skills, which requires students to have the inner capability of developing them. For example, for analytical and conceptual thinking, a person needs logical thinking, which is not specifically stated in a book. Conversely, this skill can be attained through experience. It may also be related to the tacit knowledge which reinforces the idea of effective learning through experience. Hence, these transferable skills are not the same as the technical skills which can be learned manually or from study. Generally, these findings confirm and justify the crucial need for transferable skills, as highlighted above.

By and large, most employers want employees who are going to be effective in today's changing economy. Generally, this case study supported the need to produce graduates with a "work ready" attitude, and the importance of transferable skills in producing "sustainable workers." In Malaysia, the issue of high unemployment among ICT graduates is not about shortage of supply, but more about the mismatch between the supply and demand for ICT graduates.

## Note

1. In study 1, questionnaires were distributed to ICT personnel who had a diploma or degree in ICT and were directly involved in an ICT-related occupation in private companies, government offices, and related organizations who were representative of the population of ICT graduates. A total of 300 questionnaires were distributed, of which 246 were adequate for research needs. In study 2, questionnaires were distributed to final-year ICT students who had undergone an industrial training program from three local public universities. Again, 300 questionnaires were distributed, but only 220 were suitable for the research needs. Implicitly, this survey was to elicit information on skills and competencies emphasized by employers. Instead of getting information directly from employers, this research used a reverse

technique by seeking it from employees—ICT personnel. The rationales for using this approach are the difficulty in obtaining responses from employers, particularly in large numbers; and that this research aims to obtain a more detailed opinion from employees (as there tends to be a lack of information from employees compared with employers, where most of the stated views are from employers). A possible limitation of the approach is the bias in information gathered from employees instead of employers themselves. However, this limitation is small, because of the large number of ICT personnel who have worked for more than seven years. There were 74 (30.1%) of the ICT personnel who have worked for more than seven years and this was the second largest number of the respondents. This group is the more senior managers and mature workers within the firm, and they can indeed be used as a proxy for the views of the firm on skills and competencies.

## Bibliography

Czarina, A. 2009. "Enhancing Employment through Finishing School." Presentation to CareerXcell Sdn Bhd, Malaysia (July 21).

Education Development Plan for Malaysia 2001–2010. Available at http://planopolis. ieep.unesco.org/upload/Malaysia/Malaysia Education development plan 2001-2010 summary, accessed March 6, 2010.

Knight, P. and M. Yorke. 2002. "Employability through the Curriculum." *Tertiary Education and Management* 8 (4): 261–76.

*Malay Mail* 2005a. "Unemployed Grads: Market Yourselves!." *Malay Mail* (April 2).

*Malay Mail* 2005b. "Too Many Grads in Business Administration, ICT and Engineering Disciplines." *Malay Mail* (April 11).

Mason, G., G. Williams, S. Cranmer and D. Guilde. 2003. *How Much Does Higher Education Enhance the Employability of Graduates?* Available from http://www.hefce. ac.uk/Pubs/RDreports/2003/rd1303/.2003, accessed August 4, 2009.

Warn, J. and P. Tranter. 2001. "Measuring Quality in Higher Education: A Competency Approach." *Quality Higher Education* 7 (3): 191–8.

Yorke, M. reprinted 2006. *Employability in Higher Education: What It is—What It is Not?* York: Higher Education Academy.

Yorke, M. and L. Harvey. 2005. "Graduates Attributes and Their Development." In 'Workforce Development and Higher Education: A Strategic Role for Institutional Research', *New Directions for Institutional Research*, Volume 2005, Issue 128, pages 41–58, Winter.

# Part XI
# Technology, Culture, and Society

# 11
# The Empirical Case for Taking a Technosocial Approach to Computing

*David Hakken and Maurizio Teli*

## Introduction

This chapter aims to give an empirical grounding to the argument for broadening how computing is considered. It explains why it is necessary to overcome the dominant, primarily technical perspective of academic computing studies via a technosocial view, in favor of a technosocial perspective. Along the way, it documents how a substantial portion of the problems people have in the design, implementation, and maintenance of computer-based systems that pretend to support human information practices is attributable to imbalances between and improper framings of the attention given respectively to social and technical perspectives on what it means to compute.

Understanding the specific ways in which these imbalances and improper framings are sources of problems with computing is one concern central to our current research on socially robust and enduring computing (SREC), conducted together with Vincenzo D'Andrea of the University of Trento in Italy. In this research program, articulating the specific problems manifest in the dominant technicist view, the chief goal of this chapter, is an essential preliminary to fostering more effective forms of computing. The outcome should be, in our research intention, to build systems that are truly supportive of social practices, therefore being "socially robust," able to support human actions even when the social is in stressful conditions. This kind of task requires a deep understanding of the relationship between computing and social phenomena. First of all we need a comprehensive account of the empirical evidence on extent to which, and the forms in which, the contemporary problems of computing are in actual practice connected to insufficient and improper attention to the social.

This chapter is an initial attempt at such an account. In it, we begin by expressing the sense in which we have approached "the empirical." After the survey of empirical evidence, which takes up the bulk of the chapter, a brief conclusion places this empirical reading in the context of the rest of the SREC research program, including how we think the social can be integrated practically into academic computing studies. If a way were to be found in which the social and the technical in computing were more properly balanced and framed than they are in current dominant approaches to computing artifacts and systems, as well as computer-related practices, we would be much more effective at supporting human information activities.

## From the transcendental to the empirical

At least in the popular mind, all sciences base their research programs on explicit, shared statements of what is already known, what is not yet known, and what needs to be understood next. These ideally provide the starting place for empirical science. This process is one of constructing a transcendental realm of knowledge, a corpus of hypothesis that is translated into empirically verifiable sentences that are tested through research practices, and are materialized in the output of the research process (Latour and Woolgar 1979). Beyond this, however, as the sense of what it means to "be empirical" is given widely diverse meanings in different academic and policy arenas, there is a great deal of diversity regarding how what is known, and not known, is established. Some, for example, would only consider those forms of knowledge to be "empirical" that can be demonstrated to be so via laboratory or controlled condition activities, designed to test quantitatively an explicit hypothesis on the relationship between a dependent and an independent variable. This is the direct result of modern science, which considers nature and technology as "out there," transcendent truths to be discovered via research (Latour 1999).

Via framings like these that privilege the quantitative and the formal, the academic discipline of computer science largely ignores or marginalizes phenomena difficult to treat in this manner, such as the social is often considered to be, when designing, building, testing, implementing, or maintaining computer-based systems to support human information practices. This is because, in modern science, the social is inherently immanent, unable to be grasped through an approach that focuses on discovering the transcendental. However, interestingly, the discipline turns a blind eye the relative dearth of laboratory activities in its own practices!

Instead, we will approach the term "empirical" differently and more broadly, as the immanence of "the actual" or "what really exists." Thus, our basic aim is to illuminate the extent to which what is actually known about existing computing is compatible with the following hypothesis: that the social characteristics of the ways in which the systems have been designed, implemented, and maintained, those of the use situation, are linked to the many problems manifest when computer-based systems are used. These links cannot only be perceived and identified. Furthermore, in SREC research, the strength of these links justifies rethinking how social and technical perspectives should be articulated in computing practices.

In what follows, our empirical evidence is gleaned from two primary sources. One is the practice of computing professionals, as manifest largely in what they say about this practice, when there is reason to believe that their descriptions are accurate and offered in good faith. Our second source of empirical information is the, by now, vast body of research undertaken by scholars who study computing but who do so mostly from outside academic computer science and engineering and thus are less likely to privilege the formal and quantitative *a priori* but start with identifying, in the mediation between technologies and humans, the key aspects in computing research and development.

An exhaustive compilation of the vast amount of the relevant empirical data would extend far beyond a single chapter. Only multiple scholars with deep knowledge of their own field, plus the ability to talk across disciplines— something that is in practice systematically discouraged by most contemporary academic organizations—could pull this together. Yet something along these lines is essential to the program SREC is pursuing. What follows is thus an early attempt to state what this might look like. A key step in the SREC research program is to spell out and support what we know, in the form of a set of interlocked propositions about computing and the social. Here we identify three of these, which constitute the basis for future research and theoretical reflections:

–That there have long been and continue to be problems in making computing systems that endure, and that many of these problems appear linked to relevant social factors which are being ignored or related to badly. Therefore there is a case for re-thinking computing.

–That within the academic disciplines and professional practices that have accompanied computing there has arisen an alternative tradition that acknowledges, rather than ignores, the links between computing problems and improper or inadequate attention to the social, one that does so in self-conscious opposition to the mainstream and requires legitimacy in the computing field.

–That scholars in academic disciplines and professions outside of academically institutionalized computing studies, for example computer science, have also attended to the strong but complex social correlates of computing relating to social change and computing problems. This provides an intellectual grounding tool for future research.

Systems do not endure and developers underestimate the social actors.

There is considerable evidence that computing failure is widespread and costly. According to Infonetics Research (2005), IT downtime costs large US enterprises 3.6 percent of their revenues per year, with manufacturing organizations losing 9 percent and financial services organizations losing 16 percent. Some computing failures, such as those attendant on the current international socioeconomic crisis, often have disastrous consequences for the functioning of social formations (Hakken 2003). A factor often cited in accounts of this crisis is an over-reliance on computed algorithmic models of economic risk, in particular, models that failed to incorporate the "black swan" (Taleb 2007) of possible systemic failure.

A number of other statements by computing professionals characterize the past level of computing failure as high, while also indentifying a substantial continuing risk of failure. In addressing the "dismal history of IT projects gone awry," Robert Charette (2005) presents a table, labeled the "Software Hall of Shame," of specific, notable fiascoes, then adds:

> Most IT experts agree that such failures occur far more often than they should. What's more, the failures are universally unprejudiced: they happen in every country; to large companies and small; in commercial, nonprofit, and governmental organizations; and without regard to status or reputation. The business and societal costs of these failures – in terms of wasted taxpayer and shareholder dollars as well as investments that can't be made – are now well into the billions of dollars a year.

The problem only gets worse as IT grows ubiquitous. This year, organizations and governments will spend an estimated $1 trillion on IT hardware, software, and services worldwide. Of the IT projects that are initiated, from 5 percent to 15 percent will be abandoned before or shortly after delivery as hopelessly inadequate. Many others will arrive late and over budget or require massive reworking. Few IT projects, in other words, truly succeed.

> Like many who deal with the issue of failure, Charette specifically links it to not taking the social actors seriously, and in this case not taking the steps necessary to cope with the high likelihood of failure: The biggest tragedy is that software failure is for the most part predictable and avoidable. Unfortunately, most organizations don't see preventing failure as an urgent matter, even though that view risks harming the organization and maybe even destroying it.
>
> (Charette 2005)

The failure of computer-based risk models to incorporate the "fat tail" possibility of systemic failure is another example of inattention to the social. Their pronounced technicality meant these models generally hid, rather than made manifest, dependence on overtly social theories, like the "efficient market" hypothesis. Technicality increases the chances that such presumptions, when transported from their field of origin, are shorn of caveats and qualifications (Fox 2009). The problem here is computing that effectively ignores the social, transcends its dynamics, and does so in a hidden manner, rather than incorporating potentially crucial aspects of the social into practices built on such models.

Other forms of computing do take note of the social, but do so simplistically. This practice, in J. Beckett's analysis of the computing problems of the British National Health Service (NHS) (2009), is an important source of failures. He traces many of these to the insistence on a "one size fits all" computer system that could not cope with the diverse practices of the NHS's multiple units and unit types. A recent analysis of the conversion of several of Indiana University's legacy systems to a single enterprise planning system (EPS) (Roland and Gieryn 2008) similarly points out the simplifications forced by standard, formal 'transaction cost' models, leading to underestimating difficulties. Indeed, EPSs are prone to do so, as their modeling privileges attention to things like initial customizations costs over the problems that have to be attended to as they come up. Transaction costs can only be developed for problems that one can anticipate, not, for example, for problems that only emerge when trying to migrate legacy data from one domain to another.

The range of computing forms in which these problems manifest themselves is so broad as to suggest that failure to deal at all, or at all adequately, with the social is endemic to real, existing computing. In 1994 Jonathan Grudin chose to write on "Groupware and Social Dynamics: Eight Challenges for Developers." Grudin's abstract frames the central problems of groupware failure as arising from inadequate attention to the dynamics of different kinds of social relationships:

> Computer support has focused on organizations and individuals. Groups are different. Repeated, expensive groupware failures result from not meeting the challenges in design and evaluation that arise from these differences. Many expensive failures in developing and marketing software that is designed to support groups are not due to technical problems. They result from not understanding the unique demands this class of software imposes on developers and users.
>
> (Grudin 1994: 93)

After examining the origins of groupware, describing eight specific problem areas, and describing some groupware projects that succeeded in finding better approaches to supporting work in group settings, Grudin concludes by stressing the continuing need to ask elemental social questions, such as,

> Was this software designed to support groups? Is it being used to support groups? Email and b-boards are well known, but few other groupware prototypes and products have done as well despite considerable effort. Successes exist, but progress is slow and can lead in unanticipated directions.
>
> (Grudin 1994: 93)

The kind of computing Grudin describes as so difficult to do well is generally called computer-supported cooperative work (CSCW), an area that has
been of considerable interest in computing studies for more than 20 years.
CSCW arose partly, as its name implies and as Grudin attests, in recognition
of the weaknesses of computing in dealing with a specific kind of sociality.
Yet despite lengthy and intense efforts, including Grudin's intervention,
CSCW's challenges remain the same. The title of Mark Ackerman's forthcoming article in *Human Computer Interaction* is "The Intellectual Challenge
of CSCW: The Gap between Social Requirements and Technical Feasibility."
Parts of Ackerman's abstract follow:

> Over the last 10 years, Computer-Supported Cooperative Work (CSCW)
> has identified a base set of findings. These findings are taken almost as
> assumptions within the field. In summary, they argue that human activ-
> ity is highly flexible, nuanced, and contextualized, and that computa-
> tional entities such as information transfer, roles, and policies need to be
> similarly flexible, nuanced, and contextualized. However, current systems
> cannot fully support the social world uncovered by these findings.

Why, after all the effort, is this still the case? Ackerman goes on:

> This paper argues that there is an inherent gap between the social require-
> ments of CSCW and its technical mechanisms. The *social-technical gap* is
> the divide between what we know we *must* support socially and what we
> *can* support technically. Exploring, understanding, and hopefully amelio-
> rating this social-technical gap is the central challenge for CSCW as a field
> and one of the central problems for HCI.
>
> (Ackerman n.d.: 1)

Despite differences in time, their similar articulations of the centrality of the
social to computing failure is perhaps why these two articles were among
those distributed in 2010 to participants in one of a series of US National
Science Foundation (NSF) sponsored workshops on establishing an integrated
sociotechnical research agenda in computing. Not only is the need for such an
agenda in computing recognized by the leading funder of scientific research
in computing but this funder also identifies inattention or poor attention to
the social as a main source of computing failure.

Like the NSF and the organizers of these workshops, Charette sees a research
issue in this failure to take failure seriously, an issue important to the future
of computing in general, but not just to one or two of its particular forms:
"Understanding why this attitude persists is not just an academic exercise; it
has tremendous implications for business and society" (2005). That difficulties in computing are generally associated with difficulties with the social has
long been recognized as basic to problems in system specification, articulating

exactly enough what a system is for. That is, in order to approach the building of a computer-support system as a purely technical practice, one must presume that those for whom one is building it already know precisely what it is for, they know this from the beginning, and they can express all of this with precision. Most individuals, groups, and organizations actually find this very difficult, however, partly because of the extreme difficulty of knowing in advance what practices will have to be altered to make them computable and thus have specifiable parameters. Moreover, in a domain of rapid technical change like computing, those practices that are worth making computable change rapidly. In other words, system specification must be a deeply dialogic activity, socially informed, and not merely technical. Much the same, in general, is true of the other moments in building a computer-based system to support human information practices. When not approached in this manner, of course systems fail.

## The alternative tradition in computing

Were CSCW peculiar in placing the understanding of a specific type of sociality at the center of successful computing, one might with justice argue that the history of computing demonstrates the superiority of the purely or largely technical approach to the problem of failure. However, CSCW is far from unique. While privileging the technical and trying to avoid the social, if at all possible, is characteristic of mainstream practice, this is far from being the only course that computing studies have actually followed. Rather, CSCW is heavily informed by a long and arguably noble tradition that has tried and continues to try to place the social at the center of computing (Torvalds 2001: Webster 2006). In this tradition, a primary reason why computing does not meet intended objectives, or systems become quickly outdated, is the failure to match technical with social virtuosity.

An early advocate of this position was Norbert Wiener (1894–1964), one of the founders, if not the founder, of the computing field, at least in the US. Wiener opposed the Shannon/Weaver conception of the field as an information science based on a technical definition of information as signal to noise ratio, and which systematically ignored the content that made a signal information tradition. For him, the disciplined study of information, cybernetics, should attend equally to content and form; see, for example, his *Cybernetics and Society* (1948). He also opposed von Neuman's purely formal approach to computing architecture. Rather, he placed understanding the substantive social processes that give information meaning, and thus content, at the center of cybernetics. Like Wiener, I find it hard to accept as scientific a study of information that diverts attention from the things that make it what it is, resulting in a so-called information science that marginalizes attention to what makes something information rather than noise. Wiener's MIT colleague Joseph Wiezenbaum (1976) picked up his critique of computer science as a deracinated study of information, as did Terry Winograd at Stanford (Winograd and Flores 1986).

In Europe, the critical alternative was equally vibrant. It tended to recognize attention to culture as important to successful use of computing, and knowledge of social practice/action and applied social research were integrated into computing languages (see Nygaard's work on SIMULA, 1961–5) and system development (Ehn 1988). Although sharply manifest in Nordic user participation in system development (Bjerknes and Bratteteig 1995) and American participatory design (Blomberg and Suchman 1992), this perspective had continuing broad influence.

This alternative perspective has been central to the development of human-computer interaction (HCI), regularly now a part of many departments of computer science (Giddens 1991). Its organizational forms have included *ur*-meetups like Computers for People (Hakken 1993) and the Home Brew Computing Club in Palo Alto, California, whose members critiqued the political implications of technically centralized computing (and whose activity greatly influenced the Apple machine architecture). Intellectual forms of this alternative included social informatics (Bowker et al. 1997). This alternative tradition is perhaps most manifest today in the practices characteristic of Free/Libre and/or Open Source Software (FLOSS) development, especially the way it is supposed to be developed. Controversies over individual motivation, social dynamics, and broader cultural implications are central to the diverse approaches within FLOSS; indeed, the central argument for why it is a better way to develop software is the superiority of its social arrangements. The debate over FLOSS has set the terms of the current exultation over Web 2.0 and the other forms of so-called social computing (Shirky 2008).

A good way to grasp the influence of this alternative critical tradition is to focus on the specific history of software development. One of the first approaches to this moment in computing was the Waterfall model, essentially an adaptation of the then standard general procedures for engineering any artifact. A chief characteristic of this approach is its technicist nature, including a crisp separation of activities, resulting in the need for separate job titles and consequently increased coordination among phases and people. The first description by Royce of the application of this model to software development in 1970 identifies several risks inherent in it. It is interesting to note that he includes "user involvement" as one of five measures to mitigate these risks (Royce 1970). Thus, from the beginning, there is an acknowledgment of the need to include explicitly social practices in software production, even if the way in which this might conflict with primary engineering orientations remained unexamined.

Later, the Rational Unified Process (RUP) became the mainstream, de facto standard approach in the software development industry (Kruchten 2000). RUP has both a technical component, specifying the techniques to be used in the various development activities, and a processual component, expressing the practices to be associated with managing the development process. One of the distinctive aspects of RUP, as opposed to the Waterfall model, is the articulation of the software development life cycle as a social notion, one of business decision points, whereas the Waterfall stages were connected only to technical deliverables. In RUP there is also provision for some case-by-case tailoring, albeit only within a rather strict set of rules concerning roles and responsibilities (TP 165A 2002).

Despite proclaimed success in large scale software projects, RUP-informed development has only with difficulty addressed the need for loosely coupled systems and decentralized organization characteristic of more recently developed Service Oriented Systems (Ivanyukovich et al. 2005). RUP has also been criticized for demanding excessive management effort and over-structuring the software life cycle.

Critiques like these have led to the development of several substantially different approaches to achieving coordination in software development. Agile methodologies, and more specifically Extreme Programming (EP), are probably the most popular alternatives to RUP (Beck 2004). Extreme Programming involves radical circulation of personnel among project roles, using a social means to focus the development process more effectively on the final goal, the software system. In EP, coordination is achieved socially rather than administratively. Examples of EP techniques include pair programming, daily rotation of tasks to avoid "ownership," and frequent group meetings to share achievements and issues.

While hoping to subvert mainstream approaches to coordination in software development, Extreme Programming is nevertheless still characterized by a prescriptive approach. A substantially different phenomenon is associated with FLOSS projects (Stallman 2002; Torvalds 2001). Its deviations from mainstream programming follow from an explicitly different view of the social phenomenon of property rights in software. The new forms of collaboration induced result in an array of new relationships among developers, users, software companies, and other stakeholders. The common trait of these new relationships is their emergent character, which contrasts with the controlled nature of the relationships aimed at by the approaches previously described. Even manuals proposing a somewhat more directed approach in FLOSS projects present a variety of tools and practices from which each developer is to pick and choose (Fogel 2006).

A third socially dynamic approach to coordination in software development is Participatory Design (PD) (Bødker, Kensing and Simonsen 2004; Kensing 2003). Rather than focusing mostly on coding techniques, PD pivots on trying to construct a good relationship between users and developers during software design. PD is structured to bring programmers into frequent, close contact with users and thereby break down the separation of users from the development team. In PD, this separation is seen as one of the causes for failures of software development projects as well as a source of controversies between software producers and consumers.

PD doesn't have the emergent character of FLOSS as a new way to code. In FLOSS, users and coders are often the same. Thus, both approaches aim to place users and developers in roughly symmetrical, socially equal roles. Thus, even in PD, a social interaction is intended to have a substantial impact on the outcome of the technological process. As in software development, so generally in computing: practices in which the shortcomings of a purely technical virtuosity are dealt with, by finding some way to attend to the social, recur regularly. Indeed, manifestations of substantive social attention are so common that one is justified in seeing the alternative critical tradition as necessary to computing.

### Ethnography as a way of considering the social correlates of computing

Over the past 30 years, scholars from numerous disciplines outside of academic computer science have taken up the challenge of creating broad understandings of computing. One strong and influential body of work has drawn on ethnography as a way of knowing, for example, Downey (1998); English-Lueck (2002); Escobar (1994); Forsythe (2001); Garsten (1994); Hakken (2004, 2003, 2000, 1999, 1993); Hakken, Teli, and Pisanu (2007); Haraway (1991); Helmreich (1999); Hess (2001); Ito (n.d.); Jules-Rosette (1990); Miller and Slater (1999); Pfaffenberger (1992); Turkle (1995). Computing ethnography has demonstrated the importance of the social in the design, implementation, and maintenance of computing artifacts as well as systems to support information practices, while at the same time documenting the role of computing in changing work practices, most recently in communications and media. To identify as well as account for the correlates observed, ethnographers have had to address a number of methodological questions, such as how to do translocal fieldwork and conceptualize "cybercultures and cyberspace" (the social spaces engendered by computing), or the question of how much expertise is necessary to gain entry for effective field studies. Addressing humanistic issues like the proper modes of online participation and mediated performance, as well as the ethics of electronic research (Hakken 2000), these ethnographers have, in the process, discovered considerable overlap with that of media scholars (Spitulnik 1993) and others working in fields like communication (Lindlof and Taylor 2002). In order to gain entry to produce their scholarship, these scholars took on a wide variety of roles, from HSF-sponsored researcher to participant in system building, often with the encouragement of those computing professionals described in the previous section.

Diana Forsythe's work is exemplary in these field studies of computing-in-use, as well as of the complex range of issues raised by it. In the 1980s Forsythe found herself working on a project whose "cyberspacing" of knowledge practices were very different from that described by the proponents of computing (2001). The project was to develop a computer-based expert system, a self-diagnostic computer-based tool for people who wondered if they had migraines. The system prompted the 'user' to describe her symptoms. It was programmed to repeat precisely these symptoms, whatever they were, as the proper indicators of migraine!

Further, while the system was capable of recommending a wide range of therapeutic options, it would present only the sub-set of these options selected by the physician or clinic implementing it. Thus, although of different types, both the initial set of systems and the therapeutic options presented were socially constructed selections by the expert system in use, a selectivity that occurred without the user knowing this. Thus, using the system most likely did change the quality of user knowledge, and the change was indeed related to use of computers, but the result is arguably one of greater ignorance, not more knowledge. Nonetheless, the system's design and use was legitimated by its developers as "increasing control by the patient' as well as 'promoting user confidence in the system," in order to "increase the likelihood of patient compliance" (Forsythe 2001).

This is just one way in which the ethnography of computing casts doubt on popular "cyberspace as knowledge society" tropes like those of James Bailey and Peter Drucker. In these readings, cybernauts leap to some new knowledge quantum, from normal to "hyper" knowledge. In another critical perspective, Langdon Winner called such talk "technological somnambulism" (1977). Winner placed 'knowledge revolution' computer talk firmly in the tradition of technological determinism, a worship of machines that fosters policy sleepwalking, so manifest in American thought. In a related *bon mot*, Winner describes our computer-centered society as a "Mythinformation" rather than an "Information" one (1986, 1989).

The intellectual agenda of anthropological ethnographers doing computing studies intersected with other issues of concern to the discipline, such as the presumption of rapid social change as the distinctive characteristic of the late twentieth and early twenty-first centuries. Even in standard anthropology texts, endemic change was generally linked to computing. Much current anthropology addresses how important phenomena popularly claimed to be manifest in many, potentially all existing cultures, for example globalization (Appadurai 1996) or trans-national activities (Hannerz 1996), are all that is important. As computing is now central to the reproduction of contemporary social formations, it is logically a central focus of attention for scholars of social and cultural change (Suchman et al. 1999).

As described in Hakken (2003), current ethnography both assesses the extent, character, and causes of the purported changes and theorizes their broader implications. Like Forsythe, the theorization of issues like computing and social change focuses on practice, what actually happens, as opposed to what is designed or is supposed to happen. It acknowledges that computer-supported information practices, like the spread of computer-mediated communication such as the Internet, and the universalizing affordances of processes like database construction, are generally regarded as likely to generate new practices. However, this approach enquires into whether they actually do so, and, if so, how similar the actual new practices are to those intended. In looking to explain actual deviations from intention, it also examines the similarities and differences in cultural implications between computing and related technoscience practices (Helmreich 1999, Downey 1998), and the range of variation in the relevance of national and cultural background (Forsythe 2001; Fujimura 1992; Traweek 1988). It establishes a correlation between ethnogenesis (development of new cultural forms) and actual use of computers. However, it also problematizes the simplistic, generally optimistic and dazzled conceptions of the computing and social change connection dominating contemporary popular thought.

## Conclusion

Hence, for computing to become socially robust, the field must be redefined in a way that brings the social and the technical into more equal balance. Redefining an institutionally successful academic discipline is difficult, as Thomas Kuhn attested in 1962. It is a very broad intellectual task, of which we in SREC wish to take on small but important parts. To redefine and rebalance well, one needs to be able to say what specific pieces of software lead some computing practices to be less socially robust than others. Starting to specify the role of social flaccidity in failed computing practices, one focus of this chapter, should help. Honoring the alternative tradition in computing studies, in which scholars have attempted to integrate social awareness into computing, is another part of creating an alternative self-definition of computing studies; we believe there is much to be learned from the unfortunately largely marginal experience of such efforts. Additionally, because of the centrality of computing to the reproduction of most contemporary social formations, each of the social sciences and humanities has relevant scholarship on the social correlates of computing. Even in the face of computing practices, such as social media, service-oriented computing, and HCI, in which the social is obviously of relevance, existing academic computing resists taking the social seriously. What the academic study of computing sees when it looks at itself in a mirror does not include much of the social. While such practices of self-definition at the highest level or disciplinary self-organization are doubtless more reflective than constitutive of the most important institutional roadblocks, their existence is indicative of substantial internal impediments to the rebalancing that we see as necessary.

More thorough execution of the things attempted in this chapter, including offering a social reading of failed computing, reviving a focus on previous efforts to articulate and implement socially conscious computing, drawing attention to existing research on computing in the social sciences and humanities, examining the response to current social computing, and identifying forms of self-definition which on their face marginalize the social, are foci of SREC's foundational or "pure" research program. Their more systematic execution should enable us to theoretically articulate computing development practices that are symmetrical with regard to their social and technical aspects that are, in the language we prefer, properly "intercalated" (Galison 1997). Once articulated, we believe such theories can be built into development tools and these tools used to build more socially robust and enduring computing systems to support human information practices. We look forward to being able to study what happens when these tools are used, expecting this experience to point towards more ways to make computing more socially robust and therefore enduring.

## Bibliography

Appadurai, A. 1996. *Modernity at Large: Cultural Dimensions of Globalization.* Minneapolis: University of Minnesota Press,
Ackerman, M. 2000. "The Intellectual Challenge of CSCW: The Gap between Social Requirements and Technical Feasibility." *Human-Computer Interaction*, 15: 179–203.
Beck, K. 2004. *Extreme Programming Explained: Embrace Change* (Second edition). Boston: Addison-Wesley.
Beckett, J. 2009. "System Failure?" *Guardian* (July 9).
Bell, D. 1976. *The Coming of Post-Industrial Society.* New York: Basic Books.
Bjerknes, G. and Bratteteig, T. 1995. "User Participation and Democracy: A Discussion of Scandinavian Research on System Development." *Scandinavian Journal of Information Systems*, 7(1): 73–98.
Blomberg, J. and L. Suchman. 1992. "Field Studies of Work and Co-Design." *Proceedings of PDC 92:* 101–2. Palo Alto CA: CPSR.
Bloor, D. 1991. *Knowledge and Social Imagery.* London: Routledge.
Bødker, K., F. Kensing and J. Simonsen. 2004. *Participatory IT Design. Designing for Business and Workplace Realities.* Cambridge MA: MIT Press.
Bowker, G., L. Gasser, W. Turner and S. L. Star. 1997. *Social Science, Technical Systems and Co-Operative Work.* Hillsdale, NJ: L. Erlbaum Associates.
Callon, M. and B. Latour. 1992. "Don't Throw the Baby out with the Bath Water! A Reply to Collins and Yearly." In *Science as Practice and Culture*, ed. A. Pickering Chicago: University of Chicago Press, 343–68.
Castells M. 1996. *The Rise of Network Society*, Vol. I. Oxford: Blackwell.
Charette, R. 2005. "Why Software Fails." *IEEE Spectrum* (September). Available at http://spectrum.ieee.org/computing/software/why-software-fails/1, accessed September 10, 2009.
Denning, P. J., D. E. Comer, D. Gries, M. C. Mulder, A. Tucker, A. J. Turner, and P. I. R. Young. 1989. "Computing as a Discipline." *Computer* 22 (2): 63–70 (February).
Downey, G. 1988. *The Machine in Me: An Anthropologist sits among Computer Engineers.* New York: Routledge.
Ehn, P. 1988. *Work-Oriented Design of Computer Artifacts.* Stockholm: Arbetslivcentrum.
English-Lueck, J. 2002. *Cultures@Silicon Valley* Stanford University Press.
Escobar, A. 1994. "Welcome to Cyberia: Notes on the Anthropology of Cyberculture." *Current Anthropology* 35 (3): 211–31.

Fogel, K. 2006. *Producing Open Source Software.* Sebastopol CA: O'Reilly.

Forsythe, D. E. 2001. *Studying Those Who Study Us: An Anthropologist in the World of Artificial Intelligence.* Stanford, CA: Stanford University Press.

Fox, J. 2009. *The Myth of the Rational Market.* New York: Harper Collins.

Fujimura, J. H. 1992. "Crafting Science: Standardized Packages, Boundary Objects and 'Translation.'" In *Science as Practice and Culture,* ed. A. Pickering. Chicago: University of Chicago Press.

Galison, P. 1997. *Image and Logic.* Chicago: University of Chicago Press.

Garsten, C. 1994. *Apple World.* Stockholm: Stockholm University Studies in Social Anthropology.

Giddens, A. 1991. *Modernity and Self-Identity.* Cambridge, MA: Polity.

Grudin, J. 1994. "Groupware and Social Dynamics: Eight Challenges for Developers." *Communications of the ACM* 37(1): 93–105.

Hakken, D., M. Teli and F. Pisanu. 2007. "The Internet as a Library-of-People: For a Cyberethnography of Online Groups." *Forum: Qualitative Social Research* 8 (3).

Hakken, D. 2004. "'Non-western' Studies of Cyberspace Identity Formation." *Anthroplogi Indonesia* 73: 32–9.

Hakken, D. 2003. *The Knowledge Landscapes of Cyberspace.* New York: Routledge.

Hakken, D. 2000. "Ethical Issues in the Ethnography of Cyberspace." In *Ethics and Anthropology: Facing Future Issues in Human Biology, Globalism and Cultural Property,* eds. A-M. Cantwell, E. Friedlander and M. L. Tramm. New York: New York Academy of Sciences, 170–86.

Hakken, D. 1999. "Towards and Anthropology of Cyberspace" ("Cap a una anthropolgia del ciberespai"). *Revista d'Etnologia de Catalunya* 14: 8–45. Special Issue on "Culture in Cyberspace: Towards a Digitized Society," ed. M. J. Buxo.

Hakken, D. 1993. "Culture-Centered Computing: Social Policy and Development of New Information Technology in England and the United States." *Human Organization* 50(4): 406–23.

Hakken, D. 1993. "Computing and Social Change: NewTechnology and Workplace Transformation 1980–90." *Annual Reviews in Anthropology* 22: 107–32.

Hannerz, U. 1996. *Transnational Connections: Culture, People, Places.* London: Routledge.

Haraway, D. 1991. *Simians, Cyborgs and Women—The Reinvention of Nature.* London: Free Association Books.

Helmreich, S. 1999. *Silicon Second Nature: Culturing Artifical Life in a Digital World.* Berkeley: University of California Press.

Hess, D. 2001. *Selecting Technology, Science and Medicine: Alternative Pathways in Globalization, Vol.1:* Nyskayuna, NY. Published by author.

Ito, M. n.d. "Networked Localities." Available at http://www.itofisher.com/PEOPLE/mito, accessed January 21, 2009.

Huberman, B. A. 2001. *The Laws of the Web.* Cambridge, MA: MIT Press.

Infonetics Research Report. 2005. *The Costs of Enterprise Downtime.* Available at: http://h71028.www7.hp.com/NonStopComputing/cache/426962-0-0-0-121.html, accessed January 21, 2009.

Ivanyukovich, G. R., V. Gangadharan, M. D'Andrea and M. Marchese. 2005. "Towards a Service-Oriented Development Methodology." *Transactions of the SDPS* 9 (3): 53–62.

Jules-Rosette, B. 1990. *Terminal Signs: Computers and Social Change in Africa.* Berlin; New York: Mouton de Gruyter.

Kelty, C. 2008. *Two Bits: The Cultural Significance of Free Software.* Durham NC: Duke University Press.

Kensing, F. 2003. *Methods and Practices in Participatory Design.* Copenhagen: ITU Press.

Kruchten, P. 2000. *The Rational Unified Process: An Introduction* (second edition). Boston: Addison Wesley.

Latour, B. and S. Woolgar. 1979. *Laboratory Life: The Social Construction of Scientific Facts*. Beverly Hills CA: Sage.

Latour, B. 1999. *Pandora's Hope: Essays on the Reality of Science Studies*. Cambridge MA: Harvard University Press.

Lindlof, T. R. and B. C. Taylor. 2002. *Qualitative Communication Research Methods* (second edition). Thousand Oaks, CA: Sage.

Medina, E. 2005. "Democratic Socialism, Cybernetic Socialism: Making the Chilean Economy Public." In *Making Things Public: Atmospheres of Democracy*, eds. B. Latour and P. Weibel. Cambridge: MIT Press, 708–21.

Miller, D. and D. Slater. 1999. *The Internet: An Ethnographic Approach*. Oxford: Berg.

Pfaffenberger, B. 1992. "Social Anthropology of Technology." *Annual Review of Technology* 21: 491–516.

Roland, N. and T Gieryn. 2008. "Transfer Troubles." In *Living in a Material World*, eds. T. Pinch and R. Swedberg. Cambridge: MIT Press, 375–92.

Royce, W. 1970. "Managing the Development of Large Software Systems." *Proceedings IEEE WESCON 26* (August): 1–9.

Rational Unified Process for Systems Engineering. 2002. "A Rational Software White Paper." TP 165A, 5/02.

Shackelford, R., A. McGettrick, R. Sloan, H. Topi, G. Davies, R. Kamali, J. Cross, J. Impagliazzo, R. LeBlanc and B. Lunt. 2005. "Computing Curricula." *Proceedings of the 37th SIGCSE Technical Symposium on Computer Science Education*, 2006.

Shirky, C. 2008. *Here Comes Everybody: The Power of Organizing without Organizations*. London: Allen Lane.

Spitulnik, D. 1993. "Anthropology and Mass Media." *Annual Review of Anthropology* 22.

Stallman, R. 2002. *Free Software, Free Society*. Boston: GNU Press.

Suchman, L., J. Blomberg, J. Orr and R. Trigg. 1999. "Reconstructing Technologies as Social Practice." *American Behavioral Scientists* 43 (3): 392–408.

Taleb, N. 2007. *The Black Swan: The Impact of the Highly Improbable*. New York: Random House.

Torvalds, L. 2001. *Just for Fun*. New York: Harper Business.

Traweek. S. 1988. *Beamtimes and Lifetimes: The World of High Energy Physicists*. Cambridge, MA: Harvard University Press.

Turkle, S. 1995. *Life on the Screen: Identity in the Age of the Internet*. New York: Simon and Schuster.

Weber, S. 2004. *The Success of Open Source*. Cambridge: Harvard University Press.

Webster, F. 2006. *Theories of the Information Society*. London: Routledge.

Weizenbaum, J. 1976. *Computer Power and Human Reason*. New York: W. H. Freeman.

Wikipedia. *Social Computing*. Available at: http://en.wikipedia.org/wiki/Social_computing, accessed January 24, 2009.

Winner. L. 1989. *Mythinformation in the High-Tech Era*. Cambridge MA: MIT Press.

Winner, L. 1986. *The Whale and the Reactor: A Search for Limits in an Age of High Technology*. Chicago: University of Chicago Press.

Winner. L. 1977. *Autonomous Technologies: Technics Out-of-Control as a Theme in Political Thought*. Cambridge MA: MIT Press.

Wiener, N. 1948. *Cybernetics: Or the Control and Communication in the Animal and the Machine*. Cambridge: MIT Press.

Winograd, T. and F. Flores. 1986. *Understanding Computers and Cognition*. Norwood, NJ: Ablex.

# 11.1
## The Punjab Peasant and Digital Culture

*Anjali Gera Roy*

Media theorists such as Marshall McLuhan, Mark Poster, Arjun Appadurai (McLuhan 1967, Manuel 2001; Poster 1995; Appadurai 1996) and others have challenged the Frankfurt School's denunciation of the electronic media as instruments of control by demonstrating that the digital media and satellite technologies can be democratizing, through increasing connectivity and accessibility across geographical boundaries. The debates on the new media in the new millennium have largely focused on concerns about the growing digital divide that excludes a large proportion of the global population in constructing the new mediascapes that connect global, cosmopolitan peoples. However, an increasing number of studies, such as those of Peter Manuel (2001) on Cassette Culture and Mark Poster (1995) on the Internet have successfully demonstrated how the new digital technologies can, in fact, lead to sonic democratization and to the expansion of the public sphere to include local peoples in addition to cosmopolitans. The global flows of Bhangra, a hybrid British music derived from a Punjabi harvest dance of the same name, and the convergence of multiple ethnic, caste, gendered, and sectarian subjectivities in the musical production of the *Jat* cultivator caste and the performing caste of *mirasis*, offers a classic case of the appropriation of new media and technologies by non-cosmopolitan players. Bhangra's production, circulation, and consumption serves as the textbook example of the impact of digital technologies on the politics of culture and their role in altering relations of power between and within nations, classes, castes, ethnicities, and regions.

The term cyberdemocracy, coined by Mark Poster (2001) in relation to the Internet, may be may be interpreted to refer to the utilization of all electronic communication technologies in enhancing democratic processes within a democratic republic or representative democracy. Cyberdemocracy in the Bhangra context works through the valorization of *pendu*, or rustic music, musicians and values by the cultivator caste of *Jats* and other rural communities on Bhangra websites. Though Bhangra's increased global visibility has been ascribed to its mainstreaming through global circulation networks such as satellite television, informal internet networks redeem Bhangra from its

270

reification in the music industry, by appropriating it in the formation of new global communities, which can be both boundary breaking and boundary crossing. The centrality of rustic Punjabi music to diasporic Asian, Indian, and Punjabi identity formation makes its reification in global capitalist structures irrelevant to its valorization in its fan's cultures. Though the Punjabi *Jat* might remain the object of the digital divide until content is localized, *Jats* not only access the same music through other media such as radio and cassettes, but Bhangra is also valorized for its exoticization of *Jat* values and culture. While it is true that the *Jat* voice may be heard more strongly in Bhangra music rather than in the Bhangra debates in internet chatrooms, the *Jat* commands a formidable presence on the internet through the media technologies used by diasporic computer literates who archive, annotate, appropriate, and recirculate media content. *Jat* music, synthesized with global beats to reflect diasporic realities, becomes an essential component in the reconstruction of global *punjabiyat*. Thus the Punjabi *Jats* gain a voice in the destablization of traditional hierarchical structures through their centrality in the production of Bhangra music.

The exoticization of Punjabi *Jat* rusticity in global techno-nostalgia also makes the Punjabi rustic ' "kool" among middle class urban youth. Although, the reification of valorized exotica and ethnicity in global capitalism catapults the *Jats* into new relations of power, the visibility and elevation of *Jats* comes at the cost of self-exoticization. The appropriation of rusticity in the marketing of techno-nostalgia could become a source of the further disempowerment of the marginalized, but it certainly has had the effect of globalizing the voice of the margin, albeit dictated by global agendas.

## Bibliography

Appadurai, A. 1996. *Modernity at Large: Cultural Dimensions of Globalization.* Minneapolis: University of Minnesota Press.

Manuel, P. 2001. *Cassette Culture: Popular Music and Technology in North India.* Chicago: Chicago Studies in Ethnomusicology.

McLuhan, M. 1967. *The Medium is the Massage: An Inventory of Effects.* New York: Bantam.

Poster, M. 1995. *The Second Media Age.* Cambridge, MA: Blackwell.

Poster, M. 2001. "CyberDemocracy: Internet and the Public Sphere." *What's the Matter with the Internet?* Minneapolis: University of Minnesota Press.

# 11.2
## Social Ecology of Museums in the Digital Domain

*Amareswar Galla*

The museum is universally considered as an institution in the service of society. Museums are increasingly being developed in a postcolonial environment as civic spaces for critical community engagement, to safeguard intangible heritage and cultural diversity as the common heritage of humanity. However, the application of digital technologies has largely remained in the domain of tangible heritage and mostly monocultural contexts. In fact, the accelerated pace of digital globalization and the inadequate social contextualization of technologies has become a major threat to cultural diversity of the world.

The Strategic Plan of the International Council of Museums (ICOM) advocates the use of digital technologies within its overall inclusive framework for promoting cultural diversity (http://www.icom.museum). The annual International Conference on the Inclusive Museum, developed in partnership with ICOM, has been scoping the conceptual framework of inclusion, including its digital dimensions, to further explore the complexity of museums in the present day world (http://www.onmuseums.com). Another international conference was organized in 2005 as a think-tank in Hyderabad, interrogating Author, Authenticity and Authority in the Digital Domain (http://t06.cgpublisher.com/welcome.html).

Some of the key concerns that emerge from these research conferences are directly relevant to the framing of computing. Most of the museums focus on digitization of collections and providing access as well as using digital media in the exhibitions and public programming. In these endeavors ethical concerns have emerged about the digitization of sacred materials, human remains including modified body parts, ceremonial items restricted to viewing on a gender basis, and finally the ownership of the voices embedded in the decontextualized collections from colonial times. One of the urgent challenges is to bring together the host institutions, with the collections, and the source communities from where the collections were taken. This would assist in developing a negotiated process for appropriate computing frameworks to facilitate an ethical and meaningful use of technologies in museums.

A recent postcolonial development is the urgent safeguarding of intangible heritage (UNESCO 2003a). Digital technologies are being used to document and inventorize intangible heritage elements. But then documenting intangible heritage is virtually freezing living heritage in time. The challenge is to frame the use of digital technologies within the social ecology of intangible heritage elements in the host communities. Moreover, there is very poor understanding in the computing world about the significance of heritage that is born digital and hence designated as *digital heritage*. The Charter on the Preservation of the Digital Heritage developed by UNESCO (2003b) is useful to consider in addressing this gap in the knowledge of the computing world.

Another concern is the conservation of works art born digital or digital art. Several museums and galleries have been collecting digital art. There is a whole genre of art that is promoted and funded by arts councils as digital art. There is also an increasing number of artists whose sole creative focus is digital art. These digital art works are collected in museums and accessioned. However, they can only be conserved in the format in which they are created, to retain the integrity and authenticity of the art work. But with the updating of computing formats, these art works are often upgraded in the collections databases. Even the minute change in pigments and pixels would make them not legal collections, as they are no longer the original collections.

The above-mentioned concerns are selections within a large corpus of critical engagement, dealing with digitization and digital heritage. One common concern is the ownership of intellectual and cultural property rights in the digital domain. Framing the appropriate computing to address the various concerns is urgently needed. There are many important interventions being made. However, what is needed is a critical discourse that brings the sciences of computing and museums together to negotiate the way forward. This is essential for the safeguarding of the cultural diversity of humanity.

## Bibliography

International Council of Museums. 2007. "Our Global Vision – Strategic Plan 2008–2010." Available at: http://www.icom.museum, accessed January 21, 2008.

Second International Conference on Technology, Knowledge and Society 2005. Hyderabad, India (December 12–15) http://t06.cgpublisher.com/welcome.html, accessed March 10, 2009.

The Inclusive Museum. http://onmuseums.com, accessed April 7, 2007.

UNESCO 2003b. "Convention for the Safeguarding of the Intangible Cultural Heritage." http://unesdoc.unesco.org/images/0013/001325/132540e.pdf, accessed October 18, 2009.

UNESCO 2003b. "Charter on the Preservation of the Digital Heritage." http://portal.unesco.org/ci/en/ev.php-URL_ID=13367&URL_DO=DO_TOPIC&URL_SECTION=201.html, accessed October 18, 2009.

# 11.3

## Information Technology and the Construction of Moral Reasoning, Empathy, and Affect: Crossing Time, Space, and Attitudes in Virtual Reality

*William James Stover*

The study of international relations is often an examination of foreign relations and world affairs from an ethnocentric perspective (Nossal 1998; Wendt 1999). Instructors accept their own values and then superimpose them on students. This ethnocentric analysis excludes a basic element for understanding international relations—empathy, the ability to participate in another's values, perceptions, and feelings.

Information technology helps students learn more about the world by developing their empathy through simulations and virtual visits (Stover et al. 2009). The Middle East diplomatic simulation introduces the practice of empathy, as students choose to represent countries from the region—Israel, Jordan, Egypt, Iran, Lebanon, Syria, and Palestine. They select decision-making roles within their countries: heads of state, foreign ministers, or defense ministers (Stover 2005).

Their preparation involves research for a paper based on their simulated country. Heads of state write about the goals of their country in the Middle East, foreign ministers about their history, and defense ministers about their security situation. These country teams are often assigned "executive advisors," diplomats, journalists, faculty and students from the Middle East to help participants with their research and decision-making during the simulation. Students use academic journals, news media, and Internet sites from countries they represent. This makes them look beyond the American oriented media and academic community, "visiting" sites from the Middle East made available on the website.

After completing their papers and becoming familiar with the rules, students receive a scenario that extends Middle East conflict several months into the future. They consult their own country team members, including their foreign national advisors, in a "private conference" on the website, where only the country team has access. After agreeing on their strategy, heads of state representing the country teams respond to the scenario, making moves on the Internet that can be viewed world-wide.

Students' experience has a positive affect on their understanding of the Middle East as well as their sense of empathy with countries in the region. Surveys indicate a shift in perceptions about the countries they represent. Open ended responses are even more striking, and class discussions following the simulation indicate a more complete understanding of all the countries involved in the conflict. Information technology seems to help students transcend limited ethnocentric attitudes about the Middle East.

A second project permits religious leaders in the Middle East to transcend space, borders that prevent them from talking together (Stover 2006). This "dialog of faith" joins online Jewish, Christian, and Muslim clergy from the Middle East to discuss resistance to occupation, suicide–homicide bombing, the US role in the Middle East, and the future shape of peace. The clergy reply to a series of questions, and question each others' replies (Stover 2005).

These conversations have helped participants to understand each others' concerns, values, and commitment to peace. University classes in the Middle East and North America have joined individuals from all over the world to learn from the dialog. The online conversations over so great a distance and high a barrier were in themselves an important accomplishment. Many of the participants could not visit each other because of governmental exclusion or outright hostility. Yet they could exchange viewpoints through information technology, respecting one another, listening, and being heard.

The third simulation takes students to a different time, the height of the Cold War, when the Cuban missile crisis almost led to thermonuclear conflict. Joining decision-making groups, representing the US, Russia, and Cuba, students write papers, using once classified sources, based on their country and role in the decision making process (Stover 2007).

The students receive a scenario written from their simulated country's perspective, following historical events up to the launch of the US "spy plane" that discovered the missiles. Instead of photographing them, however, the aircraft crashes in Cuba, and the pilot becomes a prisoner. Negotiations for his release are started, giving Russia time to complete the missile installation if they can cooperate with Cuba, and if the US becomes distracted with the downed pilot.

Students say they feel the tension of the Cold War by experiencing the danger and unpredictability of a different time when two countries could have destroyed the world.

Each of these projects uses information technology to help students understand international relations more completely. They are taken beyond ethnocentric attitudes, across national borders, and to a different time, where they can view the world in a more meaningful way.

The software supporting these simulations is easy to use, adaptable to various situations and time periods, and free to educational institutions at all levels. You may look at the website www.scu.edu/crs and obtain access by emailing the simulation director, wstover@scu.edu.

## Bibliography

Nossal, K. R. 1998. *The Patterns of World Politics*. Scarborough, Ontario: Prentice-Hall Canada.

Stover, W., M. Mann and M. Mankaryous. 2009. "Information Technology and the Construction of Moral Reasoning, Empathy, and Affect: Crossing Time, Space, and Attitudes in Virtual Reality." *International Journal of Science in Society* 1(1): 157–70.

Stover, W. 2007. "Simulating the Cuban Missile Crisis: Crossing Time and Space in Virtual Reality." *International Studies Perspectives* 8: 111–20.

Stover, W. 2006. "Moral Values and Student Perceptions of the Middle East: Observations on Student Learning from an Internet Dialog." *Journal of Political Science Education* 2: 73–88.

Stover, W. 2005. "Teaching and Learning Empathy: An Interactive, Online Diplomatic Simulation of Middle East Conflict." *Journal of Political Science Education* 1: 1–14.

Stover, W. 2005. "A Dialog of Faith: Reflections on Middle East Conflict from Jewish, Muslim and Christian Perspectives." *Journal of Beliefs and Values* 26 (1) (April): 65–75.

Wendt, A. 1999. *Social Theory of International Politics*. Cambridge, UK: Cambridge University Press.

# Part XII
# Digital Identities

# 12
# The Internet, Gender and Identity: Proletarianization as Selective Essentialism

*Marcus Breen*

Is it a manifestation of the state of things, when a married, middle aged white transnational male and father writes a chapter about gender and identity in an anthology about the Internet? I suspect so. The fields of gender and identity have been created, populated, colonized perhaps, and protected by all sorts of recently vivified "weirdoes" who we can celebrate. They and we are included in the life of society. The fringe is with us—the "majoritorians"—who populate the center of civilizations, who live in the suburbs, worry about our children and mortgages and find the new iterations of everyday identity almost incomprehensible. And yet that is where we are: fully embedded in a restructured social world that has cast off many of the historical constraints of otherness, even while it imposes new ones. We sit endlessly, incessantly watching, searching, typing and reading at a monitor screen. What kind of life is this? Cultural studies would argue that life should be involved in claiming a "social order of self-clarification, resulting in a heightened responsibility of fact and experience" (Peters 2006: 58). With this new sense of instrumentalism to offset the patchy performance of liberal democracy, the state of things requires that a white guy must try to meet the self-clarifying challenge.

Being burdened with the yoke of the dominant culture is not as easy as it once was. These days the Internet and all the associated communications technology generate a sense of urgency that is thoroughly dialectical. Christian Fuchs has suggested that the antagonisms of competition and cooperation are inherent in the Internet and its political economy (2008: 105). Indeed, I would suggest that the Internet is not some other, but the central characteristic of our lived, material experience, where we are constantly moving within, across and around, escaping, engaging, detesting and embracing. Where once it was possible to speak of cyberspace as some kind of ethereal, experimental other of the virtual, it is less and less like that as our lives become increasingly mediated, determined and defined by digital "stuff." It is in the constancy of this movement back and forth, through mercurial meaning structures, that the questions of cultural studies

279

are valuable. What new meanings are emerging about identity? Is it possible that this is changing the culture and if so how? What are the new facts and the new experiences that will make identity clear? Is our understanding of gender changing through the influence of the Internet? The Internet offers answers to these questions and thereby realizes a dialectical set of meanings with newfound immediacy. Given this, we can accept the challenge proposed by Sheller and Urry that "new social theory will need to develop a more dynamic conceptualization of the fluidities and mobilities that have increasingly hybridized the public and the private" in the face of Internet communications (2003: 113).

What exactly would that new cultural landscape look like? Assuming that newness is something which exists discursively, newness is always on the move to another point of realization, the hybrid cultural mash up that characterizes the Internet. It is a complex scene where the landscape is utilitarian and instrumental, static and mobile, amorphous and material, clean and dirty. It is totally theorized and becoming theoretical, while immediately applicable in a material way. And yet it evokes an "ordinary culture" (Hartley 1999: 16). It suggests that the "pixilated people" we are becoming can see our quotidian lives as newly enriched (Breen 2005: 24). And in all that complexity we inevitably suffer from seeing ourselves in ever increasing detail that creates a kind of Kierkergaadian anguish: "How far does the Truth admit of being learned?" (154).

The quotidian fragments of our lives are being so massively magnified in our pixilated world that we risk missing the truth of the entire corpus of the world we inhabit. In investigating this proposition I want to suggest a way of examining identity and gender by introducing the term "selective essentialism." In taking this term I owe a conceptual debt to Gayatri Chakravorty Spivak, who introduced the concept of "strategic essentialism" to the academic literature of race politics (1987). For Spivak, "strategic essentialism" described the way aspects of racialized ethnic identity could be reiterated by those it was used against, to reclaim those same images to contest and disrupt racist discourses. For example, in Australia "wogs" went from being a racist term for Mediterranean immigrants, to a term of affection propagated by second and third generation Mediterranean immigrants about themselves. In contrast, I want to turn around Spivak's notion of essentialism to suggest that the Internet undermines an empowering strategy to selectively reinforce the singular dimensionality of identity and gender. In this way, "selective essentialism" undoes the politics of strategic essentialism by pointing to the way the Internet produces atrophy. It creates specific types of essentialism.

This concept goes against much of the prevailing cultural work about the Internet which suggests that "rhizomes or informal networks" undo essentialisms, because of the inherent vitality of the fluidities and mobilities of life (Nakamura 2002: 128). These processes of identity within the Internet have been characterized as a new "system of relations," where the multiple identities in circulation constitute enriched possibilities in their multiple contingencies (Rossiter 2006: 167). My examination of identity and the Internet suggests that the quotidian aspects of everyday life are essentialized through and because of Internet technology, in what Jaron Lanier refers to as "lock in:" the options for diversity through experimentation are restricted by the structure of the software and the way it limits social relations (2010). The power of this pull towards selective identity formation is added to by the procedural rhetoric of persuasion that reinforces and thus limits options on the Internet (Bogost 2007). Adding to "lock in," this persuasive form fundamentally determines and limits the way reality is communicated (Šisler 2009: 278). As more content circulates on the Internet, the response is to find a space where identity is reinforced, not challenged. This is within the scope of human nature; people seek reassurance in known spaces, rather than challenges to who they are.

There is an additional way of exploring these limitations from the perspective of cultural studies. Slack and Wise describe how identity falls into a "cultural default" when there is an absence of alternative information available to challenge a point of view about an identity (2005: 167). This "unfortunate fact of our culture" is applicable to selective essentialism because of the way identity is articulated with the Internet to become a more singular perspective (Slack and Wise 2005: 167). This is in contrast with the claims that the Internet is a veritable cornucopia of social representations about lived experience. In this critical formulation, the Internet becomes a site for the narrowing of options about identity.

I want to suggest that selective essentialism can assist our analysis of the change the Internet has induced in the way we see ourselves and in the way we navigate, or are limited in navigating, the world. In doing this I acknowledge that my study of the Internet and the concept of selective essentialism is my own "particular form of essentialism" which is required in order to undertake "intellectual work and (the) historical scholarship" (Bannett 1993: 100). In order to explore essentialism I want to look at the development of sexuality, specifically pornography on the Internet and the way it has consolidated specific perspectives about sexuality.

## From single sex to multiple pornographies

Any discussion of sexuality has to acknowledge the magisterial, even overbearing, work of Michel Foucault's *The History of Sexuality*, the purpose of which was to "analyze sexuality as a historically singular form of experience" (1994a: 199). Foucault set out to show how attitudes and responses to sexuality vary across history by applying a discourse approach to the study of change over time in human behavior. To indicate how extensive the enterprise became for Foucault, and to signal the dimensions of the challenge presented by Internet pornography, it is helpful to repeat Foucault's formulation of the theoretical and practical considerations that he took into account in his study of sexuality. His research, he wrote, sought to:

> treat sexuality as a correlation of a domain of knowledge [savois], a type of normativity, and a mode of relation to the self; it means trying to decipher how, in Western societies, a complex experience is constituted from and around certain forms of behavior: an experience that conjoins a field of knowledge [connaissance] (with its own concepts, theories, diverse disciplines), a collection of rules (which differentiate the permissible from the forbidden, natural from monstrous, normal from pathological, what is decent from what is not, and so on), and a mode of relation between the individual and himself (which enable him to recognize himself as a sexual subject amid others).
>
> (1994: 200)

If Foucault offered this muddy historiographical map as a result of his efforts to describe and interpret vast amounts of research spanning hundreds, even thousands of years, then Internet pornography adds additional layers as a subset of the above. In so doing, the Internet structures the *selective essentialism* of sexuality, through pornography. My hypothesis is that the Internet is the "historical singularity" of contemporary life, and extends the history of sexuality because it is a continuation of the history of sexual behavior and human knowledge, as suggested by Foucault. This means the Internet has produced multiple new layers of theory and practice in the powerful global networked context. In terms associated with cultural studies, we can examine the multiple articulations or linkages created by the constituents of Internet pornography, such as multidisciplinary considerations of sexuality, media, communications, business, trade, psychology, physicality and pleasure. These articulations help expose the political economy of sexuality through Internet pornography.

Foucault's work offered insights into human behavior, but we need to remember that his work was done before the Internet was in everyday use. This means that while Foucault's foundational theoretical effort cannot be denied, the Internet is the new moment of "historical singularity" when new knowledge has emerged. It is in this new knowledge that the shift occurs from the historical singularity of sexuality as it was described by Foucault, to the sea of multiplicities that are revealed, represented, replayed, reproduced and reconstructed through, by and because of the Internet. In these multiple potentialities, an array of identities can be constructed. They appear as divergent concepts about gender and sexuality and are defined by overlapping yet distinct fields of interest. This idea can be understood in terms of postmodern concepts where fluidity of image and sensibility move across a landscape of simulacra (copies), which are always already in a marketplace of trade and the exchange of ideas. Nevertheless, while there are multiplicities of identity type in this postmodern landscape, the types of identity are hardening. This is how selective essentialism occurs. The Internet reinforces the producer's and the user's preference to a limited list of priorities. Rather than expanding priorities, the Internet offers selective essentialism where a "politics of knowledge" is constructed around "an ultimately uninteresting alternation of presence and absence," making "the assertion and reassertion of identity," a narrow band of experience (Said 1991: 181). In terms of human behavior, it is possible to predict that while the Internet offers so much content, it offers repetition, similarity and locked-in limits. It reinforces identity. This takes two forms: at a macro level it reinforces connection to infrastructural global information engagement, while at the micro level it reinforces the personalized value of interactivity.

The mode of communication that the Internet presents to the world is clearly complex. This is why the shift from Foucault's concept of sexuality as a historical singularity needs to be contrasted with the vastness of sexuality as it now exists, primarily through pornography on the Internet. The number of websites, videos, images, sounds and ideas associated with the millions of sexually explicit media objects in circulation incorporates hundreds of ideas about the body, desire, pleasure, pain and abuse. In their turn these ideas are linked to the intense utilization of pornography by producers and audience members through the Internet. The proliferation of knowledge about sexuality is deeply personal, as Foucault suggests, yet the circulation of vast volumes of pornography removes the certainty about sexuality as a singular experience. In this respect human history is in a very different cultural moment because the fragmented and differentiated pornographic content appears in massive amounts, making the idea of sexuality, as it was traditionally conceived, open to question. Running counter to the openness is the "lock in" factor that determines gender identity by drawing on the habituation of our social interactions with known objects. Despite the vastness of the available objects, human beings close down the options by seeking respite from the flood. By looking again and again at selected objects, we reproduce our preferences and thus determine our identity, orientation and values in relation to everything. The hardening of identity takes place to become selective essentialism: the new fundamentalism.

## Conceptual mapping

Internet pornography offers a market-based approach to cultural expression and in so doing it highlights public policy responses to unregulated media circulation. It operates within the history of the global media system, yet pushes it beyond anything that has previously emerged. My interest is in drawing a conceptual map based on the emergence of unregulated amounts of pornography that had previously been suppressed. Of particular interest to me is the social organization that can best be characterized as proletarian. By that I mean behavior that is not deemed rational or logical within the terms of Enlightenment ideals. Such behavior can now achieve 'emancipation' and appears as Internet pornography. To reiterate, Internet pornography emancipates proletarian culture by facilitating the production, distribution and consumption of private behavior, which circulates without the regulations that have previously suppressed representations of that behavior. In this sense, I define culture that is not managed by Enlightenment standards of "self-knowledge" to be proletarian. In contrast, the prevailing assumption in bourgeois liberal democracy has been the idea of knowledge that is linked to regulation of the self to produce manners, etiquette, responsibility and civility; Michel Foucault had plenty to say on this as well, in his essay "Technologies of the Self" (1994b).

My thesis owes a lot to another Frenchman, the cultural sociologist Pierre Bourdieu, who argued that the state's regulatory regime historically operates along two parallel lines of instrumentalism: (1) to suppress proletarian expressions of social life which the state considers uncivilized; (2) to act in the interests of the middle classes who are civilized by their exposure to systems of education (1986). Both of these positions are constructed out of the prevailing ideology of liberal democracies, where the educated bourgeoisie generally dominate the legal system and legislature where laws are written to determine the look and feel of regulatory regimes and the institutions that manage them. This structure of relations—the political economy—is described by the broad principles of the Enlightenment and the bourgeois revolution that produced the contemporary form of the nation-state. Explaining this unsteady relationship between proletarian behavior and bourgeois propriety is a long term goal of moral philosophy and of course, political economy. Furthermore, research on the relations between classes in the new Internet context continues the important work of observing the "vertical and horizontal" system of relationships in society (Hobsbawm 1997: 87). The use of the term proletarian here should not, however, be taken to suggest only traditional class structures and meanings, but also a mode of operation, namely the refusal to accept the bourgeois standards of decency, equality and justice.

The shift from the established system of regulated relationships to the one that applies in the Internet era mobilizes the circulation of unregulated information and knowledge. I distinguish between the two because they are constituted in different ways; information through the immediacy of a market-based transaction model, while knowledge is mediated through historical analysis and critique (Breen 1997). Pornography incorporates information and knowledge on the Internet to complicate ideas about sexuality, gender and identity. To pretend that these categories are made clearer by the Internet would be to engage in a type of falsity that walks away from the truth. The proliferation of proletarian culture—proletarianization—through pornography contains within itself selective essentialism. So, where previously Enlightenment systems of relationships could be managed by state-based regulations, the unregulated Internet has liberated proletarian culture, and in that proliferation specific horizontal functions of production and use are manifest. That is to say, audiences are highly active in selecting their identity and their gender orientation from the choices that are "freely" available to them. In the following sections I will provide some details about how the Internet became a site for these new social arrangements that make selective essentialism possible.

## Data networks

Internet pornography offers a means to study the relationship between bourgeois and proletarian social and economic formations. This can be achieved through the lens of the regulatory regimes that control data networks, which are not regulated in the way voice and public media communications (free to air radio and television) have been. Generally, data networks have been unregulated, as a function of finance and banking network requirements, which sprang in turn from computer industry demands which converged with the telephone voice transport network. The computer industry was not historically regulated by the United States Government's Federal Communications Commission (FCC). Indeed, the computer industry is regulated according to competition laws defined by the Sherman Antitrust Act around "restraint of trade" and "conspiracy" laws stemming way back to 1890 (Sherman Act). In effect, the history of communications industry regulation is the history of struggles over privatization of the public resource of telecommunications. This is the story of technology and associated media emerging to be regulated by the state, because the ownership and content of media was defined within the unsteady terms associated with "the public interest." Over many years, since the US Telecommunications Act of 1934, often in a belated sense and after technical innovation is publicly established, laws and regulations for the media and communications have been cast as being "in the public interest." The public interest theory that drives this regulatory world view has been a democratically nuanced instrument, taken up by national governments and their regulatory agencies like the FCC and the US Securities and Exchange Commission (SEC) as well as supranational organizations like the World Trade Organization (WTO), the United Nations (UN), the International Telecommunications Union (ITU) and the European Union (EU), to name a few.

In the US, data and information services have been unregulated and differentiated from the regulations that control voice telecommunications, even as data and voice moved across the network. Consequently, unregulated computer data achieved hegemony over voice communications and the public interest as ideas about the free market were applied by the voice regulator, the FCC. This unregulated computer–data domain includes the media space occupied by cable and satellite providers whose digital transmissions are pretty much in a category of unregulated non-public broadcasting and communications, and therefore outside the province of the regulatory system that dominates public speech concerns. Free speech, as radio broadcaster or "shock jock" Howard Stern showed most prominently with his move to satellite radio in 2006, operates in this unregulated space because it is mediated by subscriptions and thus "regulated" by the marketplace not by public interest.

For its part, the US Telecommunications Act of 1996 enhanced the deregulatory landscape, promoting interconnection between voice and data transport providers. In doing so, it promoted the idea that a reduction in existing regulations would increase competition and reduce the cost of telephony. In the environment of data and voice convergence, which is what the Telecommunications Act of 1996 was accommodating, the FCC promoted deregulation as the market-based solution to technical convergence. This is the world the dominant school of neo-classical economists love, a world exclusively conceived of market-based solutions. The 1996 telecommunications law diminished "the public interest" as a foundational concept that determines the role of the regulator, the FCC. Instead, public interest theory was displaced by market-based neo-classical economic theory. The ensuing struggle has been over the meaning of the public interest and what should be left to the market. This struggle actually defines the history of regulatory authorities in democracies as they seek to negotiate both the public interest and profit incentives for business. Critics of this change argue that neo-liberalism has handsomely won this struggle, giving up public interest responsibilities to the market (Mosco 2005; McChesney 2008).

The cultural change that came with the de-regulatory—some call it anti-regulationary, others anti-statist—shift in the US telecommunications sector was the realignment of the relationship between the state, the regulatory authority and the public interest (Cunningham 1992). As the government scaled down public interest regulation and left the public to the market, there was diminished protection for consumers against price increases and poor service, plus a diminished urgency about what should be done with the public network and the infrastructure to serve public needs. For example, when Vonage offered its digital IP phone service in the US in 2004, it did "not include access to traditional emergency 9-1-1 service," leading the state of Texas to sue the company. (*Consumer Affairs* 2005). As the public interest was diminished, regulations over the communications environment were characterized in neo-classical economic and ultimately populist discourse as "constraints" on the possibilities for growth and profit for the entire sector. Consequently, the state could use neither its legal authority nor it moral suasion to control what happened in the environment it deregulated; the exceptions were private legal cases against anti-trust or monopoly behavior.

(the US Department of Justice cases against Microsoft and Intel being the highest profile examples. The abuses of Enron are less commonly discussed in this context, but it is considered the "leader" in economic abuses resulting from deregulation, as suggested by the 2005 documentary *The Smartest Guys in the Room*.)

In summary, the Internet—the data network—is unregulated. This is the context where content—the ideas, images and concepts that had not previously been permitted to move unregulated along telecommunications transport systems—was freed of public interest constraints to move on the public network. It is in this context that proletarianization occurs. The US Telecommunications Act of 1996 reaffirmed that the ideology of the market became the regulator, not a set of laws seeking to determine the public interest.

## Freedom?

Lawrence Lessig and his cohorts in legal studies and the free culture movement have provided an important service through their research and advocacy around questions of corporate constraint of the Internet and the potential for creativity and shared community ownership (1999). Much of their work has been about regulation and the implications of corporatist property rights over content. Many of the people gathered around the free culture movement provoked debates about the way digital code has analogs in the legal structures which restricted the flow of information to ownership on the private network. At the same time there was little or no commentary on the emergence of unregulated content, except to welcome it. In fact, there is a consistent effort to refuse regulation of content under US claims of "free speech." Furthermore, web-based organizations such as Chilling Effect illustrate how the campaign is organized to oppose limits on the circulation of content, while the same group's efforts offer a view of how proletarianization emerges in this "free" space composed of US claims to unregulated digital rights. Claims to communication freedom inform my argument that selective essentialism is reinforced by this approach and in so doing brings proletarianization into focus.

John Perry Barlow suggested that the phrase "Information just wants to be free" characterized the way secrets cannot be kept once they enter the digital domain of the Internet (1994). The deregulatory transformation matched Barlow's libertarian political philosophy, which aimed at minimizing the role of the state and enhancing personal freedom in line with the flow of ideas on the Internet. In "freeing information," deregulation removes from the data network those standards of social interaction and behavior that reflect the Enlightenment values that are expressed in the legally and socially constructed concepts of "the public interest." In the US social and policy context, these values reflect a commitment to individualistic self-interest universally directed at the pursuit of liberty, while operating within a set of collectivist values about "being American" that feed back to individualistic self-interest. Almost every debate in civil society is about the changing boundaries that determine individual freedom versus collective interests, as John Commons, the founder of institutional economics, argued (1931). Contemporary debates about these issues as applied to the Internet are located in the context of the de-regulatory, pro-market shift, where individual freedom is the *sine qua non*. The infrequency of debate about this shift is startling. For example, Andrew Chadwick offered an otherwise comprehensive summary of global communications issues with almost no analysis of deregulation in *Internet Politics* (2006). Regulation is considered by Chadwick in relation to the UK Government's Regulation Investigatory Powers Act (RIPA) introduced in 2000, as the UK Government sought to monitor traffic over the Internet for the purpose of finding terror suspects (2006: 278–9).

The co-existence of collectivism and individualism is complex and has been analyzed and theorized within institutional economics (Greif 1994). The current context is more complex still as data networks redirect the collective impulse that is part of the history of telecommunication as a social phenomenon and replace it with individualism, which raises the weight of individual identity formation. In other words, by deregulating, the state removes its claim to collective interests—in particular, moral and ethical standards—and encourages a style of individualistic decision-making action that operates outside the historically established legal and social boundaries of the collective. In this sense, the individual subject is redefined and thus emboldened by the state, creating a mode of power as he or she moves into Internet space to become an Internet subject in the market. In this deregulated context, to claim an individual identity means that the subject who was previously defined by the explicit boundaries of law, regulation and social etiquette is no longer defined by them. As a result of this shift to individual autonomy the subject becomes newly untethered—as it were—from the behavioral regulations of the state. In such an unregulated place, the individual can seek concentrated types of stimulation that reinforce identity, gender or behavioral orientation. Pornographic material does this especially well, to make powerful new gender identities, because it draws on the emotional energy of sexuality and the subconscious mind. This newly sexualized Internet subject is defined by apparent freedom from collectivist norms and mores, emancipated to embody selective essentialism.

## Monitor space

The Internet subject is reinforced in the way the Internet delivers content to the individual. This is more acute in the delivery of pornography to the individual at "the desktop." If we did not have this shift in the orientation of unregulated data consumption by individuals—proletarianization—this would only be a discussion of the history of pornography within visual culture. To understand the significance of the desktop we need to see it as a new space generated out of our relationship with technologically mediated communication, operating within intensified global interactions that are mobile and dynamic. The power of the desktop computer continues to transition rapidly to wireless monitors, which can be viewed as a space where the individual is positioned in a unique relationship with a global digital delivery system. This system manifests itself in the monitor. The digital delivery system—the Internet—delivers data to an individual in an environment defined by transactions, and mostly driven by the market's incentive for profit. However, what users experience in or at the monitor is an artifact which is the end product of creative production. For most of us, consumption of data takes place in this private monitor space. Data is delivered to and intended for individual consumption; it is not made for masses, groups, small numbers, families or couples. Its value is in the capacity to constantly reproduce itself within the global network (Barney 2004). It is used by vast numbers of individuals who massively multiply the value of the data simply by using it to create value from all the transactions; this phenomenon is referred to as network externality.

The individual user perspective gives a social structure to the monitor space; it is primarily isolated and isolating for the individual. This is despite the fact that so much has been made of networking and connectivity. The reality is, and recent Pew Research reports bear this out, that despite commonplace high speed broadband connectivity, Americans feel more isolated from other human beings than ever before, while showing a "marked decline" in the size and diversity of the social networks with which they conduct their "core discussions" (2009). Clearly there is a lot of research to be undertaken on the impact of the monitor space and what Jody Berland described as "the unacknowledged contradictions of liberal capitalism" (2009: 19).

## Ethics: The boundary issue

Internet pornography enters the monitor space in a complex set of moves that are at once historical and social as well as deeply personal. Internet pornography incorporates cultural issues associated with a politics of the body, sex, promiscuity, legality and ultimately privacy, in a shift that Janice Radway has characterized as "intricate interdependencies," because of the way contemporary relationships are connected and dependent, and yet deeply fragmented (2000: 219). Internet pornography magnifies multilevel relationships in a new space, even while it constrains them. This is a space where micro-level decisions need to be made by the user about, as Foucault noted (above), what is "permissible and forbidden." If a pop up or phishing advertisement appears on a desktop or cell phone showing violent sexual acts against women, what is the response of the Internet subject? The answer to that question begins to define a new ethical challenge presented by Internet pornography. In other words, what are the limits to individual consumption at the desktop of private, intimate activities? For example, when a video of a Brazilian model and her boyfriend making love on a beach was distributed on YouTube, a Brazilian judge asked for its removal. Why? Because, he argued, it violates "people's fundamental rights, like privacy and intimacy" (*The Age* 2007). What or who decides on these boundaries of privacy and intimacy? If you are filmed, is there no way to regulate the circulation of your image? Answers to these questions will reveal the ethical values of the person answering them and their attitude to regulation while problematizing the taken-for-granted ideals of Enlightenment values.

## Conclusion: technological annunciation

The monitor space interpellates the individual; it creates a structure of meaning that reinforces the individual as subject (Althusser 1971). This identity is continually called out, reconstituted and reinforced through the individual's connection to a type of technological annunciation. In this context, Internet pornography reinvigorates individual identity by allowing the subject to enter a new space and become sexually unique. This new subject is linked to unregulated proletarian instincts that are "liberated" in and around data networks. There are two ways in which Internet pornography is proletarian. The first is in the flow of socially repressed and hidden proletarian sexual behavior that appears on the Internet. This content includes the representation of culture which expresses the subjugation of women, their exploitation and degradation, and the abuse of children and minorities. The second proletarian form is to be observed in the way pornography moves onto the data network to be viewed, because it has moved out from under the bourgeois regulatory shadow. When combined, these two aspects of Internet pornography form the basis of a new social movement that allows activist identity formation through selective essentialism. The exclusivist nature of the selection reduces alternative perspectives, promoting a fundamentalist narrowing of views among limited cohorts of Internet users. Consequently, otherwise hidden sexual behavior is no longer constrained by social mores and laws and in this sense it is creating changes in attitudes to sexuality and human behavior in general. Proletarianization happens through the move of otherwise unregulated sexual behavior via the Internet to the monitor space where it is willingly reproduced in the marketplace. Here it revivifies suppressed identities and genders in the limited spaces of the individuals' monitors, where it is pared down to selective essentialism.

The question is, does selective essentialism offer a fresh vision of human potential, or is it a dastardly calling out of the worst human excesses? Proletarianization, as it manifests itself on the Internet, suggests that the answers to this question will not be easy. What can at least be hoped for is that the Internet will help society achieve a type of emancipation that moves beyond the limits of our knowledge about identity and gender.

## Bibliography

*The Age* 2007. "YouTube Ordered to Block Steamy Video." *The Age* online, Associated Press. http://www.theage.com.au/news/web/youtubes-brazilian-ban/2007/01/05/11 67777256482.html#. Retrieved January 7, 2007.

Althusser, L. 1971. "Ideology and the Ideological State Apparatus (Notes towards an Investigation)." In *Lenin and Philosophy and other Essays*, ed. B. Brewster. London: New Left Books, 134–65.

Bannett, E. T. 1993. *Postcultural Theory: Critical Theory after the Marxist Paradigm*. New York: Paragon House.

Barlow, J. P. 1994. "The Economy of Ideas." *Wired*. http://www.wired.com/wired/archive/2.03/economy.ideas.html, accessed January 1, 2007.

Barney, D. 2004. *The Network Society*. Cambridge: Polity.

Berland, J. 2009. *North of Empire: Essays on the Cultural Technologies of Space*. Durham: Duke University Press.

Bogost, I. 2007. *Persuasive Games: The Expressive Power of Videogames*. Cambridge: MIT Press.

Bourdieu, P. 1986. "The Forms of Capital." In *Handbook of Theory and Research for the Sociology of Education*, ed. J. G. Richardson. New York: Greenwood Press, 241–60.

Breen, M. 1997. "Information Does Not = Knowledge: Theorizing the Political Economy of Virtuality." *Journal of Computer Mediated Communication* (December). http://jcmc.indiana.edu/vol3/issue3/breen.html, accessed April 21, 2009.

Breen, M. 2005. "US Cultural Studies: Oxymoron." *Cultural Studies Review* 11(1): 1–26.

Chadwick, A. 2006. *Internet Politics: States, Citizens and New Communication Technologies*. Oxford: Oxford University Press.

Chilling Effects. http://www.chillingeffects.org/, accessed March 21, 2008.

Commons, J. 1931. *Institutional Economics*. Available at http://www.cooperativeindividualism.org/commons_insteconomics.html, accessed February 6, 2009.

*Consumer Affairs*. 2005. "Texas Sues Vonage Over 911 Access." (22 March). Available at http://www.consumeraffairs.com/news04/2005/tx_vonage.html, accessed 6 March 2007.

Foucault, M. 1994a. "Preface to the History of Sexuality," Volume 2. In *Ethics: Subjectivity and Truth*, ed. P. Rabinow. New York: The New Press, 199–205 (originally published in English 1984).

Foucault, M. 1994b. "Technologies of the Self." In *Ethics: Subjectivity and Truth*, ed. P. Rabinow. New York: The New Press, 223–51 (originally published in English 1982).

Fuchs, C. 2008. *Internet and Society: Social Theory in the Information Age*. New York: Routledge.

Greif, A. 1994. "Cultural Beliefs and the Organization of Society: A Historical and Theoretical Reflection on Collectivist and Individualist Societies." *Journal of Political Economy* 102 (5): 912–50.

292   *The Internet, Gender and Identity*

Hobsbawm, E. 1997. "From Social History to the History of Society." In *On History*. New York: The New Press, 71–93.

Kierkegaard, S. 1946. "A Project of Thought." In *A Kierkegaard Anthology*. New York: The Modern Library, 154–164.

Lanier, J. 2009. *You are Not a Gadget: A Manifesto*. New York: Knopf.

Lessig, L. 1999. *Code and Other Laws of Cyberspace*. Reissued (2006) Code: Version 2.0 and Other Laws of Cyberspace. New York: Basic Books.

McChesney, R. 2008. *The Political Economy of Media: Enduring Issues, Emerging Dilemmas*. New York: Monthly Review Press.

Mosco, V. 2005. *The Digital Sublime: Myth, Power and Cyberspace*. Cambridge, MA: MIT Press.

Nakamura, L. 2002. *Cybertypes: Race, Ethnicity and Identity on the Internet*. New York: Routledge.

Peters, J. D. 2006. "Raymond Williams' Culture and Society as a Research Method." In *Questions of Method in Cultural Studies*, eds. M. White and J. Schwoch. Malden: Blackwell Publishing, 54–70.

Pew Research Center. 2009. *Social Isolation and Technology* (4 November). http://pewresearch.org/pubs/1398/internet-mobile-phones-impact-american-social-networks, accessed February 3, 2010.

Radway, J. 2000. "What's in a Name? Presidential Address to the American Studies Association." In *American Cultural Studies*, eds. J. Hartley and R. Pearson, New York: Oxford University Press: 218–26.

Rossiter, N. 2006. *Organized Networks: Media Theory, Creative Labour, New Institutions*. Rotterdam: NAi Publishers.

Said, E. 1991. "The Politics of Knowledge." In *Debating PC* ed. P. Berman. New York: Delta Books: 172–89 (originally published in *Raritan* XI (1) (Summer)).

Sheller, M. and J. Urry. 2003. "Mobile Transformations of 'Public' and 'Private' Life." *Theory, Culture and Society* 20 (107): 107–25.

Sherman Act, Statutory Provisions and Guidelines. http://www.justice.gov/atr/public/divisionmanual/chapter2.pdf, accessed January 27, 2010 and Restraint of illegal trade clause. http://www4.law.cornell.edu/uscode/html/uscode15/usc_sec_15_00000001000.html, accessed January 27, 2010.

Šisler, V. 2009. "Palestine in Pixels: The Holy Land, Arab-Israeli Conflict and Reality Construction in Video Games." *Middle East Journal of Culture and Communication* 2 (2): 275–92.

Slack, J. D. and J. M Wise. 2005. *Culture and Technology: A Primer*. New York: Peter Lang.

Spivak, G. C. 1987. *In Other Worlds: Essays in Cultural Politics*. London: Taylor and Francis.

# 12.1
## Virtual Glass Houses: The Process and Politics of Bisexual Identity Discussions in Online Diary Communities

*Emily D. Arthur*

Bisexual-identified people commonly experience cultural invisibility and stigmatization within lesbian, gay and heterosexual discourses and communities. These feelings of un-belonging, combined with the desire for community with like-identified people, have been cited as key reasons why bisexual-identified people have begun to create and participate in bisexual-themed online social spaces. These digital spaces, through their dedication to bi-themed content and presumably bi-identified membership, have become safe spaces for community members to engage freely in identity discussions and publish personal narratives on experiences with bisexuality and bisexual-identification. With this in mind, the Internet is emerging as an important venue in which bisexual subjects, in particular, may meet to commune as well as to share their experiences.

This study focused on bisexual-themed online journal communities as key digital spaces where sexual identity narratives are being negotiated. What is perhaps most notable about online journals or blogs is the coalescence of public and private space (Bryson et al. 2006: 799). Like journals or diaries, these spaces feature intimate narratives about the lived experiences of everyday people. Unlike journals or diaries, however, these intimate narratives are published to a publicly accessible space and, given this, are written with the expectation that this accessibility will allow for an audience not only to read but also to comment on them. In blurring the lines between public and private space in this way, Carroli (1997: 359) contends that blogs may be understood as liminal spaces "where articulations of 'self' and perceptions of 'community' collide." Online journal communities are similar to individual journals or blogs, in that participants are publishing intimate narratives to the online space with an expectation of audience interaction with the text, for example, by reading and/or responding to the posted entry. Online journal communities are differentiated from individual blogs, however, in that they are defined by a specific topic of discussion, they are able to grant membership to any number of people, and they allow for multiple authorship, so that any member, not just the founder of the community, is able to contribute narrative through the posting of an entry.

Through their text-based interactions and the contribution of bisexual-themed textual narratives, online journal community members begin to develop knowledge based on the actualities and lived experiences of bisexuality and bisexual-identification. Norms, definitions and parameters of what qualifies as bisexual begin to be created through the identification of continuities and discontinuities across the individual experiences of bisexuality that are being shared. Moreover, the texts created in the process of expressing and performing sexuality, through the sharing of stories and engagement in dialogues, form the foundation of a discourse on bisexuality based in personal experience.

It is the online community, as a direct discursive environment, that is central to the collaborative development of this alternative discourse. Its production is enabled by the technical and social configurations of the online space. Users are able to contribute content and develop knowledge through the publication of diverse narratives outlining lived experiences. The text is produced on a grassroots level by individual people, rather than by ruling social institutions. This online space acts as a contradictory site of consciousness where participants can begin to collectively reconfigure contemporary discourse on sexuality and sexual identification. In addition, these bisexual-themed online journal communities give voice to a group that has been silenced, by allowing them to weave together narratives of bisexuality through the voices of their own experiences. As Plummer (2003) points out, the narratives and stories that new sexual communities form become the foundation on which acceptance of the legitimacy of a sexual identity develops.

## Bibliography

Alexander, J. 2002. "Homo-Pages and Queer Sites: Studying the Construction and Representation of Queer Identities on the World Wide Web." *International Journal of Sexuality and Gender Studies* 7 (2/3): 85–106.

Barker, M., H. Bowles-Catton, A. Iantaffi, A. Cassidy and L Brewer. 2008. "British Bisexuality: A Snapshot of Bisexual Representations and Identities in the United Kingdom." *Journal of Bisexuality* 8 (1): 141–62.

Bereket, T. and J. Brayton. 2008. "'Bi' NO Means: Bisexuality and the Influence of Binarism on Identity." *Journal of Bisexuality* 8 (1): 51–61.

Bryson, M., L. MacIntosh, S. Jordan, and L. Hui-Ling. 2006. "Virtually Queer? Homing Devices, Mobility, and Un/Belongings." *Canadian Journal of Communication* 31: 791–814.

Carroli, Linda. 1997. "Virtual Encounters: Community or Collaboration on the Internet." *Leonardo* 30 (5): 359–63.

Deschamps, C. 2008. "Visual Scripts and Power Struggles: Bisexuality and Visibility." *Journal of Bisexuality* 8 (1): 131–9.

Pallotta-Chiarolli, M. and S. Lubowitz. 2003. "Outside Belonging: Multi-Sexual Relationships as Border Existence." *Journal of Bisexuality* 3 (1): 53–85.

Plummer, K. 2003. *Intimate Citizenship: Private Decisions and Public Dialogues*. Montreal: McGill-Queen's University Press.

Wakeford, N. 2000. "Cyberqueer." In *The Cyberculture Reader*, eds. D. Bell and B. H. Kennedy. New York: Routledge, 402–15.

Woodland, R. 2000. "Queer Spaces, Modem Boys and Pagan Statues: Gay/lesbian identity and the Construction of Cyberspace." In *The Cyberculture Reader*, eds. D. Bell and B. H. Kennedy. New York: Routledge, 416–31.

# 12.2
## UsMob: Remapping Indigenous Futures in Cyberspace

*Jan Lüdert*

Geographical remoteness, lower economic status and technological inexperience have often placed Indigenous Peoples on the "wrong side" of the digital divide. Notwithstanding these challenges, digital technologies provide a vista, from a grassroots position, that enhances their own visibility and permits the expansion of Indigenous Peoples' sphere of influence and ability to "mobilize political support in their struggles for cultural survival." The present case study of the Aboriginal Youth project UsMob (http://www. usmob.com.au/home.php) exemplifies such an indigenization of digital technologies, one that "invites kids from 'elsewhere' to come over and play on their side." UsMob underscores a potential to permeate oral history and storytelling with the help of digital technologies that broadens and engenders understandings of indigenous realities and cultures for wider audiences in general, and, crucially, for indigenous cultural futures and identities.

UsMob is a collaboration between activist lawyer and documentary filmmaker David Vadiveloo and the Arrernte Aboriginal community in central Australia. The seven-part movie series and interactive website was launched at the Adelaide Film Festival in 2005 and has since been aired by the Australian Broadcasting Corporation (ABC). The project is Australia's premier Aboriginal Youth television series and online portal that broke the mold because the community had control from beginning to end with over 70 local people employed on Australia's first Indigenous children's television series. The project is best described as an interactive television–web series with multi-path stories featuring forums, games, emails from the characters and photo galleries. It offers a unique virtual experience of one of the remotest places on earth—the Hidden Valley in Central Australia. Situated in a Town Camp (also known as a Special Purpose Lease) outside of Alice Springs, UsMob depicts the lives of the four teenagers Jacquita, Charlie, Della and Harry and their "Aboriginal community friends, taking you through the extraordinary challenges and cultural experiences that shape their lives." UsMob is accessible in English and Arrernte with subtitles. After having to apply for a permit to enter the Town Camp, users meet the four teenagers and their friends and families. "Permit holders" become "friends" and learn about and can converse with the characters via email, forum discussions and diary entries. Users in turn can interact by uploading their own video stories, thus becoming "an even more integral part of the UsMob community." The seven video episodes forwarded once a week highlight, among other topics, the difficulties of having to navigate the disappearance of Arrernte tradition with a non-Aboriginal world that seems to be encroaching into Hidden Valley. At the same time the UsMob games teach users from elsewhere how to build bush bikes and skills necessary to survive in the Australian outback. Importantly, the project, as stipulated by the community, is "not fictional" but reflects "real life and real voices" that the Arrernte want to share and convey to "viewers from around Australia and the globe" in a broader attempt to "spark an exchange" of culture, creativity and experience between "non-Indigenous and Indigenous young people." The topics that UsMob addresses are thus couched in a critical pedagogy that raises issues identified by the community, ranging widely from "Aboriginal Law, ceremony, and hunting to youth substance abuse and other Aboriginal health issues." The project remaps indigenous cultural futures by extending outward into cyberspace. Consequently UsMob, and worth emulating elsewhere, puts the authorship of digital technologies into the hands of indigenous peoples, from the local upward into a cyberspace elsewhere with people everywhere. UsMob thus instigates an imaginative dialogue toward an improved understanding between indigenous peoples and dominant society, not from the "wrong side" of a digital divide but self-determinately from their own realities and spaces. Such an invitation, I put forward, should not be passed, but followed.

## Bibliography

Australian Broadcasting Corporation (ABC) et al. 2005. "UsMob." Available at http://www.usmob.com.au/home.php, accessed May 12, 2010.

# 12.3
## Gender Structure, Gender Identity, Gender Symbolism and Information Technologies

*Verónica Sanz*

During the past two decades the underrepresentation of women in IT education and employment has become an important issue. Starting in the early 1990s, this concern arose among women computer scientists worried about the shortage of students entering and graduating in computer science, while the number of women in other university courses was increasing gradually (Pearl et. al., 1990; Frenkel, 1990; Camp 1997). Recent data (NCWIT 2009) showed that this percentage is still declining. In 2008 women earned only 18 percent of all computer science degrees in the US and the same phenomenon occurs in IT-related occupations, although women's presence in ICT study and careers has greatly increased, as has their Internet usage. In Europe, the European Union special interest group on information society technologies, within its Sixth Framework Program, funded the SIGIS (Strategies for Inclusion: Gender and the Information Society) Project, which showed that the situation in Europe is not much different from the US (Sorenson 2002; Sanz 2008).

The so-called Women into IT approach includes analysis of the different kinds of barriers that women and minorities face that prevent them from entering and staying in IT and proposals for practical interventions (Ahuja 2002; Gürer and Camp 2002; Simard, 2007). Some authors have criticized the "deficit model" involved in this approach (Henwood 1993; Zorn et al. 2007), though the causes are related to socialization and education, the problem is located in women, who are expected to "adapt" to IT by increasing their knowledge and skills and changing their attitudes to computers. Neither technologies nor men are expected to change in the same fashion.

The analytical framework developed in this study is based on the three levels of the gender system defined by Sandra Harding: identity, structure and symbolism. Harding defined gender as an ordering principle by which every society is organized. As an analytical category, Harding identified the three levels of the gender system as "a pivotal way in which humans identify themselves as persons (Individual Gender or Gender Identity), organize social relations (Gender Structure) and symbolize meaningful natural and social events and processes (Gender Symbolism)" (Harding 1986: 18).

A second goal in this study is to propose alternatives for change at each level of the gender system, based on the concept of "de-gendering" (Lorber 2000; Bath 2008).

## Gender structure and IT

At the structural level I have studied the ways in which IT-related activities are segregated by gender. As those in charge of the design phases of technology are mostly white western men, this conditions the power structure in society, and also the design and development of IT products.

De-gendering strategies at this level must analyze particular work settings and the relationship of gender to other factors, such as class or race, in concrete contexts, to understand the underlying power relations that support the segregation. Some research has been done about structural-economic and cultural differences, to explain the different situation of women in computing in countries such as Singapore, Thailand or Malaysia (Galpin 2002; Lagesen and Mellström 2004), where girls from poor families enter computer science degrees and become programmers in order to support their families.

## Gender identity and IT

In feminist constructivist theories gender identity is understood not as something that we "have" but as something that we "do" (Butler 1990), as skills contribute to the auto-conception of the self because people identify themselves very much with what they do through their professional identities. These theories have addressed how technology is implicated in the construction of subjective gender identities, by appropriating some objects or engaging in some technical activities, and how this contributes to the maintenance or transformation of particular gender performances. The most common relationship between technology and gender identity is the equivalence of masculinity with technological competence and femininity with incompetence. But, as feminist constructivists show, there exist complex and even contradictory relationships between gender identities and technology, for example, working-class masculinity is related to heavy and greasy machinery while ICTs, which are clean and light technologies, are related to white, middle class, educated young males. (Connell (1987) showed that there is not only one type of masculinity but several. However, in a concrete socio-historical context one of them is "hegemonic," and others subordinated.)

De-gendering strategies at the level of identity and IT involves exposing and stressing the contingency of the processes that relate technology to gender identities, and the relationship with other types of identities such as racial, class or sexual orientation identities. A de-gendering strategy should work towards the development and promotion of a wider range of femininities and masculinities in order to make IT more inclusive for different marginalized groups.

## Gender symbolism and IT

In gender and technology studies from a feminist constructivist perspective, it is considered that technology is not only a set of artifacts, but also has an ideological and semiotic dimension. In the case of IT, some studies have analyzed the symbolic gendered meanings of "computer cultures" like "the hacker culture" (Håpnes and Sørensen 1995).

De-gendering strategies at this level must avoid essentializing dichotomies and taking concrete paths of gendered technological meanings as being fixed. Instead, it is necessary to study cases in context to see which gender meanings are being constructed and how they change over time, to avoid generalist, oversimplified and gender essentialist affirmations. The work of Maria Lohan on the telephone is a very good example (Lohan 2001), as the telephone was first designed for and related to men's use for business. However, after a process of "appropriation" by women in rural areas to avoid their isolation, it became related to feminine sociability (although sometimes with a negative connotation as "gossip").

An important example related to IT is the "brain metaphor," which identified computers with the human mind, a metaphor that was not present from the beginning but was constructed in a concrete period of time and for specific reasons (van Oost 2000). As Faulkner's ethnographic studies show (2000), symbolisms remain, although women interact a lot with technology, and although men also use technologies in so-called feminine ways. By showing the malleability of the interpretive flexibility of artefacts and the dynamic process of their gendered meanings, these meanings can be re-shaped, leaving an open space for human agency. The contingency of these processes and the contradictions between images and real practices create a strong claim for developing and promoting alternative meanings and gender representations of IT.

## Bibliography

Ahuja, Manju K. 2002. "Women in the Information Technology Profession: a Literature Review, Synthesis and Research Agenda." *European Journal of Information Systems*, 11: 20–34.

Bath, Corinna. 2008. "De-Gendering Computational Artefacts: From Gender Analysis to Technology Design Methodologies." In *Yearbook 2008 of the Institute for Advanced Studies on Science, Technology and Society*. Munich and Vienna: Profil.

Butler, Judith. 1990. *Gender Trouble. Feminism and the Subversion of Identity*. New York: Routledge.

Camp, Tracy. 1997. "The Incredible Shrinking Pipeline." *Communications of the ACM* 40 (10): 103–10.

Connell, Robert W. 1987. *Gender and Power: Society, the Person and Sexual Politics.* Cambridge, UK: Polity Press.

Faulkner, Wendy. 2000. "The Power and the Pleasure? A Research Agenda for 'making gender stick' to Engineers." *Science, Technology, & Human Values* 25 (1): 87–119.

Frenkel, Karen A. 1990. "Women & Computing." *Communications of the ACM* 33, (11): 34–46.

Galpin, Vashti. 2002. "Women in Computing Around the World." *Inroads-SIGCSE Bulletin* 34 (2): 94–100.

Gürer, Denise and Tracy Camp. 2002. *Investigating the Incredible Shrinking Pipeline for Women in Computer Science.* Report for the National Science Foundation Project 9812016. Available at http://women.acm.org/documents/finalreport.pdf, accessed May 1, 2008.

Håpnes, Tove and Knut H. Sørensen. 1995. "Competition and Collaboration in Male Shaping of Computing: A Study of a Norwegian Hacker Culture." In *The Gender-Technology Relation: Contemporary Theory and Research*, eds. Keith Grint and Rosalind Gill. London: Taylor and Francis: 174–91.

Harding, Sandra. 1986. *The Science Question in Feminism.* Ithaca: Cornell University Press.

Henwood, F. 1993. "Establishing Gender Perspectives in Information Technology: Problems, Issues and Opportunities." In *Gendered by Design? Information Technology and Office Systems*, eds. E. Green, J. Owen and D. Pain. London, Washington: Taylor & Francis, 31–49.

Lagesen, V. and U. Mellström. 2004. "Why is Computer Science in Malaysia is a Gender Authentic Choice for Women? Gender and Technology in a Cross-cultural Perspective." In *Symposium Gender & ICT: Strategies of Inclusion.* Brussels: 2004.

Landström, Catharina. 2007. "Queering Feminist Technology Studies." *Feminist Theory* 8 (1): 7–26.

Lohan, Maria. 2001. "Men, Masculinities and 'Mundane' Technologies: The Domestic Telephone." In *Virtual Gender*, eds. Alison Adam and Eileen Green. London: Routledge, 189–206.

Lorber, Judith. 2000. "Using Gender to Undo Gender. A Feminist Degendering Movement." *Feminist Theory* 1(1): 79–95.

National Center for Women and Information Technology (NCWIT). 2009. *Women in IT: The Facts.*

Pearl, Amy, Martha E. Pollack, Eve Riskin, Becky Thomas, Elizabeth Wolf and Alice Wu. 1990. "Becoming a Computer Scientist: A Report by the ACM Committee on the Status of Women in Computing Science." *Communications of the ACM* 33 (11) (November): 47–58.

Sanz, Veronica. 2008. "Women and Computer Engineering: the Computer Science School at the Technical University of Madrid." (In Spanish), *Arbor: Ciencia Pensamiento y Cultura (Revista General del CSIC)*, CLXXXIV (733), Madrid, 905–15.

Sanz, Veronica. 2010. "Gender Studies of IT: From Equality Strategies to Epistemological Issues." In *Yearbook 2010 of the Institute for Advanced Studies in Science, Technology and Society*, eds. Arno Bammé, Günter Getzinger and Berhanrd Wiesen, Munich, Vienna.

Simard, Caroline. 2007. "Barriers to the Advancement of Technical Women: A Review of the Literature." *Report for the Anita Borg Institute for Women and Technology.* Available at http://anitaborg.org/files/womens-tech-careers-lit-reviewfinal_2007.pdf, accessed November 23, 2009.

Sørensen, Knut H. 2002. *Love, Duty and the S-Curve. An Overview of Some Current Literature on Gender and ICT*. IST-2000-26329 SIGIS Deliverable Number: D02_Part 1. Available at http://www.rcss.ed.ac.uk/sigis/public/displaydoc/full/D02_Part1, accessed April 1, 2007.

van Oost, Ellen. 2000. "Making the Computer Masculine: The Historical Roots of Gendered Representations." In *Women, Work and Computerization: Charting a Course to the Future*, eds. E. Balka and R. Smith. Dordrecht: Kluwer, 9–16.

Zorn, Isabel, Susanne Maass, Els Rommes, Carola Schirmer and Heidi Schelhowe, eds. 2007. *Gender Designs IT: Construction and De-Construction of Information Technology*. Wiesbaden: VS Verlag für Sozialwissenschaften.

# Part XIII
# Information Globalism

# 13
# Digital Capitalism and Development: The Unbearable Lightness of ICT4D

*Jan Nederveen Pieterse*

The application of information and communication technologies (ICTs) in development policies—in short, information for development or ICT4D—follows ideas of "digital divide" and "cyber apartheid." This discussion situates ICT4D in critical development studies and global political economy and argues that information for development is primarily driven by market expansion and market deepening. As the latest accumulation wave, digital capitalism generates information technology boosterism and cyber utopianism with the digital divide as its refrain. The first part of this discussion criticizes the discourses and policies of bridging the digital divide; the second section views information for development as part of a package deal in which cyber utopianism is associated, not exclusively, but primarily, with marketing digital capitalism. This is examined further in the third section on the relationship between digital capitalism and cyber utopianism of which ICT4D is a part.

The actual task of information for development is to disaggregate ICT4D and to reconsider ICT in development policy in this light. This is taken up in two sections that place ICT4D in the context of development studies and development policy. I argue that less emphasis on the Internet and more on telephone, radio and television would normalize and ground the discussion. I conclude that the ICT4D discussion should move away from development aid, NGOs and externally funded digital projects and focus on the central question of disembedding technology from capital.

## Bridging the digital divide

The digital divide, the theme of a dazzling outpouring of literature,[1] is typically portrayed in statistics, for instance "the fact that half the world population has yet to make its first telephone call, or that the density of telephone lines in Tokyo exceeds that of the entire continent of Africa" (Campbell 2001: 119). Or, Manhattan has more Internet providers than all of Africa (Fors and Moreno 2002) and 88 percent of Internet hosts are in North

America and Europe and 0.25 percent in Africa (half of which are concentrated in South Africa). With 13 percent of the world population, Africa has only 0.22 percent of landline telephone connections and less than 2 percent of global PC ownership (Ya'u 2004: 14).

The digital divide is a deeply misleading discourse: the divide is not digital but socioeconomic, but representing the divide in technical terms suggests technical solutions. It suggests digital solutions for digital problems (Warschauer 2003: 298; Cullen 2001). With the digital divide comes reasoning that correlates connectivity with development performance: "Area A is rich, integrated into market relationships, and has a lot of telephones; area B is poorer, less integrated into market relationships, and has fewer telephones: therefore, a telephone rollout will make B richer and more integrated" (Wade 2002: 450). The next step is to equate connectivity and economic development and to view ICT as key to bridging the rich–poor gap and "national 'e-readiness' as a cornerstone of capacity building ... the discourse surrounding ICT has thus become part of developmental discourse itself" (Thompson 2004: 105).

Hence follows the policy of bridging the digital divide. Since digital capitalism does not go where profit margins are low, such as rural areas and developing countries, the rationale of bridging the digital divide is that development intervention can make up for market imperfections and jump-start nonprofit connectivity.

Bridging the digital divide has become a keynote of development policy, heavily promoted by major institutions. The World Bank and its Global Information and Communication Technologies Department launched the Development Gateway, InfoDev, the Global Knowledge Partnership, the Global Development Learning Network, World Links for Development, the African Virtual University and a host of other initiatives (Luyt 2004; Thompson 2004; Wade 2002). The G8 countries launched the Digital Opportunities Task or DOT Force, which is endorsed by the United Nations Development Program (UNDP) (Shade 2003). The UNDP started the Sustainable Development Network Program and the Global Network Readiness and Resources Initiative and has teamed up with Cisco Systems to offer ICT courses in developing countries (McLaughlin 2005). The UN is involved via the World Intellectual Property Organization (WIPO). Following the 1997 Basic Telecommunications Agreement the World Trade Organization (WTO) looks further towards e-commerce (Shade 2003). The World Summit on the Information Society met in Tunis in 2005. Development cooperation in Australia, Canada, the Netherlands, Scandinavia and Switzerland, among others, sponsors digital projects in developing countries. NGO initiatives include Computer Aid International, World Computer Exchange and the International Development Research Centre (Ya'u 2004: 23).

In Thomas Friedman's book *The World Is Flat*, information technology is the key to bridging the development gap between the US and India and to bridging the rich–poor gap. "Three billion people—from India, China, and the former Soviet empire—walked onto a 'flattened playing field.' They can now 'plug and play, connect and collaborate, more directly with your kids and mine than ever before in the history of the planet'" (Friedman 2005). The combination of rising educational levels in developing countries (at a time when the American educational system is showing weaknesses) and the business strategies of multinational companies, with ICT as an enabling factor, creates economic opportunities for developing countries. On the downside is a troubling message to Americans: over the next ten years up to 11 percent of the American workforce may be outsourced (cf. Luyt 2004).

Call centers are opening from Argentina to Kenya and Russia. But are teleworking and teleservices beneficial to India and other information processing countries? They offer jobs to a new middle class segment, but already, after a few years, the attrition rate in India is 30–35 percent. "Indian staff are required to keep odd hours, adopt American accents, and have few options for career advancement" (Luyt 2004: 7). Call centers are a dependent economy geared to patrons and clients in the North to the point that Indians must adopt American names and fake identities. They are a pseudo transfer of technology which only transfers end-user capability.

> With the exception of some groups (like software programmers), it seems that most teleworkers who are predominantly women are receiving extremely low wages; and some of them work in the kind of modern-day sweatshop conditions that characterized export oriented manufacturing throughout the developing world.
>
> (De Alcantera, quoted in Ya'u 2004: 21)

At times information for development comes with an extraterrestrial optimism (e.g. Sims 2002; Friedman 2000; Alden 2003) that is oblivious to the checkered history of international development efforts. Suddenly technology becomes a development shortcut, even though this flies in the face of obvious constraints. First, "Relative to income, the divide today hardly exists" (Wade 2002: 444), so bridging the digital divide is actually about bridging income gaps, and here the evidence is that they are generally growing. Second, a major cause of growing inequality within and between societies since the 1980s is growing skill differentials and IT and digital literacy is a major part of this growing gap (Nederveen Pieterse 2004: Ch. 5; Cornia 1999). Thus, bridging the digital divide as a means to narrow inequality in effect presents the problem as a solution. Third, "the digital divide is increasing rather than decreasing" (Ya'u 2004: 24), which is plain, given the rapid changes and competitive drives in the IT field. Fourth, research suggests that "the digital divide will never be bridged: it would take Africa about 100 years to reach the 1995 level of Ireland" (Ya'u 2004). Bridging the digital divide is mopping up with the tap open. This presents

308 Digital Capitalism and Development

us with the unbearable lightness of ICT4D and the illogical nature of bridging the digital divide.

Unpacking this approach, an obvious and often discussed problem is technological fetishism.[2] Some discussions argue that connectivity should be addressed not as a technological fix but as part of a capabilities approach and in terms of social capabilities. This is true and by the same token it implies certain priorities: "Once the illiteracy problem is solved (as in Kerala, India), cheap books are a great boon, but giving illiterate people cheap books does not solve illiteracy" (Wade 2002: 443).

## ICT4D as a package deal

The wider issue is the package character of ICT4D and the interrelated nature of ICT components and the constellation that it is part of. This suggests that the *means* of bridging the digital divide contradict the very idea of bridging: "efforts to bridge the digital divide may have the effect of locking developing countries into a new form of dependency on the West. The technologies and 'regimes' (international standards governing ICTs) are designed by developed country entities for developed country conditions" (Wade 2002: 443).

From the package character of ICT4D emerges the actual task of ICT4D, which is to unpack ICT4D so its development potential can be diagnosed and possibly harnessed.

Contemporary globalization is a package deal and ICT is deeply wired into this cluster. Information technologies and microelectronics-based telecommunications since the early 1980s created the possibility of the globalization of supply: the information and communication revolution cheapened long-distance communication and enabled plant relocation and outsourcing to low wage areas. Information technology also enables providing global product information or the globalization of demand. While flexible production has come with growing research and development costs, it also comes with a shorter shelf life of products and thus pressure to expand market shares to amortize the cost of technology investments, thus generating incentives for global marketing and creating global brand recognition. With advertising, growing three times faster than trade, and the global advertising boom comes the political economy of branding and the culture of logos. Information technology is also tied up with the globalization of competition; the changing dynamics of global inter-firm competition involve inter-corporate tie ups, networking and mergers and acquisitions to manage the cost and risks of research and development and global marketing. Corporate mergers both downsize companies and seek to make brands stronger. ICT further provides the technical means for financial globalization, as in 24 hour electronic trading, which has come together with financial deregulation and "securitization," or the dilution of the separation between banking and non-banking forms of corporate finance, which have, in turn, enabled corporate globalization and the wave of corporate merger activity from the 1980s onward. One form this takes is the spread of new financial instruments such as options and derivatives.[3]

In global political economy these trends are discussed under headings such as flexible accumulation and post-Fordism, and as a mode of production. A mode of production or regime of accumulation combines systems of production (technologies and the organization of firms) and forms of regulation (political and legal regulation of business and capital). This suggests that we cannot pick and choose elements from this configuration without in effect activating and transplanting much of or the entire constellation. This is already apparent at a technical level:

> Complex ICT systems have "layers" of components—including PCs, computer hardware, telecommunications, cables, software—and decisions made about standards for one layer in one part of a large organization can easily interfere with decisions about standards for another layer made in another part of the organization. Compatibility can take years to achieve at a huge cost, by which time new incompatibilities may have arisen.
>
> (Wade 2002: 448)

What is at stake in contemporary globalization is both different national capitalisms, each of which is dynamic and in flux, and the interaction of capitalisms, which is mediated through complex layers of technology, international finance, international trade, international institutions, macroeconomic policies, knowledge systems, legal standards and proprietary arrangements. Development policy is part of the interaction of capitalisms. The terms of this interaction are generally set by hegemonic powers and institutions.

Accordingly, what matters too is with which perspective we approach these questions, from the inside (the advanced countries) looking out, or from the outside looking in (from the point of view of developing countries). The development approach suggests the latter whereas the realities of power and privilege imply the former. The trade-offs involved in investing in ICT tend to be viewed differently in developed countries than in developing countries: "it does not make sense to have hospitals connected to the Internet when there are no drugs in the hospitals, or for schools that have no chairs to be connected to the Internet" (Ya'u 2004: 26). ICT4D reworks several familiar problems in development policy, some of which are sketched below.

*Development policy is incoherent.* Surely education is more important to development than digital access and is also a condition for digital literacy. Uneven education worries organizations such as UNESCO, which calls on states to devote as much as 26 percent of their budgets to education

(Ya'u 2004:19). Yet the structural reform policies advocated by the International Monetary Fund (IMF) and World Bank require cuts in public spending, including education. It does not make sense to cut education spending and argue for ICT4D, to erode basic capabilities and advocate fancy digital capabilities.

*ICT4D implies the imposition of a development model.* According to techno-determinists the spread of technology = development. This recycles conventional modernization thinking which ranges from Enlightenment positivism (and Lenin's formula of progress as "Soviets + electricity") to postwar modernization theory. In this series, ICT4D is Modernization 2.0 (Shade 2003: 14). Second, for neoliberal economists and entrepreneurs the spread of market forces = development. Both these discourse communities make an instrumental use of information for development; what matters is technological transformation and market expansion. What these views share is at minimum development naivety, which may be both genuine and deliberate (involving not just the sociology of knowledge but the sociology of ignorance). In defining poverty as the absence of technology and market forces, they lack awareness of social development. In the process these views present the disease as the remedy and hegemony as freedom (cf. Shade 2003: 117). More precisely, what is at issue is the imposition of a development model.

Time and again technological modernization has served as a means to effect political and economic reforms. Information technology also functioned this way in western countries, making reforms seem inevitable and thus selling the Reagan and Thatcher reforms in the US and UK ("There Is No Alternative") to trade unions and labor constituencies. ICT indeed is wired in many directions.

*ICT promotion serves as a rationale for trade and investment liberalization in developing countries.* As Ya'u notes, "African countries that have undertaken the liberalization of their telecommunication sector have ended one form of monopoly—state monopolies—and found themselves saddled with a new monopoly—that of foreign investors" (2004: 19).

ICT support also undergirds changes in development institutions. ICT promotion fits the World Bank's new career as Knowledge Bank. Joe Stiglitz's theory of information asymmetry as a cause of market imperfection provides the World Bank with a rationale to improve the functioning of markets by remedying information gaps, which sidesteps critiquing markets themselves (Thompson 2004).

*ICT raises the question of appropriate technology.* It may be true that in the information economy the cost of a copy is zero (Verzola 2004), but the cost of the delivery systems—infrastructure, electricity, hardware, software and human ware—is far from zero. Questions that are seldom asked are "which technology is appropriate, are low-tech more appropriate than high-tech options, and for what are the technologies going to be used?" (Fors and

Moreno 2002: 199). A further question is how the returns on investments in ICT compare with returns on other investments?

*ICT is designed according to the requirements of the prosperous markets.* "Developing countries are placed at a growing disadvantage by the software-hardware arms race in the global market for savvy computer users ... The effect of this technological arms race is to keep widening the digital divide between the prosperous democracies and the rest of the world" (Wade 2002: 452).

*ICT privileges western content.* While ICT places the emphasis on the channels of information, in the process it privileges western content. "What does it mean that people have access to information or channels that they do not own? Citizens are provided access to channels over which they have no control. Increasingly, also, they are offered little or no real choice over content" (Ya'u 2004: 24).

*Intellectual property rights presuppose Western legal norms.* Intellectual property rights and the harmonization of patent law are a major frontier of contemporary globalization (Drahos and Braithwaite 2002; Drahos 2003). As Ngenda points out, "The international intellectual property model is a product of Western legal norms" (2005: 60). It carries the imprint of individualism and proprietary individualism such as the *droit d'auteur* (2005: 66). The incentive for reward principle has become enshrined in the World Intellectual Property Organization (WIPO) along with the view that "patent protection is an indispensable incentive to creative and inventive work" (2005: 67–8).

> Like too much of all good things, too much IP protection does not reward society. The intensification of intellectual property benefits the owners of the innovations, while society at large suffers welfare loss due to rent-seeking or monopolistic behavior of knowledge economy firms that depend on patents, copyrights, and other IP rights regimes as their source of profit.
>
> (Parayil 2005: 48)[4]

*ICT manufacturing does not necessarily add up to ICT diffusion.* Latecomer nations lack the financial resources to invest in new technologies which also presuppose a business infrastructure in soft social capital, such as appropriate institutions (Wong 2002: 168). While East Asian countries have been strong in electronics manufacturing they have been weak in services, especially financial services and knowledge-based services of the kind that use ICT. This was a factor in the 1997 Asian crisis. Disparities in ICT diffusion are significantly higher among Asian countries than among non-Asian countries; Japan and the four Asian newly industrialized economies (NIEs) rank above the norm in ICT diffusion whereas the six Asian least developed countries (LDCs) underperform, especially in Internet services (Wong 2002: 185). Thus there is a significant digital divide between the five more advanced countries of the region and the other developing Asian countries. Wong concludes that high involvement in ICT production has little or no spillover effect in ICT diffusion.

## Digital capitalism—cyber utopia

The digital divide theme is unusual because it is quite ordinary for new technology to spread unevenly, so why should digital technology be different? Now, however, cyber apartheid and information apartheid loom and, according to a flood of studies, we must get wired: schools, libraries, community centers, senior citizens in retirement homes, and the homeless to meet their information needs (e.g. Stansbury 2003; Wicks 2003).

Media reports discuss, for instance, "Ethiopia's Digital Dream" and the enthusiasm about applying IT in e-government, education and communications across the countryside, an aim that is pursued with great zeal, despite poverty, and in the hope that digital solutions can make up for the lack of infrastructure (Cross 2005). Yet, look at the fine print and we find that to implement this, the Ethiopian government and Telecommunications Corporation team up with Cisco Systems and Business Connexion of South Africa; the reporter visited the country as a guest of Cisco Systems, which prompts the question, is this Ethiopia's digital dream or that of IT corporations? This illustrates a key dilemma of ITC4D. Part of this is what is known in economics as the expert service problem: the expert who is to diagnose the problem has a stake in the solution.

The boundary between ICT4D and ICT marketing is thin. ICT4D may be a terrain in its own right but it is also part of general ICT boosterism in which ICT is the latest major wave of capital accumulation—think railroads, electricity and chemical industries in the nineteenth century and automobiles and telecommunications in the twentieth century. Each accumulation wave comes with its own boosterism: it is not sufficient for new products to be made; they must also be invested in, sold and used. They must be the talk of the town.

In the series of capital accumulation waves, the ICT wave is a special case in that it is a highly capital intensive sector that has not delivered on its promise; it has absorbed multibillion dollar investments in infrastructure (such as the Fiber Optic Link around the globe and satellite systems) that are vastly underused. ICT has been in the forefront of trans-national corporation (TNC) operations; in the 1990s typically up to a third of American TNC investments in developing and emerging markets from Mexico to Russia went to the telecom industry (Schiller 1999). It is a prime terrain of transnational mergers and acquisitions, and mega corporations such as WorldCom, Vodaphone, Viacom, MCI and Mannesmann. ICT is both a dream space of multinational capital (according to President Clinton the Internet should become a free trade zone), the spearhead of market-led development in a world-to-come of minimal regulation, and typically faces preferences for national regulation of telecoms.[5]

In Schumpeter's analysis of capitalism, new technologies and inventions are the motor of capital accumulation. This also looms large in the long wave approach to capitalism. But the cycle of emerging technologies, from trigger to inflated expectations and overinvestment, to maturity, also involves cultural changes; it is also a hype cycle.

Accumulation boosterism is an exercise in the economy of appearances, which is about conjuring up economic opportunities as much as reflecting them and in the process opens up frontiers (Tsing 2004). It is about the *aura* of innovation, the *creation* of markets, the effervescent *buzz* of entrepreneurial dynamism and expansion. The general propensity to drama in capital accumulation is enhanced in ICT because ICT *is* and is *about* the communications business. Just as broadcasters typically broadcast the gospel of broadcasting, ICT communicates the wonders of communication and preaches the ICT gospel. According to this accumulation script, ICT is essential to opening up new business opportunities, unprecedented translocal and global horizons and vast empowerment opportunities.

ICT4D is a strategic part of ICT expansion: ICT4D is digital capitalism looking South, to growing middle classes, rising educational levels, vast cheap labor pools, and yet difficult regulatory environments. It is about market expansion and converting unused capacity into business assets on the premise that new technology is the gateway to hope. And it is about the deepening of the market by pressing for liberalization, opening up spaces for competition and investment, bypassing regulations or devising new regulations that will shape the future.

One might view this as a marketing campaign for Internet service providers (Gurstein 2003), but probably more is at stake. Brendan Luyt asks, "Who benefits from the digital divide?" (2004) and identifies several beneficiaries of cyber utopianism: information capital, elites and states in the global South, the development industry, and civil society groups and NGOs. Information capital stands to gain new markets and cheap labor. "If the South increasingly assumes the role of information processor for the North and acts as a lucrative market for the new products of informational capitalism, this is not due to chance" (Luyt 2004: 5). Measures against software piracy are a significant part of its interest: "The Business Software Alliance, an organization initially established by several of the biggest names in the industry ... with the express purpose of fighting software copyright infringement, has been especially active in the developing world" (4).

For elites and states in the global South where economic development is essential to state legitimacy, ICT4D serves as another development tool. For the development industry, ICT is a strategic tool around which to fashion new public–private partnerships, matching the growing corporatization of development. Traditionally about 30 percent of World Bank disbursements have gone to infrastructure projects in transport and communication which also aid transnational capital. Civil society groups and NGOs find in ICT a low cost instrument to communicate with like-minded groups.

ICT4D is a prism in which profiles of neoliberal globalization are refracted. It stands at the crossroads of today's major forces in private, public and social spheres: telecoms, international institutions, states and civil society groups and cyber activists.

If we take a step back it is clear that cyber utopia is an unlikely project. Digital capitalism has been in the forefront of the neoliberal globalization of the past decades. The telecom industry and the dotcom economy have been central to the economic expansion of the 1980s and 1990s (Schiller 1999). For Susan Strange, telecoms were a key instance in the making of casino capitalism (1996). Although the telecom industries do not rank among the Fortune 100, they include mega conglomerates. Telecoms have been a major force in the worldwide neoliberal turn and several have also played a key part in the conservative turn.[6] As the saying goes, the media do not defend corporate capitalism, they *are* corporate capitalism. That the media are part of the problem is keenly understood in the US.

From the early 1900s on, the US has developed the world's most extensive communication infrastructure. Because of its large geography and thin population, radio, telephone and later television play a large role in American society and also information technology is more developed than anywhere else. So, should ICT be able to bridge rich–poor gaps, the US would be the leading case. Digital divide arguments have led to providing local community Internet access in schools and libraries (Menou 2001). But this has been to little or no effect, *n'en déplaise* techno determinism, public–private partnerships and silicon snake oil. Social inequality in the US has *grown* significantly, precisely since the 1980s and along with the ICT wave. In the US ICT has either been indifferent to or has contributed to increasing social inequality.

American telecoms have typically practiced "'two-tier marketing' plans, polarizing products and sales pitches to reach 'two different Americas', rich and poor ... 'Nobody puts as much effort into dual marketing as the telecommunications industry,' stated *Business Week*" (Schiller 1999: 53). This has resulted in sharply polarized provision of services, from telephone to Internet access, privileging power users: "Evidence mounted that the corporate-sponsored build-out of high capacity networks was systematically evading poor neighborhoods in order to concentrate on well-off suburban residences and business parks" (Schiller 1999: 54). Internet access among blacks and minorities in the US varies by income, so inequality is social, not digital. As Mark Warschauer notes, "just as the ubiquitous presence of other media, such as television and radio, has done nothing to overcome information inequality in the US, there is little reason to believe that the mere presence of the Internet will have a better result" (2003: 297; cf. Davis 2001; Schiller and Mosco 2001).

## ICT4D and development studies

From the point of view of development studies we can situate ICT4D at various levels. First, technology represents knowledge and capability, and forms part of a capabilities approach to development, notably the human development approach. Second, the new technologies are embedded in capital and as such they evoke development from above; most public–private partnerships around ICT are typically too technical and capital intensive in nature to be participatory. Third, technology is a means of control; witness the surveillance capabilities of ICT (such as global positioning systems aligned with cell phone signals) and the corporate campaigns against software piracy and open source software. Fourth, ICT revives the old debates on appropriate technology and dependent development (Hyder 2005; Tandon 2005).

The digital projects sponsored by foreign aid and implemented by NGOs display the usual dilemmas of alternative development; most projects are not locally owned, not sustainable and fold when the funding runs out.

ICT is wired into contemporary accelerated globalization, which in development has meant structural adjustment and rolling back the developmental state in favor of market forces. Digital divide discourse is reminiscent of previous techno fixes that stressed the need for mechanization and tractors, infrastructure development or the construction of large dams, generally prioritizing capital needs over local needs. Software, the second digital divide, involves intellectual property rights, cognitive frameworks, cultural styles and vernaculars (such as English) that raise questions of knowledge monopolies and cultural imperialism. "Human ware," the third digital divide, returns us to the basic questions of education and human development, the familiar terrain of capability and inequality. Yes, education is a leveler *if* it is available and *if* it comes with other reforms—land reform, social provisions, etc. A précis of general development implications of ICT4D is given in Table 13.1.

*Table 13.1*   ICT4D and development policy

| ICT4D | Dimension | Development |
|---|---|---|
| Technology | Capability | Human development |
|  | Embedded in capital | Development from above |
|  | As means of control | Dependent development |
| Digital divide | Technological fetishism | Development as techno fix |
| Accumulation | Neoliberal globalization | Structural adjustment |
| Development aid | NGO projects | Alternative dependency |
| Software | Intellectual property rights | Monopoly rents |
| Human ware | Education | Human development |

## ICT4D and development policy

A contributor to a discussion on the implications of technological change noted:

> Poverty is a choice the world has made. It is a political choice. The infor-mation revolution will be another instrument to implement that choice. Only a governance revolution would represent a real change. And to link the information revolution with democratization is naïve in the extreme, parallel to the current leap of faith linking democratization and open markets.
>
> (quoted in Hedley 1999: 86)

Govindan Parayil offers a more benevolent view:

> What is most urgent is to find ways to integrate informational economy with traditional economy in a fair manner such that the asymmetric relationship between the two could be overcome ... While information and communications technologies, like any other general-purpose tech-nology cluster, have the potential to benefit all, it is the unfair politi-cal economic context within which they are developed, deployed, and diffused that needs to be reformed or better reconfigured for equitable development.
>
> (2005: 49)

It is just that changing "the unfair political economic context" is a tall order. The weary succession of development decades shows that it takes a lot more than technology and capital inputs to achieve development. Development policy is a terrain of hegemony and struggle and policy compromises among hegemonic forces and institutions shape and obfuscate the terms and nature of this struggle. Hegemonic compromises introduce development fads and shibboleths, such as good governance, transparency, democracy, civil soci-ety, participation and empowerment, which, when all is said and done, usu-ally mean business as usual with fresh paint. ICT4D in most senses in which it is used is another development fad and part of the process of obfuscation. The problem is not just that many info development projects are under-funded and ill conceived, or that ICT4D is driven by corporate interests; the deeper issue is that ICT4D is a Trojan horse that locks developing countries into everlasting dependency.

The instrumental approach according to which information technology can be used and appropriated towards diverse ends and serve either utopia or dystopia is contradicted by more complex assessments of the nature of information technology such as Bruno Latour's actor-network theory (Hand and Sandywell 2002).

First, from the point of view of development policy, the emphasis on the Internet is inappropriate, reflects class bias and is inspired by commercial interests. Of course, information technology is meaningful for social movements, as in the Zapatistas' use of Internet or the Filipinos' use of cell phones in their people power interventions (Castells 1996; Léon et al. 2005) and enables "organized networks" of many kinds (Lovink 2005). Yet the Internet is principally a middle class medium; as a medium, essentially an extension of the typewriter, it presupposes literacy and the ability to absorb or create content and digital literacy. It may be termed a Starbucks approach to ICT4D.

From the point of view of development policy it would be appropriate to place more emphasis instead on television, radio and telephone. For instance in Indonesia the Internet is minuscule but radio and television are huge. Of 11 or so TV channels only one is a public channel, the others are commercial. Looking at the ordinary communication technologies grounds and normalizes our discussion. If some of the digital debates are over our heads because of novel technical and legal issues, we are all familiar with the problems of ordinary mass media: problems of ownership, unequal services and access, commercial bias and questions of content.[7] Obviously this is not a development shortcut; rather it can serve development ends only after several hurdles have been passed. Then media such as community radio allow more local input and have greater outreach and development potential than the fancy digital media.

Another question is who is the agent of information for development? Here the role of development aid and NGOs may be overplayed. The digital NGO projects display the usual characteristics of alternative development: reliance on project funding; uneven NGO unaccountability (to donors more than to communities); authoritarian or non-participatory management styles; non-replicable projects because they rely on specific capabilities and social capital, so most projects are not locally owned and not sustainable; and insufficient attention to the problems of "scaling up" (Wade 2002; Sorj 2003). While the projects run they produce alternative dependency and when the funding dries up so do most projects.

Government supported information projects with government providing inputs of content (making access to government forms and licenses available online) may be more viable than foreign aid projects, but usually fall short of their promise (Weerasinghe 2004; Gupta 2005) and turn citizens into customers (Wade 2002). India offers several good examples of developmental uses of Internet such as nonprofits that bring agricultural extension and other information to farmers in rural India,[8] for instance *e-Choupal*.[9] Several are supported by the government of India. Government supported initiatives such as *Drishtee* use a kiosk based revenue model to bring IT enabled services to the "rural masses."[10] Other projects focus on village level e-education.[11] However, as Sanjay Gupta (2005) notes, "E-governance is limited to e-government or e-services. Little participation is granted to the beneficiaries in decision-making or the design of the initiatives. Few cater to the needs of the poorest of the poor; *Drishtee*, for example, does not even consider the lowest 25 percent income-wise as its clients."

If we look to ICT4D as a new threshold in development policy rather than as another round of business as usual, then development aid on the part of bilateral or multilateral aid agencies or foundations, and also government provided services, may not be the first place to look. Realistically, the first place to look is at IT services that are provided by the private sector.

In many developing countries village phone networks such as n-Logue in India and Grameenphone in several countries (www.grameenphone.com)[12] have a considerable impact. Consider for instance the mobile phone coverage of Safaricom in Kenya (www.safaricom.co.ke). Many, often low-key private sector enterprises and entrepreneur networks (www.tie.org) use IT. There is no reason to overstate or exaggerate the significance of these initiatives; their purpose and reach are limited. But these private sector enterprises are not financially dependent on external funding, and, operating at low profit margins they have a greater reach and are more sustainable than donor or public sector projects. India may have an edge among developing countries in digital literacy (high education levels, English language, development as a national priority, decentralized state and local developmental states), yet Sanjay Gupta notes:

> Only 10,000 of the over 600,000 villages have seen some Internet-based ICT for development initiatives, most of which have important ingredients missing: social focus, community-driven, need-based and local initiative ... Their business strategy has been primarily focusing on certain types of transactions: related to land and agriculture or the provision of government services. They are mostly undertaken by the private sector with the intention of making them financially sustainable and profitable. Few have the empowerment of the socially and economically underprivileged groups as an objective ... Benefits for women in such initiatives are scarce, and little effort is made to encourage the use of services by women. Part of the problem is that most kiosks are operated by men, which discourages women from using them, given the social milieu in most parts of India. Also, content and services are more geared towards the needs of men rather than those of women. Most initiatives suffer from problems such as power cuts or lack of adequate power, and low-quality connectivity.
>
> (Gupta 2005)

But it is important to look beyond the attempts to bridge the digital divide by replicating and extending existing hardware and software technologies. Digital capitalism presents more pressing issues. Robert Wade notes that "LDC governments should not take technological and international regimes as given ... They need more representation in standard setting bodies and more support in the ICT domain for the principle that 'simple is beautiful'" (2002: 444). What matters is to shift the discussion away from the assorted applications of information technology to the technologies themselves.

The core problem that ICT4D poses is *disembedding technology from capital.* This is the real challenge of information for development, which brings us back to old questions of technology transfer and to full technology transfer rather than pseudo or adaptive transfer (Tandon 2005).

During the cold war years South Korea and Taiwan could disaggregate products and obtain their embedded technologies through reverse engineering, and by redesigning them bypass property rights and acquire intellectual property. The WTO regime of intellectual property rights and the talks on the harmonization of patent laws seek to forestall and limit these options. China follows a different avenue and uses its market power and bargaining clout to disaggregate foreign direct investment packages to obtain not just end-user capability but design technologies. But this route is not open to the smaller developing countries.

Digital capitalism poses the problem of technology dependency anew in both hardware and software. Efforts to develop appropriate IT hardware include simple computer (simputer), $100 laptop and one laptop per child (OLPC) projects. Entrepreneurs in China, India (Arifa 2002) and Brazil are developing low cost designs that may provide "Southern high-tech alternatives." Whether they compete or cooperate in these efforts is now not the most important question. Countermoves in this situation are attempts to re-embed technology in capital, as in Microsoft teaming up with OLPC and providing its operating systems at cheap rates to avoid youngsters becoming computer literate without Windows, or Microsoft making software available to Indonesia at a steep discount.

The second major frontier is software and the free and open software systems (FOSS) movement. This is of special importance because intellectual property rights are a major site of North–South negotiation and contestation. With the advanced economies increasingly losing their edge in manufacturing, services and research and development to emerging economies, owing to offshoring and outsourcing, intellectual property rights are a major remaining advantage (leaving aside the ongoing international trade talks on agriculture and textiles). In software development many corporations large and small have a stake in outflanking Microsoft monopolies and instead developing and fine tuning the Linux operating system and other open source systems, because these allow reprogramming of core codes and may thus offer greater flexibility, stability and security (Weber 2004). Governments such as Brazil and other emerging information economies increasingly use Linux in government administration, also with a view to savings (Sugar 2005). Cyber activists and other "organized networks" (Lovink 2005) are also active in this domain.

It would be a fantasy to think of a "digital Bandung" or an "IT Cancún" (similar to the walkout from the WTO talks in Cancún in November 2003 initiated by China, India, Brazil and South Africa and the Group of 21 developing countries). This assumes more policy cohesion than is now available. But there is room to strengthen this general approach and the convergence of interests of various stakeholders with a view, ultimately, to fashioning an alternative digital political economy.

We are at a major cusp in globalization. It may well be that in the next round of globalization the winners of the previous round—the US, Europe and Japan—will be placed second. The growing imbalances in production, consumption, trade deficits and financial deficits, particularly in the US, have become unsustainable. To a considerable extent courtesy of ICT, the old winners of globalization have been losing their production advantages to newcomers who combine low wages with good infrastructure, capable workers and docile labor regimes, and now this also applies to information processing services. One edge they can hold on to is intellectual property rights. As long as the advanced countries, more precisely, corporations in the advanced countries, can monopolize IPR and draw monopoly rents from IPR, they may be able to hold on to their advantage, which in other respects is slipping away. This means that ICT is not only important in its own right, it is also an arena in which at this stage the shape of globalization is being decided. The major tipping points in this arena are FOSS, TRIPS and patent laws. Here we find major corporations, governments in the global North and international institutions on one side, and most developing countries on the other. This is the real frontier of ICT4D.

## Notes

1. Google gave 6,260,000 entries for <u>digital</u> <u>divide</u> in August 2005 and over 4 million in June 2008.
2. "The digital divide is often portrayed in crassly reductive terms as a mere technological access problem that can be ostensibly addressed by providing cheap computing and communication technologies to the poor. However, the digital divide is not merely a technological problem due to the absence of connectivity or access to cyberspace. This instrumentally informed discourse on digital divide is a modernist tendency to unreflectingly categorize and compartmentalize complex sociotechnological changes into one-dimensional social problems in a bid to resolve them through simple technological fixes' (Parayil 2005: 41). Cf. Hand and Sandywell 2002.
3. Cf. Dicken 2007; Nederveen Pieterse 2004: Ch. 1.
4. On wider criticisms of copyright see Smiers 2000.
5. "It was symbolic that, in many of the world's capital cities, postal or communications ministries were physically situated near the seat of power" (Schiller 1999: 48).
6. Media tycoons such as Rupert Murdoch and Conrad Black have backed conservative politics. Rupert Murdoch funds the *Weekly Standard*, the house magazine of American neoconservatives. Silvio Berlusconi in Italy, Thaksin Shinawatra in

Thailand and Hugo Slim in Mexico emerged as entrepreneurs through the telecom and info industry.

7. The 1970s discussion on the role of the media in North–South relations, under the heading of the New International Information and Communication Order, led to a stalemate, parallel to the New International Economic Order debate. The current WSIS discussions may be a replay of these power plays (cf. Hamelink 2004; Shade 2003).

8. See http://www.indiagriline.com, www.mahindrakisanmitra.com, http://www.agriwatch.com and http://www.mssrf.org. A website that gives comprehensive information on ICT initiatives in India is http://www.bytesforall.org. I owe these examples to Sanjay Gupta (2005); cf. Arifa 2002; Chandra 2002; Singh 2002; Wade 2002; Ashraf 2004; Prestowitz 2005.

9. e-Choupal is a web-based initiative of ITC's International Business Division; it offers the farmers of India all the information, products and services they need to enhance farm productivity, improve farm-gate price realization and cut transaction costs. Farmers can access latest local and global information on weather, scientific farming practices as well as market prices at the village itself through this web portal—all in Hindi. Choupal also facilitates supply of high-quality farm inputs as well as purchase of commodities at their doorstep (http://www.echoupal.com).

10. "The services it enables include access to government programs and benefits, market related information, and private information exchanges and transactions" (http://www.drishtee.com).

11. This involves projects such as "Every Village a Knowledge Centre" (http://www.mssrf.org/special_programmes/ivrp/ivrpmain.htm). "Breaking the traditional confines of a school, Hole-in-The-Wall Education Limited takes the Learning Station to the playground, employs a unique collaborative learning approach and encourages children to explore, learn and just enjoy!" (http://www.niitholeinthewall.com). Cf. http://www.trai.gov.in.

12. "With the assistance of n-Logue and the financial support from the State Bank of India, the local strategic partnerships (LSPs) recruit local entrepreneurs to set up and run village based information kiosks. These kiosk owners are typically locally based men or women who have at least a 12th Standard education, and demonstrate the ability and motivation to run their own business. Marketed under the brand name 'Chiraag', which means enlightenment, these kiosks offer a variety of services aimed at providing benefit to rural areas while contributing to the kiosk's sustainability" (http://www.n-logue.com).

## Bibliography

Alden, C. 2003. "Let Them Eat Cyberspace: Africa, the G8 and the Digital Divide." *Millennium* 32 (3): 457–76.

Arifa, K. 2002. "Access to Information by the Socially Disadvantaged in Developing Countries with Special Reference to India." *Information Studies* 8 (3): 159–72.

Ashraf, T. 2004. "Information Technology and Public Policy: A Socio-Human Profile of Indian Digital Revolution." *The International Information and Library Review* 36 (4): 309–18.

Campbell, D. 2001. "Can the Digital Divide be Contained?" *International Labor Review* 140 (2): 119–41.

Castells, M. 1996. *The Rise of the Network Society*. Malden, MA: Blackwell.

Chandra, S. 2002. "Information in a Networked World: The Indian Perspective." *The International Information and Library Review*, 34 (3): 235–46.

Coco, A. and P. Short. 2004. "History and Habit in the Mobilization of ICT Resources." *The Information Society*, 20 (1): 39–51.

Cornia, G. A. 1999. *Liberalization, Globalization and Income Distribution*. Helsinki; UNU Wider Working Paper 157.

Cross, M. 2005. "Ethiopia's Digital Dream." *Guardian Weekly*, August 19–25: 27.

Cullen, R. 2001. "Addressing the Digital Divide." *Online Information Review*, 25 (5): 311–20.

Davis, C. M. 2001. "Information Apartheid: An Examination of the Digital Divide and Information Literacy in the United States." *PNLA Quarterly*, 65 (4): 25–7.

Dicken, P. 2003. *Global Shift: Reshaping the Global Economic Map in the 21st Century*, fourth edition. New York: Guilford Press.

Drahos, P. 2003. "The Global Intellectual Property Ratchet in the Information Age: Consequences and Costs." National University of Singapore.

Drahos, P. and J. Braithwaite. 2002. *Information Feudalism: Who Owns the Knowledge Economy?* New York: New Press.

Fors, M. and A. Moreno. 2002. "The Benefits and Obstacles of Implementing ICTs Strategies for Development from a Bottom-up Approach." *Aslib Proceedings*, 54, (3): 198–206.

Friedman, T. L. 2000. *The Lexus and the Olive Tree: Understanding Globalization*, second edition. New York: Anchor.

Friedman, T. L. 2005. *The World is Flat*. New York: Farrar Straus and Giroux.

Gupta, S. 2005. "ICTs for the Poorest of the Rural Poor." Available at http://www.globalatider.nu, accessed February 20, 2008.

Gurstein, M. 2003. "Effective Use: a Community Informatics Strategy beyond the Digital Divide." *First Monday*, 8 (12). Available at http://www.firstmonday.dk/issues/issue8 12/gurstein/, accessed December 1, 2008.

Hamelink, C. J. 2004. "Did WSIS Achieve Anything at All?" *Gazette* 66: 281–90.

Hand, M. and B. Sandywell. 2002. "E-topia as Cosmopolis or Citadel: On the Democratizing and De-democratizing Logics of the Internet." *Theory Culture and Society* 19 (1–2): 197–225.

Hedley, R. A. 1999. "The Information Age: Apartheid, Cultural Imperialism, or Global Village?" *Social Science Computer Review*, 17 (1): 78–87.

Hyder, S. 2005. "The Information Society: Measurements Biased by Capitalism and Its Intent to Control-Dependent Societies—a Critical Perspective." *The International Information and Library Review* 37 (1): 25–7.

International Institute for Communication and Development 2004. *The ICT Roundtable Process*. The Hague.

Léon, O., S. Burch and E. Tamayo G. 2005. *Communication in Movement*. Quito: Agencia Latino Americana de Información.

Lovink, G. 2005. *The Principle of Notworking: Concepts in Critical Internet Culture*. Amsterdam: HvA.

Luyt, B. 2004. "Who Benefits from the Digital Divide?" *First Monday* 9 (8). http://www.firstmonday.org/issues/issue9_8/luyt/index.html, accessed May 1, 2009.

McLaughlin, L. 2005. "Cisco Systems, the United Nations, and the Corporatization of Development." Available at incom-l@communicado.info, accessed July 21, 2009.

Menou, M. J. 2001. "The Global Digital Divide: Beyond HICTeria." *Aslib Proceedings* 53 (4): 112–14.

Nederveen Pieterse, J. 2001. *Development Theory: Deconstructions/Reconstructions*. London: Sage.

Nederveen Pieterse, J. 2004. *Globalization or Empire?* New York: Routledge.

Ngenda, A. 2005. "The Nature of the International Intellectual Property System: Universal Norms and Values or Western Chauvinism?" *Information and Communications Technology Law* 14 (1): 59–79.

Parayil, G. 2005. "The Digital Divide and Increasing Returns: Contradictions of Informational Capitalism." *The Information Society* 21: 41–51.

Prestowitz, C. 2005. *Three Billion New Capitalists: The Great Shift of Wealth and Power to the East.* New York: Basic Books.

Schiller, D. 1999. *Digital Capitalism: Networking the Global Market System.* Cambridge, MA: MIT Press.

Schiller, D. and V. Mosco (eds.) 2001. *Continental Order? Integrating North America for Cyber-Capitalism.* Boulder, CO: Rowman & Littlefield.

Shade, L. R. 2003. "Here Comes the Dot Force? The New Cavalry for Equity!" *Gazette* 65 (2): 107–20.

Sims, M. 2002. "A Digital Dividend?" *Intermedia* 30 (1): 28–30.

Singh, J. 2002. "From Atoms to Bits: Consequences of the Emerging Digital Divide in India." *The International Information and Library Review* 34 (2): 187–200.

Smiers, J. 2000. "The Abolition of Copyright: Better for Artists, Third World Countries and the Public Domain." *Gazette* 62 (5): 379–406.

Sorj, B. 2003. *Confronting Inequality in the Information Society.* Sao Paolo: UNESCO Brazil. Available at brazil@digitaldivide.com, accessed December 1, 2008.

Stansbury, M. 2003. "Access, Skills, Economic Opportunities, and Democratic Participation: Connecting Four Facets of the Digital Divide through Research." *Canadian Journal of Information and Library Science* 27 (3): 142–3.

Strange, S. 1996. *The Retreat of the State: the Diffusion of Power in the World Economy.* Cambridge: Cambridge University Press.

Sugar, D. 2005. "The Free Software Challenge in Latin America." *Countercurrents.org.* Available at http://www.countercurrents.org/gl-sugar030605.htm, accessed March 6, 2008.

Thompson, M. 2004. "Discourse, 'Development' and the 'Digital Divide': ICT and the World Bank." *Review of African Political Economy* 31 (99): 103–23.

Tsing, A. 2004. "Inside the Economy of Appearances." In *The Blackwell Cultural Economy Reader*, eds. A. Amin and N. Thrift. Oxford: Blackwell, 83–100.

Verzola, R. 2004. *Towards a Political Economy of Information: Studies on the Information Economy*, second edition. Quezon City: Foundation for Nationalist Studies.

Wade, R. H. 2002. "Bridging the Digital Divide: New Route to Development or New Form of Dependency?" *Global Governance* 8: 443–66.

Warschauer, M. 2003. "Dissecting the 'Digital Divide': a Case Study in Egypt." *The Information Society* 19 (4): 297–304.

Weber, S. 2004. *The Success of Open Source.* Cambridge, MA: Harvard.

Weerasinghe, S. 2004. "Revolution Within the Revolution: The Sri Lankan Attempt to Bridge the Digital Divide through e-governance." *The International Information and Library Review* 36 (4): 319–32.

Wicks, D. A. 2003. "Building Bridges for Seniors: Older Adults and the Digital Divide." *Canadian Journal of Information and Library Science* 27 (3): 146–57.

Wong, P.-K. 2002. "ICT Production and Diffusion in Asia: Digital Dividends or Digital Divide?" *Information Economics and Policy* 14 (2): 167–87.

Ya'u, Y. Z. 2004. "The New imperialism and Africa in the Global Electronic Village." *Review of African Political Economy* 31 (99): 11–29.

# 13.1

## Information and Communication Technologies for Least Developed Countries: A Case Study of the Republic of Malawi

*Robert M. Bichler*

Information and communication technologies (ICTs) are claimed to be a central engine for societal progress and prosperity and therefore in the last 15 years an increasing body of literature dealing with the interrelation between ICTs and development has been emerging (cf. e.g. Mansell and Wehn 1998; Braga et al. 2000; Okpaku 2003; Wilson 2004; Unwin 2009). So far, only the Western world has benefited from these technologies, while developing countries especially are facing the challenge that the already existing tremendous gap between them and the high-income economies in the West may still widen. In this paper the results of a research project in Malawi, which aims to identify strategies for closing this gap, are briefly summarized.

The study combined quantitative and qualitative methods. The primary data were drawn from a survey of users in eight Internet cafés, in the two major Malawian cities Lilongwe and Blantyre in November 2007. Lilongwe is the administrative capital whereas Blantyre functions as the unofficial economic capital. Internet cafés were chosen because the Internet penetration in Malawi in 2007 was only 0.4 percent of the total population and hence Internet cafés seemed to be the ideal place to find out about the Internet habits of those who actually use the Internet. Usually the Internet cafés were not very well equipped, providing neither headsets nor web cams. I personally handed out 270 questionnaires to Internet café visitors and also acted as a client in the selected Internet cafés, which enabled different observations concerning the cafés' equipment and the nature of the clients. In addition expert interviews with Malawian decision-makers from governmental institutions, universities, business companies and NGOs were carried out to gain a broader picture.

### Central findings

The study shows that at the moment technical difficulties are the main barrier for the diffusion of ICTs, especially the Internet, in Malawi. The fact that the country is not yet connected with a fiber cable leads to very high costs combined with an extremely low bandwidth and this again results in the exclusion of the majority of the Malawian citizens.

In 2007 a dial-up connection with 33 kbit/s, for example, cost about US$30 per month, whereas a faster WiMAX[1] connection with 64 kbit/s started around US$200 per month. Additionally, one had to pay between US$70 and US$100 to Malawi Telecommunications Limited (MTL) for the telephone line, which made the line costs about three times more expensive than the service costs (Nyirenda 2007; Personal Interview). Taking into account that for 2006 the GNI per capita[2] was calculated to be US$170[3] it becomes obvious that the majority of the population is financially excluded from Internet use. Neoliberal market reforms, which included the privatization of MTL and the opening up of the Malawian telecommunications market to foreign investors, did not really improve the situation. It is also questionable if the installation of the fiber cable between Malawi and Mozambique by Alcatel-Lucent will contribute to the sustainable development of the Malawian telecommunications landscape. The data-signalling rate will definitely increase, but the costs will probably remain high as a result of the amount of return the company aims to make on its investment.

The very high costs result in problematic user demographics: the typical Malawian Internet user, aged between 26 and 40 years, is quite old, taking into consideration that the life expectancy at birth is 43 years, and the average income of Malawian Internet users is superior or high; for example, 18 percent of the respondents had a net income of between US$300 and US$500 per month and 21 percent had an income higher than US$500 per month.

The education level of Malawian Internet users is also far above average, for example, 33 percent of the respondents completed secondary school and 19 percent graduated from university. Generally the low number of users (55,029 in the year 2006; 139,500 users in 2008)[4] and the limited availability of services resulting from the poor infrastructure remain immense challenges for the political decision-makers. Using the Internet for business purposes (56.7%) is very popular, whereas up until now the possible fields of application are limited to business communication and conducting market research online. E-commerce, in the form of online shopping, is a marginal phenomenon (10.6%) as Malawians don't have credit cards and lack confidence.

Furthermore the study shows a great interest in education (Category "Research for School/University": 47.8%; Category "Using Internet for Training and Educational Purposes": 29.8%) and health related issues (18.8%), which opens up the possibility of adopting new Internet based services in the form of e-learning and e-health applications. The results of the survey indicate that such services would be well received by Malawian users and thus would contribute to a sustainable use of ICTs.

## Notes

1. Worldwide Interoperability for Microwave Access.
2. Gross national income per capita.
3. Data taken from the World Bank: http://devdata.worldbank.org/external/CPProfile.asp?CCODE=MWI&PTYPE=CP.
4. Data taken from http://www.internetworldstats.com.

## Bibliography

Braga, C. P. A., C. Kenny, C. Qiang, D. Crisafulli, D. D. Martino, R. Eskinazi, R. Schware and W. Kerr-Smith. 2000. "The Networking Revolution: Opportunities and Challenges for Developing Countries" (working paper). Washington, DC: World Bank Group. Retrieved April 2007 from: http://www.infodev.org/library/working.html.

Mansell, R. and U. Wehn (eds.) 1998. *Knowledge Societies: Information Technology for Sustainable Development.* Oxford: Oxford University Press.

Nyirenda, P. B. (Coordinator). 2007. "Malawi Sustainable Development Network Program" (SDNP). Blantyre, November 20.

Okpaku, J. (ed.) 2003. *Information and Communication Technologies for African Development: An Assessment of Progress and Challenges Ahead.* New York: United Nations ICT Task Force.

Unwin, T. (ed.) 2009. *ICT4D. Information and Communication Technology for Development.* Cambridge, New York, Melbourne: Cambridge University Press.

Wilson, E. J. III. 2004. *The Information Revolution in Developing Countries.* Cambridge, MA: MIT Press.

# 13.2
## Dot.Com Marriages in India: Examining the Changing Patterns of the Arranged Marriage Market in India

*Mili Kalia*

Traditionally, finding marriageable singletons in India was considered the prerogative of a few members within a community network. Later, this was augmented by the use of matrimonial newspaper advertisements (Gist 1953). Nowadays, Indians entrust this process to a new system of "virtual matrimonials." This enables Indians to find that "perfect someone" in, homes, cafés or Internet parlors. Individuals and families can create web accounts of available brides and grooms, with practically no financial liabilities. They can compile a list of favorite profiles, post photographs and videos, and arrange virtual rendezvous via online video conferences and chats, and so on.

The logic of virtual matrimonials is derived from the practices of the evolving capitalist market economy, which works through and in conjunction with global technological development. Todd Wolfson (2005) refers to this type of system as "Cyber Capitalism," which is a "contemporary regime of accumulation" that relies on "new information and communication technologies." New technologies can affect the nature and process of accumulation by inculcating a need for flexibility, effectiveness and predictability. Accumulation, in the context of this study, could refer to material and non-material resources. However, its impact, in this context is most palpable in one's ability to accumulate information, contacts and knowledge about the pool of singletons worldwide. Wolfson (2005) further suggests that Cyber Capitalism also, "affects the cultural logic of its accompanying activities," such as the basis of finding a desirable mate.

The search for marriageable mates in this virtual system suggests a change in the mate-search. With millions of Indians online, it is imperative to explore how this shift may or may not transform some of the most important social institutions in India.

The data for this study were collected in 2006, with the aid of a computing language designed specifically to extract information from 2705 randomly selected profiles from one of the most popular matrimonial websites catering to Indians. Probabilistic analysis was conducted to decipher the predicted odds of most likely choices under the given circumstances, such as stated background, gender, authorship, religion and age. Specifically, I tried to look at how important caste was in relation to economic status, such as educational background. The following discussion highlights just one of the most important findings of this study.[1]

Traditional systems of finding a mate focused on ascribed markers such as caste or sub-caste backgrounds. These markers were considered reliable indicators of cultural and economic similarities, familiarity and successes (Dumont 1981). Recent scholars have questioned the efficacy of caste to determine the social boundaries and interactions, as in recent years new boundaries are being formed based on economic status, such as education level (Srinivas 1956, 1959; Caldwell, Reddy & Caldwell 1983; Corwin 1977; Marriott 1960). To what extent then does this hold true for single men and women who are trying to find a mate, using advanced global technologies which are giving them more choices, information and knowledge about available singletons all over the world?

Findings suggest that the preceding theoretical proposition holds true for most single men on this website (log odds of mentioning caste = $-0.621$***[2]). However, even though profiles of single women did signal economic markers such as educational background (49%), their main focus was on their caste backgrounds (51%). It would be a fallacy to state that this asymmetrical signaling pattern is indicative of women and men tending to let go or hold on to various aspects of their tradition, that is caste background. In fact, this finding is indicative of a complex cultural conundrum that most affluent educated Indian women may find themselves in today. This finding may be symptomatic of a society that is coming to accept women as being able or needing to "do it all" (see Hochschild 2003). A marriageable woman thus needs to be able to be, at the same time, sensitive to cultural sentimentalities within her home and educated enough to find her place among her peers beyond this private domain.

Let us remember that these actors are not passive amid these social expectations. On the contrary, I suspect that women may be using both caste and class to express their own desires and "personhood," a personhood that values tradition but simultaneously is reflexive, able to understand and re-appropriate the nature and purpose of tradition contextually within the confines of what is accepted by the larger society. Men, on the other hand, seem to be comfortable with their role as the primary provider. Again, this is not to say that they readily accept it. I do suspect that men, like women, feel the pressures to prove their worth. But, their marriageability can only be shown through their economic achievements, giving them less "space" to mold their personhood in the infinite space of virtual matrimonials.

Virtual matrimonials in India are one of the many representations of cyber capitalism. This is global in its intention but "privately reflexive" in its outcome. It is global because of its logic of efficiency and flexibility, necessitating the participation of many actors. Concomitantly, it is also a "privately reflexive" system because these global technological and market influences are reappropriated in distinct ways by actors. This may be contingent on the social, cultural, psychic, historical and economic structure of the country or community in question (Appadurai 1996). In this case it is a unique amalgamation of Indian socio-cultural structures along with global technological growth. The Indian case of Cyber Capitalism thus produces a system of mate search that is asymmetrical; requiring women to show that they are multifaceted and men to show that they can be the primary providers for their families.

## Notes

1. Please note, the actual study is more nuanced with analysis using a three actor model. The analysis here is based on only two actor interactions.
2. Significant at 0.001 level.

## Bibliography

Appadurai, A. 1996. *Modernity at Large: Cultural Dimensions of Globalization.* Minneapolis: University of Minnesota Press.

Caldwell, J. C., P. H. Reddy, and P. Caldwell. 1983. "The Causes of Marriage Change in South India." *Population Studies* 37 (3): 343–61.

Corwin, L. A. 1977. "Caste, Class and the Love Marriage: Social Change in India." *Journal of Marriage and the Family* 139: 823–31.

Dumont, L. 1981. *Homo Hierarchicus: The Caste System and its Implications.* Chicago: University of Chicago Press.

Gist, N. P. 1953–54. "Mate Selection and Masscommunication in India." *The Public Opinion Quarterly* 4: 481–95.

Hochschild, A. R. 2003. *The Commercialization of Intimate Life: Notes from Home and Work.* Berkeley CA: University of California Press.

Marriott, M. 1960. "Caste Ranking and Community Structure in Five Regions of India and Pakistan." Poona, India: Deccan College Monograph Series 23.

Srinivas, M. N. 1956. "A Note on Sanskritization and Westernization." *Far Eastern Quarterly* 15: 481–96.

Srinivas, M. N. 1959. "The Dominant Caste in Rampura." *American Anthropologist* 61: 1–16.

Wolfson, T. 2005. "Mediated Democracy: Activism and the Promise of Politics in Cyber Capitalism." Presented to the *American Anthropological Association,* Washington, DC (December).

# 13.3
## Bridging the Digital Divide in Developing Countries: A Case Study of Bangladesh and Kuwait

*Charles C. Chiemeke*

The key trend in ICT is that the nature of international flow of fees, and the resulting accumulations of net capital flow of patents, licenses and royalty payments, are asymmetrically distributed between the industrialized and developing countries (Bartel and Chiemeke 2008). Such a structure of flow and distribution is entirely consistent with the view that technology flows are externalized in markets and have a "North" and "South" dimension (Copy/South Dossier 2006), which may result in technological upgrading of the "South."

Three groups of countries, the technological innovators, the technology adopters and adapters, and the technologically excluded, depict the current nature of the technological divide in the world economies (Sachs 2000). Most middle and low income countries produce relatively little knowledge; they are net importers of information capital and widely regarded as the technologically excluded economies. Although the digital divide gap in developing countries is narrowing, in some countries in South Asia and the Middle East region the gap seems to be widening.

### The concept of the digital divide

The digital divide is referred to as a gap between those who have access to information technology (IT) and those who do not (Rice and Katz 2003). This geo-economic phenomenon, in recent times, has received a great deal of attention from researchers and policy analysts (Sachs 2000; Wheeler, Dasgupta and Lall 2001; Corrocher and Ordanini 2002; Akbar 2001). There has been a widespread urgency to understand the digital divide because of the recognizable and obvious role of IT as an important source of a country's economic growth.

Interestingly, the term digital divide has been further broadened to include all aspects of digital inequality: technical means (hardware, software and connectivity); autonomy (location of access, freedom of use); use patterns (purposes of the Internet uses); skills (ability to use the Internet effectively);

and social support networks (access to advice from more experienced users). This framework helps us to better understand the complexity surrounding this issue and affirms the increasing awareness in developing countries of the extreme disparities in the conditions of production and use of "creative information" across the globe.

## Bridging the digital divide in Bangladesh

Bangladesh is regarded as one of the developing countries where the digital divide gap is widening. There is acute backwardness in many communities, particularly among the women of Bangladesh, in the use of IT. Access to tele-communications infrastructure is quite deficient, with only about 0.5 percent of people having access to fixed telephone lines.

According to Md. Shahid Uddin Akbar, CEO of womenBD.com, Bangladesh is facing an acute crisis in numbers of skilled computer users as a result of literacy problems, given that information on the net is designed in advanced technology, which requires adequate knowledge from the user. He further stated that women in Bangladesh have limited access to the Internet because of lack of education, fear of technology use, economic conditions, lack of English language skills and family responsibilities. They also lack the social aspects for utilizing Internet facilities from outside and within the working environment and the professions.

A recent essay by Rashid (2006), however, finds an exponential diffusion of cellular markets in such emerging countries as Bangladesh. In other words, cellular telephony in Bangladesh is expected to dominate other technology in bridging the digital divide. Accordingly, key major initiatives in bringing cellular connectivity to rural Bangladesh include the Village Phone (VP) concept of Grameen Telecom, which provides loans to women borrowers for phone equipment and subscriptions to cellular services, and the Emerging Market Handset (EMH) program initiative launched in February 2005 by the GSM Association to provide low cost handsets. The VP concept has provided access to telecommunications facilities for more than 60 million people living in the rural areas of Bangladesh, while the EMH program has the potential to add more than 100 million new global connections per year.

## Bridging the digital divide in Kuwait

A recent report has found that there are 3.54 million Internet users in the Arab world, as of 2001 (Ajeeb.com) and there will be over 20 million users possibly by the end of 2010. More importantly, most Internet users in the Arab world live in the Gulf Cooperation Council (GCC) countries, which include Kuwait, Qatar, Bahrain, Oman, United Arab Emirates and Saudi Arabia.

Similarly, differences exist in the percentages of penetration across countries in the GCC. But, Kuwait, with a 9 percent penetration rate, is among those that have high percentages of their populations online. Furthermore, the new Digital Opportunity Index (DOI),[1] created from the set of internationally agreed core information and communication technology (ICT) indicators, shows that Kuwait's rank moved from 49th in 2005 to 60th in 2006 (Allam 2007). The DOI includes indicators that show the opportunities for a country's citizens to use ICTs, indicators that represent the infrastructure needed by any country to use ICTs, and indicators that show the extent of ICT utilization within the country. Evidence suggests that issues concerning freedom of information explain one of the reasons for the low progress in Kuwait. Lack of a stable political environment in Kuwait, an important precondition to ensure free access to information in the society, remains one of the key challenges facing the country as it builds an information society.

## Note

1. The DOI is a standard tool that governments, operators, development agencies, researchers and others can use to measure the digital divide and compare ICT performance within and across countries (Allam 2007).

## Bibliography

Akbar, M. S. U. 2001. "Bridging Digital Divide: Electronic Medical Information Systems Bangladesh Aspect." Presented to the Regional Conference on Medical Librarianship: Building the Virtual Health Sciences, Tehran, November 11–13.

Allam, A. 2007. "Managing Knowledge in the 21st Century and the Roadmap to Sustainability." In *World Sustainable Development Outlook 2007: Knowledge Management and Sustainable Development in the 21st Century*, ed. A. Allam. Sheffield UK: Greenleaf.

Almeida, P. R. de. 1995. "The Political Economy of Intellectual Property Protection: Technological Protectionism and Transfer of Revenue among Nations." *International Journal of Technology Management* 10 (2/3): 214–29.

Bartels, F and C.C. Chiemeke. 2008. "The Structure of Patent, Licensing and Royalty Fees Flow: The Global Role Of Multinational Enterprises." *Journal of World Universities Forum.*

Bertot, J. 2003. "The Multiple Dimensions of the Digital Divide: More than the Technology 'Haves' and 'Have Nots'." *Government Information Quarterly* 20 (2): 185–91.

Braga, P. and A. Carlos. 1989. "The Economics of Intellectual Property Rights and the GATT: A View from the South." *Vanderbilt Journal of Transnational Law* 22: 243–64.

Bu-Hulaiga, I. A. 2001. "Challenges of e-Business in GCC: Digital Divide & Economic Reform." Available at https://www.itu.int/arabinternet2001/documents/pdf/document17.pdf, accessed February 11, 2010.

Copy/South Dossier 2006. "Issues in the Economics, Politics, and Ideology of Copyright in the Global South." A paper produced by Copy/South Research Group.

Available at: https://www.kent.ac.uk/law/copysouth/V2-documents/v2-pdf-1.pdf, accessed March, 2006.

Corrocher, N. and A. Ordanini. 2002. "Measuring the Digital Divide: A Framework for the Analysis of Cross-Country Differences." *Journal of Information Technology* 17 (1): 9–19.

Haythornthwaite, C. 2006. "Digital Divide—Social Barriers On and Offline." Paper presented at the *International Conference Cyberworld Unlimited* (February 9–11) Bielefeld, Germany.

Ikiara, M. M. 2003. "Foreign Direct Investment (FDI), Technology Transfer, and Poverty Alleviation: Africa's Hopes and Dilemma." *African Technology Policy Studies Network* (ATPS) Special Paper series no. 16.

Khalil, T. 2000. *Management of Technology: The Key to Competitiveness and Wealth Creation.* Boston: McGraw-Hill.

Kumar, N. 1996. "Foreign Direct Investments and Technology Transfers in Development: A Perspective on Recent Literature." *Institute for New Technologies* (INTECH) Discussion Paper 9606. Maastricht: United Nations University.

Norris, P. 2001. *Digital Divide: Civic Engagement, Information Poverty, and the Internet.* Cambridge: Cambridge University Press.

Rashid, S. M. Mamun Ar. 2006. "Bridging the Digital Divide—Paradigmatic Evolution of Bangladesh as the Microcosm of Emerging Economies." Available at http://www.itu.int/osg/spu/youngminds/2006/essays/essay-rashid-mamun.pdf, accessed February 15, 2010.

Rice, R. and J. Katz. 2003. "Comparing Internet and Mobile Phone Usage: Digital Divides of Usage, Adoption, and Dropouts." *Telecommunications Policy* 27 (8–9): 597–623.

Sachs, J. 2000. "Technological Divide." *The Economist* (June 22).

Techatassanasoontorn, A and R. J. Kauffman. 2006. "Is There a Global Digital Divide for Digital Wireless Phone Technologies?" Special Issue, *Journal of the Association for Information Systems.*

Wakasugi, R. and B. Ito. 2005. "The Effects of Stronger Intellectual Property Rights on Technology Transfer: Evidence from Japanese Firm-Level." KUMQRP Discussion Paper Series DP2005-009/12: Keio University.

Wheeler, D., S. Dasgupta and S. Lall. 2001. "Policy Reform, Economic Growth, and the Digital Divide: An Econometric Analysis." World Bank Development Research Group, Infrastructure and Environment. Policy Research Working Paper 2567 (March).

# Part XIV
# Reading Machines

# 14
# Do E-Book Readers Understand Digital Documents?

*Jean-Claude Guédon*

## Introduction

The question raised in the title is ambiguous at best, absurd at worst. The word "reader" obviously creates ambiguity but here we are talking about devices, and not about humans. We are talking about devices that "read" digital documents in such a way that humans can then interact symbolically with them, and even read them if indeed this is their intention. The absurdity is linked to the word "understand": how can a machine "understand" a document? Indeed, machines do not understand documents in the usual sense of the word; however, structured and constrained by their design as they are, they do treat and apprehend documents in particular ways. Furthermore, these modes of apprehension can be viewed as the technical translation of how designers or engineers understand documents: they refer to the ways in which the same engineers relate to documents, how they access elements of their meaning and how these documents ought to relate to each other, if at all. In short, a reading device incorporates a vision of what a document is and how it "lives" among humans. This opens the possibility of partial or even total misreadings, and this observation begins to explain the title and its slightly provocative and even cryptic form.

In short, the title does not try to anthropomorphize the e-book reader; it acts as a kind of wake-up call by foregrounding a simple, but fundamental, fact: reading is inescapably tied to some technology, and this is as true of cuneiform writing as it is of contemporary digital devices. This also means that reading mutates as technologies evolve. In other words, the relationship of humans to documents shifts, which then affects the relationship among human beings whenever they rely on some communicational technology. For example, if we were studying documents around the invention of the printing press, we might ask how Gutenberg presses understood texts, but saying this would be little more than raising the well-known debate about "print culture," which arose in the wake of Elizabeth Eisenstein's study on the consequences of the printing press.[1]

It also happens that I write these lines even as a very intense campaign to promote Apple's iPad is assaulting the ears and eyes of information technology (IT) consumers. Often presented as a new hybrid object, part phone, part tablet, part computer, part e-book reader, part game platform, the iPad provides a fitting, almost providential, backdrop to what follows—namely an attempt at making some reasonably educated guesses about the future evolution of reading machines. As we shall see, they encompass much more than mere e-book readers as illustrated by Amazon's Kindle, for example.

This essay builds on an earlier text where the prehistory of the bicycle was used to revisit the evolution of e-book readers, and it does so with a deeper purpose in mind: how do technical objects, documents and people relate to each other and to one another?[2] In other words, what kinds of technical, documentary and social ecologies are emerging as a result of the gradual digitizing of everything and the radical recasting of our techno-documentary environment?

This essay actually tries to cover two different sets of questions: on the one hand, it tries to peer over the horizon and imagine what the future of the e-book readers might look like. But this effort is actually secondary; more interesting is the attempt to look at wider issues. For example, what does the combination machine-digital document portend for the status of knowledge? What does it mean for the very fabric of human societies?

Peering into the future is a risky business, generally best avoided by academics, but, in this case, we are helped by the fact that fault lines are both present and already visible. Focusing on the e-book reader as a technical category actually helps identify these fault lines: while they do not define the future, they bring out some of the issues that will affect its structure.

## Recalling some lessons from bicycle history

In my earlier text, I made use of the history of the bicycle to foreground an issue that I felt was being neglected in the discussions surrounding the e-book readers: the place of users. What roles do they play in the design of technical objects? A pioneer text in the sociology of technology sought to show how the bicycle stabilized in essentially one form out of a complex context that was crowded by a wide variety of vehicles.[3] In one way or another, these objects, sometime quite strange in appearance or even in name (witness the "boneshaker"), were trying to respond to needs, desires, aspirations, dreams of different categories of people: athletes wanted speed and did not mind showing off their acrobatic skills as well; children wanted toys but their parents wanted these toys to be safe. Women also wanted to ride, but, as with horsemanship, clothing had to be adapted to the task, and safety remained part of the equation. Meanwhile, workers wanted a cheap, fast and secure way to go to work. And so on. Although incomplete, the list brings to light a set of disparate and even contradictory functions that no object knew how to integrate. It is this simple detail that explains the variety of vehicles that, from our perspective, form the prehistory of the bicycle. In short, tension between contradictory desires resulted in technical diversity.

If we take safety and speed to be the two most fundamental requirements of a bicycle yet to be created, it becomes apparent that one particular innovation, the tire, made a huge difference: it increased comfort, certainly did not decrease safety, and it increased speed. These characteristics of the tire effectively reconciled the wishes of different sets of customers to merge them into a larger, and more viable, market. Instead of facing a bewildering array of "pre-bicycles," a general bicycle "template" could emerge at last. It included a number of basic features, and these could be tweaked for a particular usage over another: for example, more speed with less safety thanks to narrower tires. This also led to other details being generalized: the pedals and chain propelling the rear wheel became the rule; the frame was made slightly differently for men and for women to accommodate sex-related clothing constraints; front and rear wheels sported equal sizes; and steering was done with the handlebar fixed to the front wheel. But the tire remained the key innovation because it erased the fault line that placed safety on one side, and speed on the other.

## E-book readers vs. digital readers

By contrast with the bicycle, the e-book readers offers the somewhat puzzling appearance of an object that has not essentially changed in the last 20 years, yet keeps on presenting itself as the harbinger of the future. When e-book readers were only the stuff of science fiction[4] they already looked like e-book readers. The first attempts at commercializing e-book readers, and that was about a dozen years ago, reiterated the same patterns for the device. In short, e-book readers have not fundamentally changed in outward appearance for the last 30 years: it is still the size of a small paperback, and about as heavy. As expected, it sports a screen presented as the generic new page. The only significant change came when newer e-book readers began to rely on a technology called E Ink. They are sold with the argument that reflected light, which e-ink screens use, makes them feel more like a printed page. In fact, E Ink tries so much to imitate printed paper that the entire family of e-book readers appears to harbor only one ambition: replacing the printed codex. E-book readers essentially want to substitute themselves to the printed codex while avoiding disturbing as little as possible the financial, legal and institutional contexts of the book trade. Therein lie its technical essence and the main source of its design. Amusingly, exactly as Gutenberg's dream was to create a believable manuscript *Ersatz*, the e-book reader based on the E Ink technology dreams of replacing the printed codex.

The claims that are made on behalf of various, yet similar, e-book readers are quite monotonous, but they raise a deeper question: these devices clearly have not yet stabilized, and consequently their shared "look and feel" should appear puzzling: if this "family air" should correspond to a stable plateau, it would mean that the evolution of e-book readers does not follow the pattern that bicycles present. The stable appearance of e-book readers would suggest that no market fault-lines exist and that the technical solution presently offered for sale can claim a template status. Are all users really happy with e-book readers? But then, why have the sales moved slowly and with difficulty? Though growing significantly in recent years, sales do not overwhelm. Clearly, despite all the efforts of Amazon and Sony and other companies to promote their kind of e-book readers, these little machines have not yet triggered massive purchases by a broad front of users. The appeal of the e-book reader has not been as spectacularly compelling as that of the cellular (or mobile); neither has it given rise to a quick succession of innovations that tend to support the rapid spread of objects once the right "template" is identified. The collective wisdom of crowds, to the extent that it can be trusted, remains skeptical with regard to e-book readers. Only niche groups seem to take to them with some interest, if not enthusiasm. It appears, therefore, that the question may not have been put in quite the right way.

In short, is considering the e-book reader as an isolated family of technical objects the right approach? If the answer is negative, the e-book reader as we know cannot be a template. By looking at e-book readers as if they had to be (and behave) like codices, but only better, are we not being tricked into an unsuitable context? By insisting on treating the e-book reader as a family of technical objects, are we not missing the real set of objects we should be looking at? And might not this elusive set display a much greater variety of forms?

In my earlier essay (2009), this led me to propose a different hypothesis: with digitization, the world of documents is being changed in profound ways; in any case, it includes much more than the digital equivalent of printed texts. Furthermore, the digital world cannot be treated as if it consisted of a series of isolated silos storing images, sounds or videos, with each requiring different technologies to be accessed. On the contrary, and as is well known, texts, images, sound and video, as well as their elements related to their structure, can now be "written" with a single language, that of positive or negative bits. This means that the separate categories of documents we have grown familiar with no longer make natural sense. A text can incorporate a video, it can be transformed into sounds that people can listen to, it can incorporate a musical background, and so on. The boundaries between various document types vanish when these documents can be "read" with a single technology, but that is not obvious so long as we insist on treating these documents as different from each other. In short, the marriage of texts and images that print tacitly resisted for a very long time has bequeathed a legacy which although anomalous, if viewed over the last few thousands of years, has become the dominant vision in the last few centuries; we are so familiar with texts existing by themselves, in splendid isolation so to speak, that we tend to start our analyses from this peculiar perspective, even though, over the *"longue durée,"* it looks more like an exception than the rule. When the Akkadians transformed writing from symbolic images of things to symbolic images of sounds, they obviously did not banish images, and documents continued to mix glyphs with images all the way to Gutenberg. In fact, even after Gutenberg, much ingenuity was spent trying to bring back the lost paradise of magnificently illuminated manuscripts. Curiously, and in a patently reactionary manner, e-book readers work in the contrary direction: they try to keep us in the world of primitive print, with images reduced to a few shades of gray. When based on E Ink technology, no video is possible. Furthermore, copyright issues block creative souls with daunting financial and procedural obstacles. This is the reason that moving into a fully digital world, as was recently pointed out, will probably coincide with the creating of new images, sounds, videos. Existing resources, on the other hand, because they are too hard to use legally, will simply tend to be left unused.[5]

The bicycle and its history help remind us that sometimes objects multiply in such a profusion of forms that, although functionally related, they do not appear to belong to the same family. As a result, users and customers puzzle about the probable evolution of what, to them, appears as no more than a disorderly crowd of technical objects. In fact, this is exactly what happens with information technologies, as the following reaction to Apple's iPad illustrates:

> The iPhone filled many voids all at once. But the iPad doesn't fill an immediately apparent need, except maybe some springtime enthusiasm represented through rampant consumerism. It's basically a big iPhone that isn't a phone. Using it, I kept asking myself: Why do I need this?[6]

Other comments on the iPad reinforce this point: how many of these objects should I carry about? Between the smartphone, the laptop, the netbook and the tablet, the hapless user of IT tools (not to say "gadgets") feels as confused as did potential users of vehicles propelled by human muscle in the second half of the nineteenth century. Before the bicycle stabilized, any individual might have felt that he or she needed to buy several machines, each fulfilling a particular function. In the case of the proto-bicycles, we benefit from some retrospective sense that all these objects were connected. However, in the case of information technologies, looking forward without the benefit of a good "template," it is much more difficult to recognize such connections. Our IT devices appear as an unstructured collection of gadgets.

One way to project some order on this confused situation is to go back to basics and remember that all that can be "written" with 0s and 1s is a document. This immediately reveals that computers, game consoles, mobile phones, tablets, netbooks and so on are all machines that can "read" various digital documents, and transform them into symbols or representations that humans can deal with. It also brings to light a bewildering variety of digital readers that begin to look like the jungle of pre-bicycle objects. E-book readers no longer constitute a separate family of objects, only a functional niche within a wider collection of gadgets that have yet to recognize themselves as members of a single family, and the diverse collection of IT objects can be easily compared to pre-bicycle vehicles:

1. The e-book reader. It tries to substitute itself for the codex while supporting and even increasing the control of publishers over the texts they own. While designed for ease of reading, they really seek to check on transactions and limit them. For example, lending a text to a friend often involves lending the whole machine. For example, stored formats may be machine-specific. The refresh rate of E Ink screens is too slow to accommodate video and interactive contents. Color is missing in most e-book readers. Its connectivity can go from nothing to limited possibilities of linking to computers, the Internet or some telephone network. When the latter possibility is present, it is mainly to facilitate impulse buying, as is clearly the case with Amazon's Kindle. Some annotations and linkages across documents are allowed on some machines.

2. The computer is really a virtual machine and it therefore fulfills functions that are limited mainly by the software that runs on it. Its portability and ease of handling have improved. Many people still find it difficult to read on a computer screen, but the screen can be improved. The computer can connect to almost everything and the documents it processes can be modified in almost unlimited fashion.

3. The game console is built around powerful graphic cards that allow fast rendering and intense interactivity. They are very portable and can be used in a variety of ways, alone, or in combination with various kinds of screens, including large television screens. The connectivity of consoles emerged later and is generally limited to Wi-Fi. As a rule, game consoles do not allow tampering with the game software. In fact much is done to make copying very difficult as a way to enforce copyright laws. In this regard, they behave very much like e-book readers with proprietary formats.

4. Digital televisions do read digital documents, but they do so within the constraints of a program, a time-dependent sequence of documents over which the viewer has no control, or, at best, very limited control. Television can receive its signal from a variety of sources, including the air, cable distribution, satellite dishes and now Internet connections. However, it continues to work within a broadcast philosophy, whereby a single document is distributed to vast numbers of people with little or no possibility of modification or retro-action. Time-shifting of programs only underscore the rigidity of the device. However, the Internet, by providing

a novel way to access sources of entertainment, documentaries and movies, offers the promise of deeply transforming the television set by redefining it as an interface to a library of documents, bringing it closer to an e-book reader. The advent of 3-D screens brings back the intrinsic importance of the screen as interface between machine and humans.

5. Digital radios largely follow the footsteps off digital television, using essentially the same means of diffusion. Internet radios provide access to thousands of stations from all over the world, and it helps us imagine the future of television.

6. Electronic frames are optimized to "read" color photographs and display them in various kinds of sequences. They require good color capability, similar to that of a high-definition television. They generally have no connectivity at all despite all the obvious possibilities such a functionality would provide. These tools have very limited functions, as they were conceived as an electronically enhanced frame and nothing else. They incorporate a range of functions that is even narrower than those of the e-book reader.

7. The smartphone offers ever increasing entries into various forms of connectivity, most of them related to telephone networks, but a growing number include Wi-Fi connectivity as well. They include more and processing power and the capacity to take pictures and videos, bringing them closer to computers. However, because their screens are small, human readers tend to limit their reading to consulting or browsing. No wonder that social networks such as Twitter are very successful on such devices: the small screen fits perfectly with the 140-character limit of this network.

8. For the moment, digital cameras and camcorders are little more than the direct transposition of their analog ancestors. They are conceived as stand-alone machines that generally do not connect to anything but a computer. As a rule, these devices do not allow modifying the recorded document directly, but rather require the use of a computer. Their screens are small and used only to check that the recorded document is acceptable.

9. Digital sound recorders have existed for a long time, but they are not in wide use as stand-alone objects. Rather, they tend to be incorporated within machines with other functions, for example, mobile phones, video cameras and some laptop computers.

10. Digital devices that store various kinds of documents, most notably music and films, have been widely distributed in the form of CD-players, DVD-players and their successors. They actually play the same role as e-book readers, and are designed to stimulate the sale of source materials and control their use. The recent additions of connectivity to some of these machines (such as the iPod) through various kinds of networks ranging from Wi-Fi to mobile telephone is meant to facilitate impulse

purchases. In fact, e-book readers transpose to the printed text functions originally invented for the music industry.

This rapid survey is admittedly incomplete, but it helps restore a sense of how varied these devices are. Although all "read" digital documents, they are sufficiently specialized to appear disconnected from each other. No "template" links them together. In fact, the wide variety of objects works against the emergence of a template: it is hard to conceive of its very possibility. Only after a template emerges does the family resemblance become obvious.

Notwithstanding the absence of a template, our survey of digital readers allows us to identify a number of common points:

1. The question of the screen is the most obvious. Its size is of the essence: small sizes preclude sustained reading and limit the viewing of video materials. Larger-size screens commonly encountered on computers tend to be very good for working on a word processor or on a spreadsheet, but their suitability for sustained reading remains disputed. The printed page and E Ink work through reflected light, and our eyes seem reasonably comfortable with that approach; reading from a back-lit LCD (liquid crystal display) screen may be more difficult on the eye. Also, reflected light depends on ambient light, unlike the back-lit screen which competes with ambient light. If e-book readers advertising often depict reading on a beach, it is because, subliminally, people know that laptops are very hard to use in such circumstances. In dark surroundings, on the other hand, an E Ink screen is useless.

Color is also important: the history of computers shows how irresistible it can be. The Apple McIntosh was first offered with only a black-and-white screen. In comparison with color PCs, they looked drab despite a superior, icon-based, interface. Microsoft improved the interface with Windows, and Apple introduced color.

The refreshing rate is another important element among screen characteristics. Without a sufficiently fast refreshing rate, videos and interactive games become impossible. In short, the characteristics of the screen constrain the range of functions that a particular device can meet, and thus define its uses.

2. Connectivity also defines the functions of digital readers. Basically, connectivity is divided between phone networks and the Internet, leaving aside the special case of television or radio that are linked to broadcast distribution systems (e.g. cable or satellite). When connected, digital readers are often limited to one kind of connectivity, often with only one company available. This is particularly true of telephones where, in fact, the device-as-gadget is used to lock in customers with a given network for a couple of years, by which time the gadget will be or look obsolete

anyway. It also applies to some e-book readers, such as Amazon's Kindle. From the user's standpoint, the most desirable form of connectivity is the Internet, of course, as it is by far the most open.

3.  Whether we can manipulate documents, transform them, link them to other documents, is also crucial. Game consoles, for example, forbid this possibility, or limit it severely. For their part, e-book readers grant varying degrees of freedom, ranging from limited annotations and forms of highlighting to the capacity of making abundant notes, linking them and texts together, creating hyperlinks of all kinds, including links to external documents in the web or in various "clouds." The ease with which these actions can be carried out is obviously crucial. Is the keyboard of a sufficient size, or, if it is embedded in a tactile screen, is it sufficiently responsive to our needs? Typing with two thumbs or banging on a tactile screen may be good enough to "communicate" through Twitter, but it is probably not be the best tool to write a relatively long text.

4.  Portability of the device is certainly another important issue. Here, obviously, compromises have to be made, as legibility requires size, and therefore, weight. An office computer and a smartphone together define the extreme points of a range of possibilities within which an optimal combination remains to be identified.

5.  Finally, the issue of openness must not be forgotten, although it is routinely ignored in the discussions of IT devices. Companies in particular often avoid discussing this issue because closed technical solutions are often part of strategies designed to lock-in customers, which also brings us back to the question of company-specific connectivity. If a machine can only read one type of document that, furthermore, is proprietary, then the consumer makes herself a prisoner of the company that sells the machine. This is an old strategy that stands at the polar opposite of open standards. For example, computer printers try to lock us in with their cartridges, and so do game consoles or a number of e-book readers. Obviously, an ideal machine would adhere to totally open formats.

While the future of technical objects remains opaque, the preceding remarks begin to point to the kinds of forces that will shape the stable templates yet to come. In the case of digital readers, the crucial issues hinge around screen characteristics, openness in a generalized way (open connectivity, open formats, open to human modifications of documents) and portability. Given the enormous interest generated by the recent offering of the Apple iPad, it is useful to review the characteristics of this particular machine from the perspective developed here.

## Is the iPad a possible template for e-readers?

Judging from its initial commercial success, the iPad is a fascinating object. At the very least, it demonstrates effective gadget consumerism. At the same time, we should recall the quotation where puzzlement was expressed about the possible uses of such an object.[7] However, and from the perspective of this text, the iPad is interesting in another way: might it not display in exemplary fashion a kind of sleepwalking progression toward a possible digital reader template? Consider the following points:

1. The iPad's portability, although less than a mobile phone, is better than a laptop computer and is in the same range as a game console or an e-book reader.
2. Its screen, although not in the reflected-light category, is high-definition, thus allowing for relatively comfortable reading in the right ambient light, along with an excellent presentation of color photographs and videos. In any case, it shows that Apple engineers were sensitive to the need for various screen functions, as were outlined earlier.
3. The iPad's connectivity can accommodate telephone networks (3G) as well as the Internet. However, unlike a computer, it cannot accommodate an Ethernet connection and it has no USB ports. As a result, it seems to have been pushed far in the direction of mobile phones. Whether it is too far remains to be seen.
4. With the iPad, documents can be processed to some extent: a keyboard is embedded in the tactile screen, and an external keyboard can be added to the machine. However, the tactile keyboard is not as responsive as a real keyboard, and lugging a keyboard along contradicts the portability principle. The machine is clearly conceived as a stand-alone device.
5. The openness appears great if one considers the number of applications available; however, they are limited because Apple controls which applications can be legitimately ported to the iPad and, indeed, it is part of the device's business plan. Even the battery of the iPad can be changed only by Apple, and this example is emblematic of the whole philosophy surrounding this object.
6. It has no camera and so texts can be illustrated only through images stored either internally or in the web.

The conclusion of this rapid analysis is that the iPad is a tool for casual readers, for example readers of popular magazines. It aims at the idle surfer looking for quick, factual, information, while keeping up with a variety of social networks. The possibility of using e-mail and of phoning makes it a possible (if somewhat bulky) substitute for a smartphone. More importantly, it facilitates shopping, in particular shopping for music, but also books. In this capacity, it incorporates the functions of e-book readers, but also of more recent portable digital players, which all limit the individual to consumer roles. In short, the iPad does not appear to qualify as the "template" that the bicycle once was, but it certainly points in a direction that must be examined carefully, because it does incorporate efforts toward some sort of market convergence. However, the iPad must be seen as a failure if it is judged from its ability to bridge disjointed markets. The market convergence it displays is limited to constructing a new kind of synthetic consumer ranging from casual browsing to superficial social connections.

Its main deficiencies are its screen and its closed structure. The back-lit LCD screen, on top of consuming more power and therefore reducing the autonomy of the machine, is, as pointed out above, sensitive to the quality of ambient light. A sensor does allow the machine to adapt the iPad's screen luminosity to ambient light, but there are limits to what can be done in this fashion. In view of present and coming developments in screen technology (such as the screens from a small firm called Pixel Xi),[8] the iPad appears conservative. It is a digital reader that understands perus-ing, consulting, browsing, but does not seem to understand the needs of serious reading. It provides some computer-like capacity but its embedded keyboard will quickly constrain anyone trying to make full and sustained use of a computer. Trying to write this text on an iPad would be difficult; for example, how should the iPad be positioned to read and write at the same time? Flat, vertical or in between? And if the adopted position is not flat, how does the iPad stay at the right angle without the help of some external stand?

The iPad could easily belong to the toy category. Indeed, it is interesting to note that the iPad is often displayed in advertising as something you curl up with in an armchair, as if marketers treated the iPad as some kind of magazine, only richer in possibilities. Perhaps the iPad should be called an e-magazine reader.

The closed nature of the iPad may look reassuring to docile consumers looking for an appliance, but it will frustrate users that insist on their own autonomy and on treating a computer like a computer, a machine endowed with an almost infinite range of potential functions. Indeed, such criticisms of the iPad have been aired over the Internet and reveal that a significant proportion of users chafe at the idea of artificial constraints.[9]

Where the iPad is probably playing its greatest role is in pointing in the direction of tablets as a possible approximation of a coming template. Hybrids between the portable computer, the netcomputer and the smart-phone, tablets have yet to define their own stable niche. The iPad, because of its temporary success (at least, this is my prediction), is providing strong visibility and a degree of legitimacy for the category itself, if not yet its precise definition.[10] By the same token, it is contributing some elements of this definition by showing that a new combination of functions and characteristics is in order and must be developed. In short, there is a strong probability that the template of the digital readers of the near future will be closer in appearance and functions to the iPad than they will be to e-book readers. And this is because, essentially, they will "understand" digital docu-ments better.

## How should digital readers understand digital documents?

The ten different varieties of digital readers listed earlier differ mainly by the nature of their screen (when they sport one), and by the way they relate to computers. All of them are enhanced as soon as they connect to a computer. Even the digital camera makes more sense once we realize that snapping the picture is just the first step in a number of operations involving the treatment of the raw picture by a suitable image-processing tool. It becomes quickly obvious that much more can be done after the picture is snapped than through taking the original shot. One of the most essential elements of digital documents is that they lend themselves to reworking, reinterpretation and repositioning within various ecologies of documents. Even the lowly digital frame incorporates minimal computer functions with its ability to shuffle the order in which pictures are displayed. Through a computer, digital documents can be linked together in various ways, for example through the http protocol, tagging or suitable meta-data. Comments can be added. In short, unlike the model of the printed book, digital documents enjoy a degree of fluidity, of liquidity even, to echo the "liquid publishing" phrase,[11] which is essential to their existence. This explains why the page images of printed documents that we can now access thanks to Google's "generosity" are but a foil: Google's apparent gift does include some limited, but real benefits, but they are minuscule in comparison with all that can be done, could be done, will be done through "digital reading." One only need think about a digital snapshot locked into a fixed, untouchable, format and compare it to the same picture amenable to treatment with PhotoShop or Gimp to understand the potential that accompanies any digital document. It is precisely the locus where Google intends to anchor its monopoly.

Being connected to computers and to the Internet is fundamental if the full potential of digital documents is to be unleashed. But connectivity also brings up the issue of "openness" that has been invoked earlier. The Internet operates on the basis of open standards, and connecting to it requires conforming to them.

Openness also refers to the formats of digital documents. If information is locked up by a format that limits "reading" to human eyes (for example a page image), it allows doing little more than parchment or paper, even though it is digital. Such technical choices often reflect a concern for proprietary control over some dimension of the digital documents. For example, Google's decision to release only page images of the works it digitizes is coherent with the decision to build a monopoly over indexing, but it does not satisfy all the wishes of readers and scholars.

Digital readers, we can now conclude, tend to understand digital documents rather well, but they do so from a perspective that does not necessarily favor the user or even the creator. E-book readers, for example, tend to understand digitized books as publishers or their distributors (such as

Amazon) would. Amazon's Kindle, for example, is optimized to control the digital document by tying it to the device, while facilitating impulse buying. In this regard, it comes close to a game console. Each kind of device is optimized for a specific function: sustained reading in one case, powerful graphic capability for exciting interactivity in another. But control over the document is common to both types of devices.

The ability to understand the perspective of a particular stakeholder all too well is obviously a factor in creating and maintaining the fragmentation of the markets; but it also accounts for the relative lack of success of the solutions brought forth by various companies. If digital readers respond too closely to the specific interests of the recording industry, the movie industry, the publishers' world and so on it is not surprising to see a proliferation of devices that find it difficult to recognize each other as essentially part of a larger, over-arching family. It is not difficult either to see why a 'template' is not emerging easily. It is not difficult to see why the market finds it difficult to grow. It is not difficult to understand why e-book readers are still niche objects.

## Conclusion: The shape of a possible template

Given all that has been said, what possible shape could (should?) a template take? The answer would probably be as follows:

1. Digital documents are too dependent on computer-like capacity to leave this element out of the equation. A digital reader should include a computer or at least some significant computing capability, as is the case of Apple's iPhones and iPads. And a digital reader should be able to connect at least to the Internet. Telephone networks are a possible plus, but they will remain very constraining, trying to lock in users. Their presence may signal the desire to reduce users to the role of passive consumers.
2. The screen issue is crucial and not entirely solved. To go back to the comparison with the bicycle, it is probably the element closest to the tire. Ideally, such a screen should be able to switch reflected light and backlighting. For the moment, tactile input is viewed as an advantage, but serious users may see this functionality as more of a fad than a trend.
3. The tablet form appears very useful for many consulting and browsing tasks on the go, but the reader should not be limited to this form. Laptop computers that, through a clever rotation of their screen, can transform themselves into a tablet are likely to incorporate an important element of the template we are seeking. In particular, this feature responds to the need of inputting data in a variety of ways without having to carry on an external keyboard. Perhaps, voice recognition will also be part of future devices when this technology is perfected.

4. The digital reader template should also accept various forms of input and production: writing, linking, taking photos and videos, and working with images, videos and so on should all be available.

5. The desired digital reader template should be built on free tools (free software in particular) so as not to depend on any private company for the access, treatment and enjoyment of digital documents. The appearance of software such as Google's Android designed as (relatively) free software is a signal that this request is being heard. The digital reader template should also favor open formats, including a built-in and user-friendly capacity to produce structured XML documents; XML now stands as the foundation for web publishing and for preservation, and it is used by most serious and important publishers.

6. Portability is also of the essence. The reference points here, as verified by centuries of use, are the smaller codex formats used in the print world. If the digital reader weighs about a pound and is about the size of an octavo or duodecimo (roughly the size of a regular paperback book nowadays), it will broadly respond to people's expectations. In fact, most e-book readers have spontaneously (and unsurprisingly) adopted these codex dimensions and weights. By contrast, the iPad screen size corresponds more to a magazine format, thus revealing some elements of its commercial positioning, but also some of its niche limitations. The iPad may well be a little too large.

If we look around the vast collection of diverse digital readers already available, the template just adumbrated is best approximated not by the iPad, but by a small computer that has been developed mainly for schoolchildren in the Third World: the "One Laptop per Child" (OLPC) that was Nicholas Negroponte's brainchild; its portability, ruggedness, screen characteristics, magnificent connectivity, openness and so on all come very close to what is needed for a template. Perhaps a bit more tweaking here and there could transform this device conceived to help education in poorer countries into a magnificent tool for everyone, thus allowing for the lowering of production costs and increasing its availability for all. In particular, one OLPC model that was abandoned at the end of 2009, the OLPC-XO-2[12], was supposed to incorporate the kind of dual screen, back-lit as well as reflected light, that was pioneered by Mary Lou Jepsen before she launched her own company Pixel Qi[13]. It may be that the digital reader "template" is already almost among us, but we do not yet recognize it. The next five years should bring this issue to rest. Then the history of e-book readers will look very different.

## Notes

1. "Print culture" is an expression that emerged with the famous book by Elizabeth L. Eisenstein, *The Printing press as an Agent of Change: Communication and Cultural*

*Transformations in Early-Modern Europe*, 2 vols. (New York: Cambridge University Press, 1979). Many of its main theses have been challenged by Adrian Johns' *The Nature of the Book: Print and Knowledge in the Making* (Chicago: University of Chicago press, 1998). See also *Agent of Change: Print Culture Studies after Elizabeth L. Eisenstein*, eds. Sabrina A. Baron, Eric N. Lindquist and Eleanor F. Shevlin (Boston, University of Massachusetts Press, 2007).

2. Jean-Claude Guédon, "What Can Technology Teach Us about Texts? (and Texts about Technology?)" in Timothy W. Luke and Jeremy Hunsinger (eds.), *Putting Knowledge to Work and Letting Information Play* (Blacksburg, Virginia: CDDC, Virginia Tech, 2009), pp. 55–75.

3. Pinch, T. J. and Bijker, W. (1984). "The Social Construction of Facts and Artefacts: Or How the Sociology of Science and the Sociology of Technology might Benefit Each Other." *Social Studies of Science*, 14(3) (1984): 399–441.

4. As exemplified by the BBC series derived from Douglas Adams' cult novel, *The Hitchhiker's Guide to the Galaxy*.

5. Marc Aronson, "The End of History (Books)." *The New York Times*, April 3, 2010, Opinion section. http://www.nytimes.com/2010/04/03/opinion/03aronson.html, accessed August 21, 2010.

6. Kent Anderson, "The Scholarly iPad—First Impressions, Relevance for Publishers." http://scholarlykitchen.sspnet.org/2010/04/05/the-obligatory-ipad-post-first-impressions-relevance-for-scholarlypublishers/, accessed August 21, 2010.

7. See note 6.

8. See http://www.pixelqi.com/, accessed August 21, 2010.

9. See, for example, http://www.pcworld.com/article/188073/apple_ipad_reviews_the_critics_weigh_in.html, accessed August 21, 2010.

10. Steven Levy, 'How the Tablet Will Change the World," *Wired* (April 2010), available online at http://www.wired.com/magazine/2010/03/ff_tablet_levy/, accessed June 21, 2010.

11. For a particular instantiation of the liquid Pub concept, see http://liquidpub.org/, accessed August 21, 2010.

12. See http://www.h-online.com/open/news/item/OLPC-XO-2-cancelled-but-XO-1-75-planned-849585.html, accessed June 10, 2009.

13. See http://www.pcworld.com/article/200376/lowpower_pixel_qi_displays_sell_out_in_a_day.html, accessed June 10, 2009.

# 14.1

## In Praise of Paper: Cultural Prejudice and the Electronic Book Market in Spain

*José Luis González-Quirós*

As history has clearly shown us, the introduction of any new technology sparks negative reactions—often because errors are made or promises take too long to materialize, among other reasons. The case of dedicated reading devices and how they have been received in Spain is marked by a number of rather peculiar features, however. Among them is a group of arguments to which I would like to call attention: arguments attempting to show that reading, and culture itself, could incur certain risks if the dedicated devices come to be very widely accepted and/or if new forms of reading arise that have the potential to ruin in-print publishing businesses.

These objectors have sought a basis for the background argument, so to speak, in the existence of certain problems with on-screen reading, deliberately ignoring the fact that electronic ink technology causes no eye fatigue whatsoever. Their arguments went beyond the physical experience, however, to suggest reading deficiencies of a vague nature—a completely unfounded and indigestible hodgepodge of physiological and semiotic fluff, as if the understanding, sensitivity, reflection and critical spirit that make reading a noble act had appeared in the world thanks to Gutenberg.

The introduction of technologies is always a bit awkward but, in the case of e-books in Spain, that process of trial and error to find the perfect or almost perfect form is put forth as proof that there are supposedly irreparable defects in the new systems. This is not the first time that has happened among us, however.

In the hands of the detractors, the lack of a definitive or universal text format, the lack of standards of compatibility between devices, and the different policies of digital publishers become proof that we have before us a type of device that offers no obvious advantages and will not achieve success. Through feignedly independent and disinterested ploys of confusion, minor problems with legibility become a definitive argument against the convenience and advantages of the new devices.

Besides the fact that this serves to defend traditional publishers, who are tremendously reluctant about any change, the heart of the matter lies in the book being confused with an object, and publishing being reduced to pagination and mere typography. This concept is entirely out of place in the digital world.

In the digital world, books will again be what they truly are, an author's ideas, set forth in the author's own way, and delivered to the author's readers and any other human beings who may wish to drink of them. A book is not a pile of ink-stained pages but a collection of arguments, metaphors and discourses. In the digital world, publishing will find its place with tremendous clarity because it will not be limited to producing copies of a particular composition in pages and types; rather, it will be complementing the book with everything that will enrich the reading of it and its comprehension, with what can illuminate its meaning and its influence in every moment (González-Quirós and Gherab-Martín 2009). The classics will be revived because there will always be scholars willing to publish them, and there will be no need for the backing of a commercial agent who calculates the cost of printing, marketing and distributing this work that will then forever belong to the agent, and not to the author.

Curiously, e-readers are also criticized for their lack of interactivity, although it has been said recently that such interactivity is one of the things that apparently alters the pleasure of the text, to use the Barthesian expression.

In an attempt to disregard the tremendous opportunities the new devices open up for publishing, in the proper, non-commercial sense of the term, their alleged problems with supporting glossaries, citations and notes is pointed out as one of their limitations. In reality, what will happen is that the digital environment will afford unprecedented opportunities for erudition and bring us into something infinitely more rich and powerful than any paper platform would be able to support. All these opportunities do not yet exist, obviously, but nor were the first trains very clean, nor did they have air conditioning, and nor could they travel at 350 km per hour. The day came, however, when they put the mule teams out of business forever in transporting goods and people.

The perspectives that are opening up may be frightening, of course, because we see that a building that has stood for several centuries is crumbling. But that should not cause us to denounce technologies that are solving problems with an efficacy we could not even imagine only 15 or 20 years ago.

We must shake off the fear that authorship will vanish and that it will be impossible for professional writers to support themselves with their publications. What will vanish, of course, are the numerous intermediaries whose contribution to culture is rather dubious, for they have made products expensive and have spawned all manner of artificial fads to increase their sales. Attempting to safeguard a significant portion of the old set-up and its associated commerce through censorship and control procedures such as digital rights management (DRM) systems represents an error of principle. The primary right of authors is the right to be read, and author remuneration will only increase in the new digital environment, which can and must strive for much less meddling than there is in the print environment.

It will be some time before we see what definitive form all these new opportunities consolidate, but I long to see the hired mourners, who cease not to proclaim evils and defend their pathetic causes, disappear as soon as possible, and for their own good. The advantage of the open market—and the market for ideas is exceptionally open—is that, in the end, it always puts into effect whatever forms the public prefers, for many can be fooled some of the time, and some can be fooled forever, but you cannot fool everyone indefinitely. The detractors may shout themselves hoarse, but reading devices are here to stay. They are an essential part of a project that, although still not perfectly defined, is very clear in its general outline, as is typical of everything new, everything that breaks with age-old routines and traditions. This project is the universal digital library, of which we have spoken, which will be very easily accessible, almost perfect and extremely inexpensive.

## Bibliography

González Quirós J. L. and K. Gherab Martín. 2009. *The New Temple of Knowledge: Towards the Universal Digital Library.* Australia: Common Ground Publishing.

# 14.2
## Why E-Readers Will Not Gain Widespread Popularity

*José Antonio Millán*

I believe that the average reader will not be able to enjoy the advantages of a dedicated e-reader—convenience and access to a great number of works, among others—because the advantages of these devices are buried beneath their drawbacks.

This hypothesis is based on my personal experience with e-readers, which began in 1999 and has continued since then with the testing of various devices for the Spanish newspaper *El País* and for my blog *Libros y bitios*.[1] I have also conducted surveys and done qualitative research with user groups for the report *La lectura en España* (Reading in Spain).[2] There are many e-readers available for purchase but, in almost all these devices, the underlying electronic paper technology is almost identical. The hardware controls, interface, and text display software, however, can vary from one model to another. I will address the problems in successive sections.

### Hardware

Some devices have a control that is only for turning pages forwards and backwards, while, in others, this function is included in a single, complex control that is also used to navigate the menus. Some incorporate number keys for choosing from different options; finally, some have a QWERTY keyboard (Kindle) or are activated using a stylus that allows writing (iLiad).

So, there is no standard design, and users have no *a priori* knowledge of how to operate the device.

### Interface

In general, the interface is very poorly designed. Because these are devices that can store hundreds of works, one might think that the manufacturers' first concern would be making it easy to browse the internal library. That is not the case, however, as on many models, navigation is by file name only.

Access to text highlighting, if the device has this feature, and other commands, is usually equally unsatisfactory.

## Text marking

On many models, one cannot make marks; other models allow only five or six marks per work. Page marking may require several not very intuitive keystrokes. Some models such as Kindle have no pagination, which makes it difficult to reference a page in a work, as may be necessary in teaching, for instance. These were the elements that resulted in students refusing to use these devices in the first pilot studies that were conducted (Mintz 2009).

## Searching for a word, or copying an excerpt

In many models, this cannot be done. These devices, which are designed to operate with the most flexible of raw materials—the digital text—cannot make the most of its advantages.

## Typography

Even though font size can be increased in e-readers, which is a distinct advantage they have over traditional books, their typically small screen complicates this feature. The text has to flow to accommodate the changes.

Consecutive paragraphs and the simple dialogues of a novel are easily read, but quotations in indented text probably will be lost. Genres such as poetry and theater that are typographically more complex may lose the layout.

## e-Babel

Despite the growth of ePub, different formats are still proliferating; among them, PDF, which is widely used by publishers because it more closely resembles the printed page. Unfortunately, as a "crystallized" format, PDF is the one that is most problematic in terms of displaying it on the screen. Conversion is not always possible, and, even starting with the same file, different e-books will give different results.

The best solution for the purchasing public may turn out to be strange, as well. Some publishers are going so far as to offer three formats for each work: PDF optimized for six-inch screens, PDF for eight-inch screens, and ePub.

## DRM (Digital Rights Management)

There are obstacles in place that make it difficult for works to be read with other devices or to be lent to other people, and this markedly distorts the perception of what one has acquired. Nicholson Baker's definition is probably the best: "They are closed clumps of digital code that only one purchaser can own. A copy of a Kindle book dies with its possessor" (Baker 2009).

## Reading

Reading on an e-ink e-reader is not better than reading on a back-lit, high-definition screen, for example, the iPhone screen.

   Moreover, with electronic ink technology, it is difficult to memorize what one reads because of the blackout that occurs with every page turn (Testard and Bettayeb 2009).

## Printing

One cannot print from most models.

## Instruction manuals

E-reader manuals are typically just as deficient as those for any other electronic device.

## Scarcity of works

Outside the English-speaking realm, there is a scarcity of works available to read with an e-reader. The number of works offered in countries such as France and Spain, however, is expected to increase considerably in 2010.

## Private documents

One of the most publicized features of e-readers is that they can be used to read private documents, for example, publishers are able to read original manuscripts. In all cases, especially with PDFs, this requires a conversion task; some users do not know how or are unwilling to undertake this.

## Price

E-readers are expensive. We might think that digital books are a good option, financially, because they are cheaper than books in print. But how many works would one have to purchase to recoup the cost of an e-reader? In the US, where the selling price of digital works is low, this figure was calculated to be 61 books (Arends 2008). In Spain one would have to purchase more, because the savings per book would be less.

## Conclusion

Where is the niche for electronic readers, then? As a Spanish manufacturer and importer's advertising stated recently, this is "a device designed not only for those who are fond of the latest technologies but also for those who love design." Yes the e-reading public will remain first and foremost among gadget lovers, even if many of the problems we have described are not resolved, and some will not be easy to resolve. Meanwhile, people will be reading comfortably on improved, non-dedicated devices such as *smart phones* and *tablets*.

As the otherwise astonishing prospects for digital reading continue to evolve, will electronic ink e-books turn out to be little more than an anecdote?

## Notes

1. http://jamillan.com/librosybitios/blog/.
2. José Antonio Millán (ed.), *La lectura en España. Informe 2008. Leer para aprender.* [Reading in Spain: 2008 Report. Reading to Learn.] Madrid, Federación de Gremios de Editores de España y Fundación Germán Sánchez Ruipérez, 2009. Online edition: http://www.lalectura.es/2008/.

## Bibliography

Arends, B. 2008. "Can Amazon's Kindle Save You Money?" *Wall Street Journal* (24 June). Available at: http://online.wsj.com/article/SB121431458215899767.html, accessed July 23, 2009.

Baker, N. 2009. "A New Page. Can the Kindle Really Improve on the Book?" n *New Yorker* (23 August). Available at http://www.newyorker.com/reporting/2009/08/03/090803fa_fact_baker?currentPage=all, accessed January 12, 2010.

Jessica Mintz, J. 2009. "Students Unready to Trade Texts for Kindle." *Boston.com* 14 (October). Available at http://www.boston.com/business/technology/articles/2009/10/14/students_unready_to_trade_texts_for_kindle/, accessed August 1, 2010.

Testard-Vaillant, P. and K. Bettayeb. 2009. "La lecture change nos cerveaux aussi." dossier de *Science et Vie* (September). Available at http://pvevent1.immanens.com/fr/pvPage2.asp?puc=2232&pa=2&nu=1, accessed December 15, 2010.

# 14.3
# Hyperactive: The Digi-Novel

*John W. Warren*

The past 12 months have been marked by significant media attention to e-books, numerous e-reader devices have entered the market, and e-book sales have grown. Still, most e-books today are merely a "picture of a book"—a book that has been digitized but adds little value besides portability, improved search and access (Warren 2009).

Despite the often personal, evocative relationship many of us have with printed books, electronic texts can provide benefits not possible with a printed volume. *Inanimate Alice* (www.inanimatealice.com) provides an example of a digital narrative that would not exist in the same way in printed form. Written and directed by writer Kate Pullinger and digital artist Chris Joseph, this "digital novel" combines text, audio, video, special effects and gaming to explore a form of storytelling in which the reader is converted into an active participant. As readers progress through each episode—four are available, a fifth is expected soon, and ten episodes are planned—their participation level increases. Alice, the character, learns to make games on a fictitious device, which the reader learns to play; gaming, in essence, provides and underscores Alice's emotional journey (see Figure 14.3.1). The player/device, as well as the player/reader, is part of the story (Pullinger 2009).

According to Kate Pullinger, *Inanimate Alice* was conceived and created in digital form from the outset, arising both from a desire to explore new digital tools, as well as because the digital form seemed the best way to tell the story. The use of music, images and sound and video effects seem to fit organically into the story instead of feeling tacked on. The form also provided a great opportunity for collaboration, between the writer and digital artist, and subsequently with the audience. It has had an unexpectedly wide appeal, among elementary and secondary school students, educators and university students. A range of electronic curriculum and reader-created stories ensued almost naturally, without a strong push from the authors or producer, but certainly not with their opposition. Dedicated sites have sprung up from educators and schools in Australia, Great Britain and the

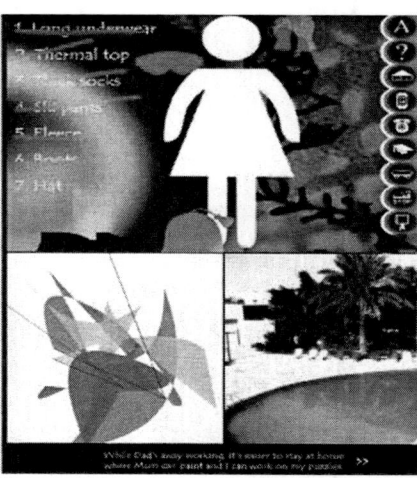

*Figure 14.3.1    Inanimate Alice*

US, and many students have created their own episode five to continue the series (Laccetti 2007; Pullinger 2009).

*Inanimate Alice* also offers challenges, foremost of which is that the income stream from the project has presented (and presents) a conundrum. The digital novel sprung up to tell a story, not to fit a business model, and although it has been embraced by many fans, this hasn't translated into revenue. Creating these stories also presents a challenge in deciding how much text the multimedia can support and how the various elements work together. Another challenge was the need gradually to bring the reader up to speed because of inexperience with form. The games within the story get more complicated as *Inanimate Alice* progresses; it was impractical to expect the reader to dive headlong into complicated gaming scenarios. Nevertheless, the key challenge is to find a new way of storytelling that fits into the commercial market.

While there is little consensus on a definition of digital storytelling or new media writing, the use of a computer or electronic device such as an iPhone is the unique component of both writing and reading digital stories. Multimedia, including text, audio, video and interactivity/gaming, can be blended in ways impossible in printed books. Hypertext provides alternative construction, concept and characterization, for fiction and nonfiction, perfectly suited to the online or e-book form (Warren 2009). Combining microcontent and social media to create distributed conversations has been called Web 2.0 storytelling (Alexander and Levine 2008). Interactive stories may employ the reader's use of an avatar to become a character that navigates through and interacts with the story (Pressman 2008). In the future, the e-books that grab and hold our attention will be those that embrace interconnection, multimedia, interactivity and customization, bringing us into a new space beyond the printed word.

## Note

This work is adapted from Warren (2010) which was not yet published at the time of submission.

## Bibliography

Alexander, Bryan and Alan Levine 2008. "Web 2.0 Storytelling: Emergence of a New Genre." *Educause Review* (November/December). Accessed October 4, 2009: http://net.educause.edu/ir/library/pdf/ERM0865.pdf.

Laccetti, Jess, 2007. "Inanimate Alice Pedagogy Pack: Lesson Plans and Education Resource Pack." Accessed October 4, 2009: http://www.inanimatealice.com/education/.

Pressman, Jessica. 2008. "Navigating Electronic Literature." Electronic Literature: New Horizons for the Literary (February). Accessed November 4, 2009: http://newhorizons.eliterature.org/essay.php?id=14.

Pullinger, Kate, 2009. Phone interview, September 7. Accessed October 1, 2009: http://www.katepullinger.com/digital.html.

Pullinger, Kate, and Chris Joseph, 2009. *Inanimate Alice*, episodes 1–4, accessed 1 October, 2009: http://www.inanimatealice.com/)

Warren, John W., 2009. "Innovation and the Future of E-Books." *The International Journal of the Book* 6 (1): 83–94. Accessed October 10, 2009: http://ijb.cgpublisher.com/product/pub.27/prod.273; also at http://www.rand.org/pubs/reprints/RP1385/).

Warren, John W., 2010. "The Progression of Digital Publishing: Innovation and the E-volution of E-books." *The International Journal of the Book*, 7, (4): 37–53. Accessed October 2, 2010: http://ijb.cgpublisher.com/product/pub.27/prod.369; also at http://www.rand.org/pubs/reprints/RP1411/.

# Conclusion

# Mapping Emerging Digital Spaces in Contemporary Society

*Karim Gherab-Martín*

It is nearly impossible to find a common thread running through all the chapters in this book. For instance, although there are many authors who think that ICTs favor decentralization, democratization, development, and the flattening of hierarchy among human beings (see Chapters 2.1, 4.2 and 11.1), others believe there is no ground for these claims (see Chapters 1 and 13). However, there seems to be a "family resemblance," as Wittgenstein would say, among the chapters, and it would be rather pretentious to try to predict how digital technology will shape social structures. There is no doubt that new material means of production and spreading information will radically transform social, economic, and labor relations—indeed, this is already happening (see Chapters 10, 10.1, 10.2, and 10.3)—but the Marxist characterization of history does not necessarily lead to a strict technological determinism. In other words, the fate of the Internet society is still unwritten—it is not predetermined.

More than 40 years ago Heilbroner (1967) distinguished between two kinds of technological determinism: hard technological determinism states that the fate of society hinges on technological changes alone, while soft technological determinism states that it hinges on a combination of technological, political, social, and cultural causes. I lean towards the soft version of technological determinism; this way I can avoid having to speak as an oracle here in this conclusion. Furthermore, I am certain that readers have already drawn their own conclusions without having to read mine.

Nevertheless, I believe I have something to offer to readers who are willing to re-read this book from a new perspective based on the notion of *borders*. Just as cartographic projections do not preserve the metrical properties of the physical contours of countries and continents, the socioeconomic and cultural projection of the physical world over the new digital scenario also suffers marked distortions. The transformation of borders and contours (political, social, commercial, cultural, scientific, and so on) is not intrinsic to ICTs, but there is no doubt that these tools radically enhance changes, which forces governments and businesses to invest in education, training, and acquisition of new skills (see Chapters 2.2 and 10, 10.1, 10.2, and 10.3). All these changes trump the socioeconomic transformation that began to take place in the fifteenth century with the invention and universalization of the printing press.

As a result of globalization, borders are no longer defined only by their territorial and geopolitical criteria. For instance, the borders of the Schengen zone are not the same as those of the Eurozone, and the boundaries of the Eurozone differ from those of the European Union as a whole. The same is true for the US dollar, the influence (and use) of which extends beyond US borders. Formal organizations such as the Organisation for Economic Co-Operation and Development (OECD) and informal ones like the BRIC (Brazil, Russia, India, and China) economies are other examples of this kind. In the military realm there are similar organizations such as the North Atlantic Treaty Organization (NATO), which represent political spaces that are not defined by geographical criteria alone.

Thus, we can speak of "variable geometries" where borders can be moved and reshaped. Political, social, cultural, scientific, economic, and even military borders can be redefined and related to the five freedoms of circulation of goods, capitals, people, services, and knowledge. Furthermore, there are several notions of borders, which will all change as a result of an increase in openness and exposure to digital technologies and the Internet. Above all, the Internet fosters the circulation of knowledge, of course, and does so in a host of (sometimes) unpredictable ways in what has come to be known as the "convergence culture" (see Chapters 2 and 9.1).

The physical dimension of borders, including maritime spaces, should therefore be re-explored in the light of recent developments in the digital dimension. Some physical barriers become meaningless as a result of the Internet, which may cross state lines providing information globally and enhancing virtual regions of knowledge in accordance with the metaphor of the "global village"—the spreading of ICTs and transportation, transnational corporations, and global trade, which makes the notion of physical borders almost irrelevant (see Chapters 13.1 and 13.3). Ideas are transformed into objects in the physical world and the digital world, which allows them to be manipulated, exchanged, and marketed. Copying and distributing a physical object can be a costly process. In the digital environment, however, any object can be copied and distributed at virtually no cost. Also, it has been shown that most innovations originate not with knowledge producers but with knowledge users and distributors—an effect that is multiplied exponentially in the digital environment where hyperlinks mean that digital artifacts (software) are only a "click" away for any potential user.

The constant reuse of digital content shapes the borders of the objects themselves. The appearance of new kinds of open licenses, such as Copyleft and Creative Commons (see Chapter 6.1), for example, allows content to be reused and builds up a new system that is much more flexible than the traditional dichotomy between copyright and the public domain (see Chapters 6, 6.2, 8, and 8.3). This enables new industries to emerge based on the reuse of previous products and contents (see Chapters 3.1, 3.2, and 8). In the book industry, for example, physical platforms and business models/processes are changing, but content does not change (see Chapters 4 and 14, 14.1, 14.2, and 14.3). Thus although the "borders" of paper books and e-books are quite different, they bear the same ideas (see Chapter 3.2). Some authors argue that e-readers, as they are currently designed, have a negative impact on e-book readership (see Chapters 14 and 14.2). Others have exactly the opposite opinion (see Chapters 14.1 and 14.3). And some authors argue that we should shape the borders of digital objects from a techno-social perspective, overcoming the dominant and primarily technical view (see Chapter 13.1).

The notion of borders also applies to the limits of the disciplines themselves. New ways of mapping borders in addition to geographical, political, and historical ones can be developed. Digital technologies allow for many disciplines to merge, and it is likely many innovations will appear in the intersection of such disciplines. Ideas and creative endeavors from different disciplines will be linked, overlapped, combined, and remixed in many ways, thereby configuring new cultural and artistic disciplines that will push forward those borders (see Chapter 8). For instance, what are the borders between science and digital technologies? Which borders are "pushed forward" as a result of the emergence of the so-called digital humanities? How will initiatives as OpenCourseWare[1] change education worldwide?

Internet users are creating and sharing digital resources they value—information, ideas, and creations such as documents, pictures, videos, music, and software—across time and space (See Chapter 2.1). Web 2.0 is characterized by the fact that producers and consumers are indistinguishable—users play both roles at once. Alvin Toffler (1980) coined the term "prosumers" to describe this user phenomenon in the post-industrial society. In summary, the meaning of "producers" and "consumers" is changing to the extent that the borders between them are steadily vanishing. In the same manner, and to some extent as a result of this change, it is increasingly difficult to distinguish between professionals and expert amateurs (see Chapters 7 and 10). The borders of expertise are less and less distinct, especially now that the technological means for creating and spreading digital objects have been somehow democratized. Physical objects cannot be sent via the Internet, of course, but they can be successfully advertised at almost no cost.

The open nature of ICTs has sparked a growing debate about the "end of privacy" (see Chapter 5). The boundaries between public and private information are more blurred than ever (see Chapter 12.1). Surveillance is growing constantly (see Chapter 6, 6.1, 6.2, and 6.3), so we can wonder where the borders of privacy are.

There are also gender borders (see Chapters 12 and 12.3). Vandana Shiva has argued that western cultures have separated "production" and "creation" into different, bounded spaces, in which creativity and production became the monopoly of men and recreation and reproduction the tasks of women (see Chapter 6). In fact, all these changes that have been brought about by ICTs are transforming the borders of personal and communal identities (see Chapters 4.1, 7.1, 7.2, 7.3, and 12.2), as well as notions of citizen integration (see Chapters 1.1, 1.2, 4.3, and 13.2). The dilution of privacy, the variety of hobbies among amateurs, and virtual spaces such as SecondLife[2] are turning our mono-identity in the physical world into a multiplicity of identities on the Internet (see Chapter 4). In turn, these identity transformations involve aesthetic changes in the digital realm (see Chapters 9 and 9.1). Inversely, new technologies also enrich our reflections on our real physical environment (see Chapters 1.3, 2.3, 9.2, 9.3, 11.2 and 11.3).

Over the last decade, researchers have analyzed the peculiarities of particular sectors and industries, their internal processes, and the changes that have taken place as byproducts of ICTs. It is time to study the borders that separate all these analyses to find the missing factor that is common to all of them. It is not likely, however, that digital technologies and the Internet will clearly identify the fate of all sectors of society. Therefore, our search for a common factor may be futile. The many nuances of geography, ethnography, culture, socioeconomics, techno-science, and gender, among others, which enrich local contexts, also make this task far more difficult. For the time being we must be content to accept a kind of "family resemblance" that underlies almost all the chapters included in this book.

## Notes

1. A well-known initiative is the Massachusetts Institute of Technology's http://ocw.mit.edu/index.htm, accessed May 15, 2010.
2. http://secondlife.com/, accessed May 15, 2010.

## Bibliography

Heilbroner, Robert L. 1967. "Do Machines Make History?" *Technology and Culture*, 8, July: 335–45.
Toffler, Alvin. 1980. *The Third Wave*. New York: Bantam Books.

# Index

Abbate, J., 90
Abdul-Raouf, D., 97
Aboriginal Youth project, 296
abortion debate, 163–4
academic practices, 157–69, 177–8
academic publishing, 61–70, 182, 186,
    189–90, 201–3
access, 5, 90–1, 175–6, 270–2, 305–6,
    314, 330–2
Ackerman, M., 260
actuarial practices, 117–18
Adajian, T., 166–7
added value, 185–9, 191–2
Adobe, 33, 215
Adorno, T., 239
advertising, 290, 308
aesthetics, 6, 207–17, 223–4
Africa, 5, 175–6, 305–6, 310, 312, 318
agency, 17–18, 91–4, 198, 229
Agger, B., 90
Aglietta, M., 239
Agro-Bio, 199
Ahuja, M. K., 298
Akbar, M. S. U., 330–1
Alcatel-Lucent, 324
Alden, C., 307
Alexander, B., 360
Alexander, J., 29
Alexander, P., 166
Aliar, R., 5
Allagui, I., 4, 27
Al Qaeda, 91–2
alternative computing tradition, 261–4
Althusser, L., 290
Altman, L. K., 191
Álvarez, F. G., 38
amateur production, 232–4, 237, 368
amateur reporting, 4–5, 38, 45, 123–4
Amazon, 75, 181–2, 186, 338, 340, 342,
    345, 349
analytical perspectives, 13–16
Anderson, B., 75
Anderson, C., 237–8
Anderson, G., 56

Anderson, R. I., 19
Andhra Pradesh, 172–3
Andrejevic, M., 114
Angelo, N., 4
antitrust, 285–6
Appadurai, A., 265, 270, 329
Apple, 216, 262, 338, 341, 344–6, 349
April 6 movement, 97–8
Arabic websites, 4, 26–7
Arends, B., 357
Aristotle, 38, 158–61, 166, 168–9
Aronowitz, S., 239
Art Beat, 6, 223–4
Arthur, E. D., 7
art/life boundary, 219–20
ArtsConnectEd, 56–7
arXiv, 190, 201–3
assemblage of surveillance, 108–9, 112,
    116, 126–7
Atari Democrats, 247–8
*Atlantic Monthly*, 46–7
Atton, C., 99
Australia, 7, 296–7
authorship, 353–4
Autodesk, 215
autonomy, 63, 92, 135, 139, 142–3, 236,
    238, 288, 347
avatars, 39, 360
averaging, 210–11

Baba, H., 75
Backlund, P., 188
Bailey, J., 265
Ballabio, E., 103
Ball, K., 114
Bangladesh, 7, 40, 331
Bannett, E. T., 281
Barbrook, R., 214
Barlow, J. P., 89, 287
Barney, D., 289
Bartels, F., 330
Bartels, L. M., 247
Basque Country, 4, 29–31
Bateson, G., 211

374    *Index*